Pharmaceutical Formulation
Development of Peptides
and Proteins

Pharmaceutical Formulation Development of Peptides and Proteins

edited by
Sven Frokjaer and Lars Hovgaard

CRC PRESS

Boca Raton   London   New York   Washington, D.C.

FIRST INDIAN REPRINT 2009

Library of Congress Cataloging-in-Publication Data
Catalog record is available from the Library of Congress

Visit the CRC Press Web site at www.crcpress.com

© 2000 CRC Press

No claim to original U.S. Government works
International Standard Book Number 0-748-40745-6

Printed and bound in India by Nutech Photolithographers

FOR SALE IN SOUTH ASIA ONLY

Contents

Figures

Tables

Contributors

Michael J. Akers is with Akers Consulting and Training Services, Indianapolis.

S. Dean Allison is at the University of Colorado Health Science Centre, Denver.

Miroslav Baudyš is at the Biomedical Polymers Research Building, Salt Lake City.

Ronald T. Borchardt is at the University of Kansas, Lawrence.

Jens Brange is with Brange Consult, Klampenborg, Denmark.

Helle Brøndsted is at the Royal Danish School of Pharmacy, Copenhagen, Denmark.

John F. Carpenter is at the University of Colorado Health Science Centre, Denver.

Michael R. DeFelippis is with Eli Lilly & Co., Indianapolis.

Nanni Din is with Novo Nordisk A/S, Bagsvaerd, Denmark.

Jan Engberg is at the Royal Danish School of Pharmacy, Copenhagen, Denmark.

Karen Fich is with H. Lundbeck A/S, Valby, Denmark.

Gitte Juel Friis is at the Royal Danish School of Pharmacy, Copenhagen, Denmark.

Sven Frokjaer is at the Royal Danish School of Pharmacy, Copenhagen, Denmark.

Chimianlall Goolcharran is at the University of Kansas, Lawrence.

Lars Hovgaard is at the Royal Danish School of Pharmacy, Copenhagen, Denmark.

Mette Ingemann is with LCO Pharmaceutical Products A/S, Ballerun, Denmark.

Mehrnaz Khossravi is at the University of Kansas, Lawrence.

Sung Wan Kim is at the Biomedical Polymers Research Building, Salt Lake City.

Lotte Kreilgaard is with Immunex Corporation, Seattle.

Deirdre Mannion is with the Danish Medicines Agency, Brønshøj, Denmark.

Bernard A. Moss is with PolyPeptide Laboratories, Hillerød, Denmark.

Theodore W. Randolph is at the University of Colorado, Boulder.

Lars Skriver is with L & K Biosciences, Vedbaek, Denmark.

Preface

The rapid advances in recombinant DNA technology and the increasing availability of peptides and proteins with therapeutic potential are a challenge for pharmaceutical scientists who have to formulate these compounds as products with optimal therapeutic effects and shelf lives. Conventional drug formulation has the same focus but, due to the unique structures of peptide and protein molecules, formulation of these compounds is more complex and challenging. The therapeutic application of peptides and proteins is limited by several problems, such as lack of physical and chemical stability, and the lack of optimal physicochemical properties for adequate transport across various biomembranes. Thus, the pharmaceutical scientists are faced with the challenge of formulating these drugs into safe, stable, and efficacious drug delivery systems.

This book focuses on general pharmaceutical development aspects of peptides and proteins rather than on design and therapeutic possibilities of advanced drug delivery systems. However, as pharmaceutical formulation is an interdisciplinary science, we believe that it is important to have a basic knowledge of related disciplines, i.e. peptide synthesis, recombinant DNA technology, and protein purification technology, as well as chemical and pharmaceutical aspects of a registration file, in order to understand and to be able to communicate with other disciplines integrated in the development process of biotechnological drugs.

This book is intended use for as a textbook in courses for pharmaceutical students at both undergraduate and graduate levels, and as a reference book for pharmaceutical scientists involved in peptide and protein formulation. The first three chapters deal with peptide synthesis, the basics in recombinant DNA technology, and protein purification. Chapter 4 provides an overview of analytical methods used for characterization of therapeutic proteins from both chemical and physical viewpoints. The stability aspect of drugs is a key concern for all formulation scientists. Chapter 5 addresses the most prevalent chemical pathways for degradation of peptides and proteins observed during production and storage. The physical stability of proteins is covered in Chapter 6. A general description of factors which can lead to protein aggregation as well as different approaches for physical stabilization of protein drugs is reviewed.

Chapters 7 to 11 discuss the various formulation principles for peptides and proteins. The challenge in formulation of parenteral suspensions is the topic of Chapter 7, which provides ideas and principles for pharmaceutical scientists involved in the development of parenteral peptide or protein suspensions. In Chapter 8, the emphasis is on the approaches used to solve formulation problems for peptides and proteins in solution. The chapter gives a practical guidance to formulation scientists working on solutions of peptides and proteins, including packaging and manufacturing aspects. Problems related to lyophilization of peptides and proteins are reviewed in Chapter 9, with a focus on the critical characteristics for obtaining stable dried protein products. A number of formulation principles for non-parenteral administration of peptides and proteins are covered in Chapter 10, including a case study related to immobilized enzyme intended for local effect in the GI tract. Examples of the use of the prodrug and analogue approach in optimization of peptide and protein drug delivery are given in Chapter 11, and in Chapter 12 the essential chemical and pharmaceutical documentation required in an application for a marketing authorization for a medicinal product in Europe is outlined, with a focus on biotechnological products.

The editors wish to thank all contributors for their valuable contributions that made this book possible. It is our hope that this book as a textbook and a reference for pharmaceutical scientists will contribute to the understanding, and possibly also to solutions, of formulation problems in the exciting world of pharmaceutical biotechnology.

Sven Frokjaer and Lars Hovgaard
Copenhagen

Peptide Synthesis

BERNARD A. MOSS

PolyPeptide Laboratories A/S, Hillerød, Denmark

1.1   Introduction

Peptides, which here include polypeptides and proteins, are biopolymers derived from the serial condensation of various natural amino acid monomers. Although more than 500 different amino acids occur in nature, those incorporated biosynthetically, i.e. ribosomally, into the growing peptide chain are selected mainly from only 20 which are genetically coded. These elementary amino acids have the general formula $RCH(NH_2)COOH$ wherein the amino function is located at C2, the α-carbon atom, giving the terminology 2- or α-amino acids. These molecules (except for glycine) are chiral with an asymmetric or stereocentre at C2, and are defined as having the S absolute configuration (in cysteine, however, the configuration is defined as R). Another, more traditional nomenclature uses the prefixes L and D, which relate all the (2S)-amino acids to (S)-2,3-dihydroxy-propanol (L-glyceraldehyde); all 20 genetically coded amino acids, with the exception of glycine, are L-enantiomers.

By convention, the amino end, or N-terminal, of the peptide chain is written on the left, while the carboxyl end, or C-terminal, is on the right. The chain incorporating the peptide bond is called the main chain and this provides a common core structure in peptides. To every third atom of the peptide chain is a substituent R-group or side chain: H- in glycine; alkyl or aryl groups in alanine, valine, leucine, isoleucine, phenylalanine; pyrrolidino (imino) group in proline; alcoholic or phenolic hydroxy-containing groups in serine, threonine, tyrosine; carboxy-containing groups in aspartic acid, glutamic acid and their carboxamide counterparts asparagine, glutamine; primary, higher order or heterocyclic amino-containing groups in lysine, arginine (guanidino), histidine (imidazole), tryptophan (indolyl); mercapto (thiol)- or sulphide-containing groups in cysteine, methionine. These side-chains give amino acid residues hydrophilic or hydrophobic properties: neutral/

hydrophobic/aliphatic in glycine, alanine, valine, isoleucine, leucine and methionine; neutral/ hydrophobic/aromatic in phenylalanine, tyrosine and tryptophan; neutral/hydrophilic in serine, threonine, asparagine and glutamine; acidic/hydrophilic in aspartic acid and glutamic acid; basic/hydrophilic in histidine, lysine and arginine. It is the particular combination of side-chains that gives each peptide its characteristic set of properties, with each functional group or ring system undergoing its own typical set of interactions and reactions.

Of all classes of bioactive macromolecules, peptides exhibit the widest structural and functional variation and are integral to the regulation and maintenance of all biological processes. For example, peptides can have regulatory roles as enzymes, antibodies, hormones, kinins, cytokines, neuropeptides/neurotransmitters that influence physiological functions as diverse as growth, reproduction, digestion, neurotransmission, blood pressure, inflammation, infection/immunity, cancer, and so on. As well as being characterized by high specificity and potency these diverse biopolymers can also undergo rapid metabolism, properties that are necessary for efficient, flexible physiological regulation. Several biological secretions and host defence/offence chemicals also comprise peptides, and include toxins (some cytotoxins are useful against human tumour cell lines), antimicrobial and antifungal agents, as well as hormones, neuropeptides and opioids.

The wide-ranging biological activities of peptides make them ideal starting points in the search for new therapeutic drugs, and research for this purpose has accelerated in recent years. Much of this increase in research activity is due not only to the remarkable progress in our understanding of molecular biology, immunology and enzymology, but also, importantly, to the dynamic progress in the synthetic technologies (chemical, and recombinant DNA/genetic engineering methods) crucial to rendering the peptides accessible for study, and the availability of sensitive bioassays, including appropriate cell types and cloned receptors. Equally important is the development of decisive methods for studying peptide and protein structure–activity relationships, the analytical technologies, such as NMR spectroscopy and mass spectroscopy, and not least the development of improved purification methods.

Small to moderately sized peptides such as bradykinins, enkephalins, gonadorelin (GnRH or LHRH), oxytocin and vasopressin and related peptides (agonist/antagonist analogues) have been chemically synthesized for medical purposes for several years. Longer peptides such as insulin, adrenocorticotropic hormones, calcitonin and secretin are also synthesized for medicinal therapy. The penetration of new peptide-based drugs into the therapeutic market is poised to advance rapidly in the next decade because of this vigorous research and development activity.

1.2 Chemical synthesis of peptides

The process of the serial condensation of the amino acid monomers into peptides essentially involves: (i) interaction of the α-carboxylic acid function of one monomer with the α-amino function of the next in a coupling reaction, (ii) loss of a water molecule, and (iii) formation of an amide link (the–CONH–peptide bond), to give a chain of covalently linked amino acid residues (Figure 1.1). The only other type of covalent link between amino acids in peptides is the disulphide bond formed between two cysteine residues in the same or different peptide chains by mild oxidation of their thiol side-chains. As more amino acids are condensed a peptide chain of any sequence and length is obtained.

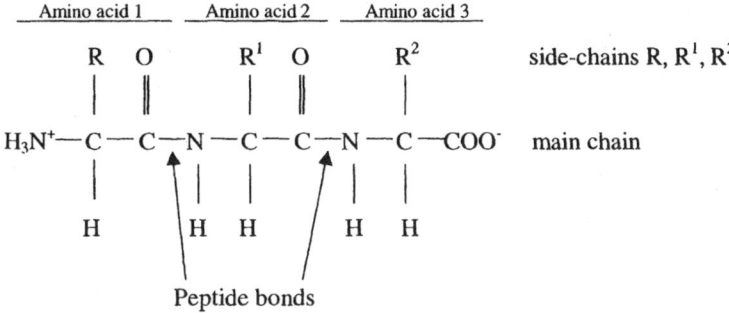

Figure 1.1 Structural representation of an amino acid and a peptide chain (a tripeptide).

Peptides can range in size and complexity from simple dimers containing two amino acid residues (dipeptides), through small peptides containing fewer than 10, to polypeptides containing from about 10 to around 50, and eventually to larger polypeptides (proteins) containing 50 to 5000 or more amino acid residues.

Most early research on the chemical synthesis of peptides had as its objective the preparation of compounds identical with naturally occurring ones. For this purpose methods were needed to enable the orderly linkage of the L-amino acids (and glycine) into peptide chains of predetermined sequence and length. To achieve peptide bond formation activated amino acid (or peptide) precursors were necessary. Since no satisfactory procedure was found to activate the α-amino component, reactive derivatives of the α-carboxyl group were developed. The essential reaction in the chemical synthesis of a peptide thus involved the acylation of the amino group of one amino acid by the carboxyl group of another, culminating in peptide bond formation. In this reaction the activated carboxyl function was prevented from acylating its own α-amino group by temporarily attaching to it a chemical blocking group. This protecting group was subsequently removed from the nascent peptide chain in readiness for elongation in stepwise fashion with another α-amino protected, α-carboxyl activated amino acid. The presence of diverse side-chain functional groups in the various amino acids, and the need to maintain the chiral integrity of the α-carbon stereocentre during coupling, complicated the peptide synthesis process. Suitable side-chain protection or the judicious selection of reaction conditions for less problematic groups, and careful choice of activating group and coupling reaction were necessary to assure syntheses of very high fidelity and efficiency, while avoiding potential side-reactions that generated unwanted by-products. Of the 20 common amino acids, only alanine and leucine appear to be generally free from specific side-reactions involving side-chains, and C-terminal glycine and proline are essentially free from racemization in the coupling process, making them good coupling points in peptide segment coupling strategies.

The basic method for the rational chemical synthesis of peptides has in essence now been solved. The principles and practice of peptide synthesis have been well documented; some papers are given in the reference section both as recommended reading and as sources for original references.

1.2.1 *Solution and solid phase peptide synthesis*

Two general synthetic approaches are available to peptide manufacturers: the classical organic synthesis methods in solution phase that have evolved since the turn of the twentieth century, and the solid phase alternatives established and elaborated since 1959. These approaches are not mutually exclusive. Solution phase synthesis has application in laboratory-scale research, but is particularly useful in the large-scale manufacturing of peptides (from tens of grams to tens of kilograms or higher), generally up to 40 residues in length. Although it is more cost-effective and environmentally sound than the solid phase method, it is both time-consuming and labour intensive, especially in the early stages of development. This is due to the need for process optimization (reaction conditions, yields and purification procedures for essentially all intermediates as well as finished product) and validation to assure the final identity and quality of product. In contrast, peptides for research purposes are more accessible and rapidly available by the solid phase approach, which retains and extends chemistry proven in solution phase (protection schemes, reagents). The fundamental premise of this method is that amino acids can be assembled stepwise into a peptide of any desired sequence by a series of addition cycles involving deprotection/coupling while one end of the growing chain is anchored to a polymeric support (usually insoluble, but soluble polymers can provide advantages in solubilizing difficult sequences). All the reactions involved in the synthesis can be driven to completion by the use of excess soluble reagents, which together with unwanted reaction by-products are removable by simple washing and filtration. Laborious optimizations of reaction conditions and purifications at intermediate stages in the synthesis are effectively minimized or circumvented. The method is also readily amenable to automation due to the speed and simplicity of the repetitive steps, which are carried out in a single reaction vessel at ambient temperature. The technique lends itself to batch and continuous flow synthesis ranging in scale from sub-mg to tens or hundreds of grams or even to kilograms; moreover, rapid synthesis on polymer pins or single beads has had a profound impact on drug discovery and design, particularly in the field of combinatorial chemistry and the synthesis of peptide libraries (Novabiochem, 1997, 1997/98). Once the desired sequence of amino acids has been obtained, a reagent is used to cleave the chain from the support, concomitant (if desired at this stage) with liberation of protecting groups, under conditions of minimal damage to the crude product. The type of chemistry used in the original linkage of the C-terminal residue to the polymeric support and the type of cleavage reagent selected will determine whether a C-terminal acid, amide, or other functional type will be present in the cleaved peptide. Finally, any remaining protecting groups are appropriately removed and the crude product purified and characterized to assure it has the correct identity.

For peptide synthesis the C-terminal carboxylic acid is activated by conversion to an acylating agent using the acid halide (e.g. chloride) method, the acid azide method, the anhydride method (as symmetrical and mixed anhydrides) or the active ester method (Bodanszky, 1993), and used preformed or *in situ* to react with the α-amino group for peptide bond formation. *N,N'*-dicyclohexylcarbodiimide (DCC) is the archetype *in situ* coupling reagent and may be used alone or in combination with 1-hydroxybenzotriazole (HOBt), 3-hydroxy-1,2,3-benzotriazin-3(4H)-one, N-hydroxysuccinimide, other succinimides or oximes. Examples of active esters are *p*-nitrophenol, pentafluorophenyl, and N-hydrosuccinimidyl, while the most successful mixed anhydrides are those generated with the help of alkyl chloroformates (e.g. *iso*butylchloroformate). As well as having an active ester function, HOBt suppresses racemization of chiral centres. Other racemization suppression additives, such as cupric chloride, may also be included in coupling reactions under special circumstances. The newer phosphonium-based coupling reagents HBTU, 2-(1H)-benzotriazole-1-yl-1,3,3-tetramethyluronium hexafluorophosphate, and TBTU, 2-(1H)-benzotriazole-1-yl-1,3,3-tetramethyluronium tetrafluoroborate, designed for use in both solution phase and solid phase peptide synthesis enable smooth, efficient couplings with very low racemization; added HOBt improves the process.

Protecting groups for both amino and carboxyl groups, as well as for the side-chain functional groups of the various amino acids, have been developed (Novabiochem, 1997/98); variants continue to be designed and perfected in an evolving process. Such protecting groups must be easily introduced into the amino acid or peptide, be able to protect the functional group under conditions of peptide bond formation, be selectively removed according to the stage of synthesis, and leave the nascent peptide undamaged under conditions of removal. Functional groups, therefore, are modified with a combination of temporary and semi-permanent protecting groups. Many such protecting groups are known and are commercially available (Greene and Wuts, 1991; Novabiochem, 1997/98). The α-amino function is temporarily protected by an acid sensitive group (e.g. N-α-*tert*-butyloxycarbonyl (Boc), 1-methyl-1-(4-biphenyl)ethoxycarbonyl (Bpoc), 1-adamantyloxycarbonyl (Adoc), 2-nitrophenylsulfenyl (Nps)), or a base sensitive group (e.g. 9-fluorenyl methyloxycarbonyl (Fmoc), trifluoroacetyl (Tfa), or a hydrogenation (e.g. benzyloxycarbonyl (Cbz or Z)), or a photolytically sensitive group, or by groups transformable into labile protecting groups. Cbz, Boc and Fmoc are the most widely used and commercially viable α-amino protecting groups. Side-chain amino groups are similarly protected, but it is usual to choose those with the property of orthogonality wherein one protecting group is retained on the peptide, i.e. is more permanent, while another is selectively removed. The most favoured C-terminal and side-chain carboxy protecting groups include benzyl, methyl, 9-fluorenylmethyl, ethyl, and allyl. These are also chosen to have orthogonality with the other protecting groups in the synthesis as desired. Although side-chain protection of cysteine, aspartic and glutamic acids and lysine during syntheses is mandatory, not all sensitive amino acids require side-chain protection every time, but each synthesis requires an informed decision based on the length and sequence of the target peptide and other considerations. Side-chain derivatives of all sensitive residues (with many different protecting groups, and which are compatible with Boc or Fmoc chemistry) are available as the need arises and these should satisfy any synthesis strategy. Either Boc chemistry with benzyl (Bzl)-based side-chain protection strategy or Fmoc chemistry with *tert*-butyl (*t*-Bu)-based side-chain protection is generally used in solid phase peptide synthesis, while for solution phase synthesis Boc and Cbz chemistry are preferred. The ready availability of reagents, knowledge of their properties, reaction conditions and accumulated peptide synthesis experience allows many strategies to be adopted in the synthesis of different peptides and even of the same peptide. Strategies may range from having full protection in solid phase synthesis to minimal protection in solution phase. Other important developments in peptide synthesis are the application of proteases (enabling minimal protection) for the formation of the peptide bond in aqueous, organic–aqueous or organic media, and the enzymatic manipulation of protecting groups when used (Jakubke, 1987; Glass, 1987). A commonly used strategy for efficiently elongating peptides and procuring the target peptide in purer form is adoption of convergent synthesis. This process refers to the synthesis of two or more peptide intermediates of desired length (segments), their purification or partial purification, and their coupling under such conditions that the target peptide has minimal or no racemization at chiral centres. Convergent synthesis may be used to couple segments serially in any suitable order. The process of obtaining a chemically synthesized peptide for therapeutic application involves design of the peptide, its chemical synthesis, and an evaluation–purification cycle to demonstrate its efficacy in an experimental situation (Grant, 1992).

Despite considerable progress in peptide synthesis methodologies, each synthesis inevitably is not perfect so that the formation of by-products, and hence the need for purification of both intermediates and final products, is a continuing challenge to be overcome in order to secure pharmaceutical peptides, especially those demanding greater than 98% purity.

1.2.2   *Large-scale peptide synthesis*

Industrial-scale peptide synthesis is governed by different factors to those in research. For economical reasons the number of chemical steps is kept to a minimum and the multi-step synthesis is simultaneously commenced at different starting points. Thus the method of choice uses the concept of convergent synthesis with minimal protection and segment condensation. After synthesis the peptides are subjected to purification, drying and characterization. The dry, purified peptides are often used directly as active ingredients in final pharmaceutical formulations. As such they are subject to strict quality control measures that impose limits on the amounts of related impurities, decomposition products, residual solvents/reagents (including heavy metals) and even the choice of counter-ions.

An example of the bulk solution phase synthesis of a peptide is the cyclic nonapeptide oxytocin. The first synthesis of this hormone together with the synthesis of the related hormone vasopressin by du Vigneaud in the early 1950s laid the foundation of the solution phase approach; oxytocin was made wholly stepwise in the C → N direction. Since these pioneering studies, many different strategies have been used to make these molecules and literally thousands of analogues. Oxytocin is produced and secreted by the posterior lobe of the mammalian pituitary gland. It was formerly isolated for investigational and clinical use from this source, but nowadays is produced by chemical synthesis as a freeze-dried powder that can be formulated as solutions for injection. The pharmacopoeial specification for the HPLC purity of synthetic oxytocin for human use is not less than 95%, with no single related impurity more than 1.5%, and its oxytocic activity not less than 560 international units (IU) per mg of peptide, whereas that for veterinary purposes is less stringent, with ~90% and 300 IU, respectively, being acceptable.

Figure 1.2 depicts one way of representing the oxytocin molecule. The amino-terminal is a half-cystine (cysteine, residue 1) paired by intramolecular disulphide linkage to another half-cystine (cysteine, residue 6) to form a hexapeptide ring which is attached to a carboxyl-terminal tripeptide tail ending with glycine amide (residue 9). For full biological activity both ring- and tail-structures are required.

Scheme 1.1 gives one of the many possible synthesis strategies that can be used. This approach uses segment condensation. Briefly, manufacture in kg scale was based on the simultaneous stepwise production of two segments by conventional solution phase

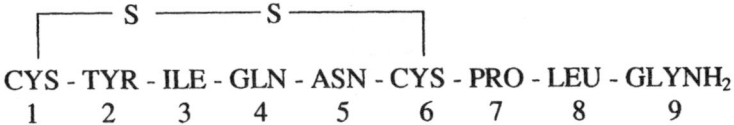

Figure 1.2 Structural representation of oxytocin.

Scheme 1.1 Synthesis strategy for bulk oxytocin manufacture

A   Stepwise synthesis of two segments, Boc1-4OH and Boc5-9NH$_2$, both with [Acm] protected cysteines (A 4 + 5 strategy; other schemes might use a different combination of segments and different cysteine protection)

B   Conversion of Boc5-9NH$_2$ to 5-9NH$_2$ by acidolytic Boc removal

C   Coupling of Boc1-4OH and 5-9NH$_2$ to produce the protected intermediate Boc[diAcm]1-9NH$_2$, i.e. Boc[diAcm]oxytocin

D   Acidolytic Boc removal from Boc[diAcm]oxytocin to obtain linear H[diAcm]oxytocin

E   Ring closure of linear H[diAcm]oxytocin, involving [Acm] removal with concomitant disulphide bridging of the two cysteines

---

methods: a tetrapeptide and a pentapeptide. This approach uses Boc and Cbz chemistry for N-terminal protection of amino acids/peptides, in conjunction with minimal side-chain protection. The phenolic hydroxyl group of tyrosine and the primary carboxamide side-chains of glutamine and asparagine were unprotected, whereas the thiols of the two cysteines were [Acm] protected. The N-terminal tetrapeptide Boc1-4OH (with a free carboxyl end) and the C-terminal pentapeptide 5-9NH$_2$ (with an amidated carboxyl end) were first prepared via active esters generated by HOSu/DCC and HOBt/DCC. Commercially available glycine-amide and Cbz protected leucine and proline were used in segment 5-9NH$_2$. A sequence of couplings, precipitations, extractions, deprotections (catalytic hydrogenations), filtrations/centrifugations of the intermediates with industrial equipment, such as 200 l reactors or pressure vessels, 75–100 cm diameter vacuum filters or centrifuges, was used in work-up. Catalyst was removed after hydrogenation by pressure filtration over a cellulose bed. Reactions were monitored by analytical reverse phase HPLC and thin layer chromatography to assure completeness of reactions (< 0.1% of the limiting species present). The remaining amino acids of the sequence were Boc derivatives and when appropriate the Boc group was removed by acidolysis with HCl/acetic acid or TFA/acetic acid. A similar series of reaction monitorings and work-up treatments, as well as process reverse phase HPLC on C-18 (10–20 l resin beds; kg scale), was used to obtain the intermediates. Control of the quality of the key intermediates Boc1-4OH, 5-9NH$_2$, Boc (diAcm)1-9NH$_2$ and H (diAcm)1-9NH$_2$ are crucial to achieving final product purities of 98% or more. The linear nonapeptide molecule was assembled by condensation of the two segments using TBTU as coupling reagent. A problem in segment condensation is racemization at the non-protected carboxyl-terminal amino acid of the segments being coupled; here, formation of the D-Gln^4-nonapeptide. In this case TBTU was chosen for *in situ* formation of HOBt active ester because of its high coupling efficiency, usually with few or no undesired side-reactions. Following isolation of the linear nonapeptide by process scale reverse phase HPLC on C-18, the intramolecular disulphide bridge was installed by methanolic iodine treatment of the [Acm] protected cysteine thiols. The oxytocin so formed was subsequently purified by a combination of ultrafiltration/reverse osmosis with a suitable molecular size cut-off membrane and chromatography (anion exchange and C-18 process reverse phase HPLC).

The oxytocin manufactured according to Scheme 1.1 could contain a variety of impurities. These may result from the synthesis process as peptide-related substances and as residual process contaminants (solvents and chemicals used during synthesis and purification).

Peptide-related impurities can hypothetically contaminate the final product because of their presence or generation in various stages of the manufacturing process, then being carried adventitiously through the final purification steps. Analytical reverse-phase HPLC is used to monitor the progression and product purity at each step in the peptide synthesis. Provided the HPLC system is suitable for its task and is properly controlled, this gives good assurance that deviations from the planned synthesis can be detected early and appropriately corrected.

The main oxytocin-related peptide impurities that could be expected to arise during the various synthesis processes are as listed below and summarized in Figure 1.3.

1   Peptides containing diastereoisomers, i.e. peptides where one or more amino acids are racemized.

2   Peptides altered by deamidation, e.g. deamidation of glycine amide, asparagine and glutamine, during manufacture or storage of the finished product.

3   Peptides altered by other side-chain modifications, e.g. chemical changes involving asparagine, glutamine, tyrosine and glycine amide, and perhaps cysteine, e.g. by oxidation to cysteic acid.

4   Peptides generated by incomplete deprotection of N-terminal and side-chain protecting groups, or incomplete ring closure of the deprotected cysteines.

5   Peptides containing dimers or higher oligomers generated by interchain disulphide bonds involving cysteine during ring closure or subsequent steps, and/or peptide aggregation.

6   Peptides lacking one or more amino acids, or containing extra amino acids from the synthesis itself or originating from amino acid/peptide intermediates.

7   Peptides altered by sequence inversion (involving glycine as a third residue).

The persistence of solvents and chemicals used in the synthesis, particularly from the final segment condensation and ring closure steps, could contribute to a decrease in product purity. Such impurities are avoided or minimized by the purification steps, which involve process reverse-phase HPLC, ion-exchange chromatography, precipitations/extractions, evaporations, ultrafiltration/reverse osmosis and freeze-drying.

Based on the synthesis strategy used and experience from other peptide researchers using actual amino acid/peptide couplings and model systems, it is possible to surmise the outcome of oxytocin synthesis concerning the generation of potential impurities. These are outlined in Figure 1.3. These possibilities were taken into account and, where appropriate, reference samples were synthesized by the solid phase method to check whether such derivatives were actually present in the final product. Although laborious and time-consuming, the elucidation of related peptide impurities during development of an industrial process helps to avoid them, thereby leading to a purer end-product.

Early development work showed a purity of only 75% for the crude oxytocin after ring closure, with a leading peak up to 10% of total and trailing peaks up to 25% (polymers 10%, others 15%). Following purification the purity increased to 93%, with 5% front peaks, 2% back peaks and no polymers. From comparative analytical data involving the

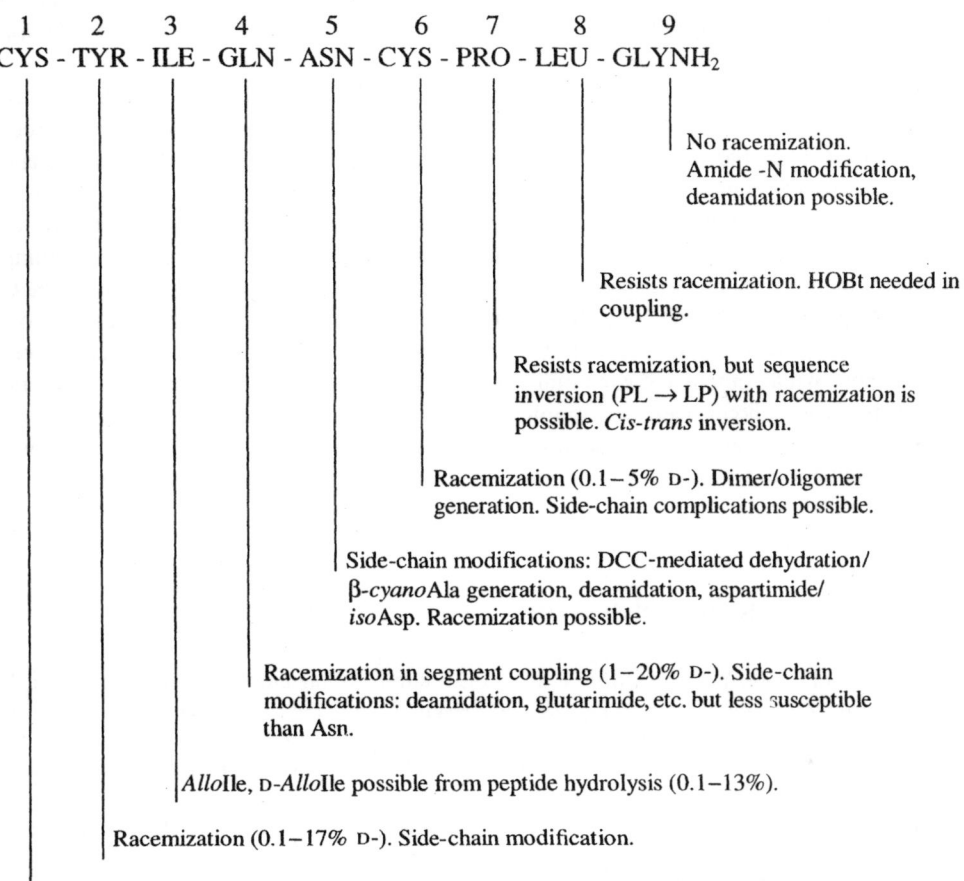

```
        1       2       3       4       5       6       7       8       9
      CYS  -  TYR  -  ILE  -  GLN  -  ASN  -  CYS  -  PRO  -  LEU  -  GLYNH₂
```

No racemization.
Amide -N modification,
deamidation possible.

Resists racemization. HOBt needed in
coupling.

Resists racemization, but sequence
inversion (PL → LP) with racemization is
possible. *Cis-trans* inversion.

Racemization (0.1–5% D-). Dimer/oligomer
generation. Side-chain complications possible.

Side-chain modifications: DCC-mediated dehydration/
β-*cyano*Ala generation, deamidation, aspartimide/
*iso*Asp. Racemization possible.

Racemization in segment coupling (1–20% D-). Side-chain
modifications: deamidation, glutarimide, etc. but less susceptible
than Asn.

*Allo*Ile, D-*Allo*Ile possible from peptide hydrolysis (0.1–13%).

Racemization (0.1–17% D-). Side-chain modification.

Racemization (0.1–5% D-). Dimer/oligomer generation. Side-chain complications
possible.

Figure 1.3 Outline of possible peptide by-products of oxytocin synthesis.

prepared references, using HPLC, capillary zone electrophoresis, amino acid analyses, sequence analyses and mass spectral analysis, several of the possible impurities were excluded. In fact the impurities were deduced as arising from two separate types of reaction: (i) racemization of L- to D-glutamine during segment coupling, (ii) removal of Boc from cysteine (Acm)-containing peptides with unexpected Acm cleavage/transfer side-reactions. From this knowledge coupling reaction conditions were modified such that racemization was reduced to less than 5% and was essentially removed by chromatographic purification (reverse phase C-18 HPLC and ion exchange chromatography). Although more complex to solve, conditions for controlling the generation and removal of the (Acm)-containing oxytocin impurities were identified. In these reactions the tyrosine phenolic group, and the carboxamide groups of glutamine, asparagine and glycine–amide were targets for the liberated Acm carbo-cation. By changing reaction conditions in Boc removal to include scavengers, the Tyr[2] (Acm)oxytocin was eliminated and the other (Acm) derivatives were reduced to about 1% in total, giving oxytocin with a purity of 99%. The lability of (Acm) under the acidolytic conditions generally used for Boc removal in peptide synthesis was quite unexpected, but similar phenomena have been reported by others (Mendelson *et al.*, 1990; Lamthanh *et al.*, 1993).

1.3 Concluding remarks

The application of native peptides as therapeutic drugs is hampered by several disadvantages, including their proteolytic instability, polar nature (making penetrating membranes and the blood–brain barrier problematic), and rapid elimination from the recipient. Therefore, the direction of medicinal chemists in peptide design and synthesis is to modify the peptide structure into bioactive compounds with improved properties. Structure variations of the native peptides, which includes cyclization, can range from changing the amino acid side-chains or the peptide main chain into peptide analogues, and backbone mimetics, to peptide mimetics wherein only the electronic and steric aspects of the parent peptide which proved to be essential for activity are retained. Such modifications can render the peptide more stable, potent and even orally active. For example, simple chemical modification of the peptide bond itself, or completely replacing it with a surrogate, could protect the peptide against enzymatic degradation, increasing the peptide's half-life and bioactivity. Modern peptide-based drug design is undoubtedly ushering in a challenging new era for the synthetic chemist.

References and additional sources

BENOITON, N.L., LEE, Y.C. and CHEN, F.M.F., 1993, Racemisation during aminolysis of activated esters, *International Journal of Peptide and Protein Research*, **41**, 512–516.

BENOITON, N.L., LEE, Y.C., STEINAUR, R. and CHEN, F.M.F., 1992, Studies on sensitivity to racemisation of activated residues in couplings of N-benzyloxycarbonyldipeptides. *International Journal of Peptide and Protein Research*, **40**, 559–566.

BODANSZKY, M., 1993, *Peptide Chemistry, a Practical Textbook*, 2nd edn, Berlin: Springer-Verlag.

BODANSZKY, M. and BODANSZKY, A., 1994, *The Practice of Peptide Synthesis*, 2nd edn, Berlin: Springer-Verlag.

BODANSZKY, M. and MARTINEZ, J., 1983, Side reactions in peptide synthesis. *The Peptides, Analysis, Synthesis, Biology*, Vol. 5, edited by E. GROSS and J. MEINHOFER, pp. 111–216, New York: Academic Press.

DUNN, B.M. and PENNINGTON, M.W., 1996, *Peptide Analysis Protocols*, Vol. 36, New York: Eaton Publishing.

GLASS, J.D., 1987, Enzymatic manipulation of protecting groups in peptide synthesis. *The Peptides, Analysis, Synthesis, Biology*, Vol. 9, edited by S. UDENFRIEND and J. MEINHOFER, pp. 167–184, New York: Academic Press.

GRANT, G.A., 1992, *Synthetic Peptides, a User's Guide*, New York: W.H. Freeman and Company.

GREENE, T.H. and WUTS, P.G.M., 1991, *Protective Groups in Organic Synthesis*, 2nd edn, New York: Wiley.

GROSS, E. and MEIENHOFER, J., 1979, Major methods of peptide bond formation. *The Peptides, Analysis, Synthesis, Biology*, Vol. 1, pp. 1–435, New York: Academic Press.

1981, Protection of functional groups in peptide synthesis. *The Peptides, Analysis, Synthesis, Biology*, Vol. 3, pp. 1–379, New York: Academic Press.

JAKUBKE, H.-D., 1987, Enzymatic peptide synthesis. *The Peptides, Analysis, Synthesis, Biology*, Vol. 9, pp. 103–165, New York: Academic Press.

LAMTHANH, H., ROUMESTAND, C., DEPRUN, C. and MENEZ, A., 1993, Side reaction during the deprotection of (S-acetamidomethyl-cysteine in a peptide with a high serine and threonine content, *International Journal of Peptide and Protein Research*, **41**, 85–95.

MENDELSON, W.L., TICKNER, A.M., HOLMES, M.M. and LANTOS, I, 1990, Efficient solution phase synthesis of (1-(β-mercapto-β, β-cyclopentamethylenepropionic acid)-2-(O-ethyl-D-tyrosine-4-valine-9-desglycine)arginine vasopressin. *International Journal of Peptide and Protein Research*, **35**, 249–257.

MIYAZAWA, T., OTOMATSU, T., YAMADA, T. and KUWATA, S., 1992, Racemization studies in peptide synthesis through the separation of protected epimeric peptides by reversed-phase high performance liquid chromatography. *International Journal of Peptide and Protein Research*, **39**, 229–236.

NOVABIOCHEM, 1997, *The Combinatorial Chemistry Catalog*, solid phase organic synthesis notes, S1-S46, Calbiochem-Novabiochem AG, Laufelfingen, Switzerland.

NOVABIOCHEM, 1997/98, *Catalog and Peptide Synthesis Handbook*, synthesis notes, S1-S88, Calbiochem-Novabiochem AG, Laufelingen, Switzerland.

PENNINGTON, M.W. and DUNN, B.M., 1996, *Peptide Synthesis Protocols*, Vol. 35, New York: Eaton Publishing.

SEPETOV, N.F., KRYMSKY, M.A., OVCHINNIKOV, M.V., BESPALOVA, Z.D., ISAKOVA, O.L., SOUCEK, M. and LEBL, M., 1991, Rearrangement, racemization and decomposition of peptides in aqueous solution. *Peptide Research*, **4**, 308–313.

2

Basics in Recombinant DNA Technology

NANNI DIN[1] AND JAN ENGBERG[2]

[1]*Department of Molecular Genetics, Novo Nordisk A/S, Bagsvaerd, Denmark*
[2]*Department of Pharmacology, The Royal Danish School of Pharmacy, Copenhagen, Denmark*

2.1   Introduction

Gene technology is one of several terms used to describe the methods which have evolved as a result of research on the structure and function of genes. Other terms are 'recombinant DNA (rDNA) technology', 'genetic engineering' and 'gene splicing'. Gene technology involves taking genetic material from one source and recombining it *in vitro* with genetic material from another source followed by introduction of the recombined material into a host cell.

By the aid of genetic engineering it is possible to introduce a gene coding for a desired protein into a biological environment where this protein can be produced in large quantities. This possibility has been one of the main reasons for the interest of the pharmaceutical industry in applying genetic engineering methods. Using these methods, manufacturers can in principle produce unlimited quantities of biologically active proteins and peptides, as has already been convincingly demonstrated in many cases. In 1982, the first approved human therapeutics produced by the aid of gene technology appeared. This landmark drug was insulin, a peptide hormone used to treat diabetes. At the present time (1999), more than 20 different peptides or proteins produced by gene technology methods have been approved as human therapeutics, and more are certain to follow. In contrast to insulin, for which a production method based on extraction from the pancreas of pig or ox had existed for many years, some of the other approved gene technology products have never been in regular use as drugs before, because they were impossible to produce in sufficient quantities. Thus, one of the advantages of gene technology is the possibility of producing substances that have hitherto been so elusive as to make their use as pharmaceuticals remote.

Another advantage of gene technology is that it allows introductions of modifications into proteins at desired positions. The modifications may be a few changes in the amino acid sequence, made by introducing specific mutations in the gene encoding the protein. Such changes may lead to desirable altered physicochemical characteristics, e.g. new solubility properties or a different physiological half-life of the active polypeptide. However, more profound changes giving rise to novel activities are also possible.

This chapter aims at describing the basic tools and concepts of gene technology as used in pharmaceutical research and production, and sums up the current status and future trends for recombinant protein therapeutics.

2.2   General methods in gene technology

2.2.1   *DNA cloning tools*

At the beginning of the 1970s, many new methods of significant importance for recombinant DNA technology were developed. Since then the methods have improved and new ones have emerged, but the principles are still based on the same tools: DNA modifying enzymes, DNA vectors and host cells (Sambrook *et al.*, 1989).

A key factor in the breakthrough of the 1970s was the discovery of two types of enzymes, the restriction enzymes and the DNA ligases. Restriction enzymes are endonucleases that recognize and cleave DNA at specific sequences, typically 4–8 base pairs (bp) long. As a result, DNA fragments of defined sizes are generated, some of which may contain a single complete gene. Restriction endonucleases can be isolated from a variety of microorganisms, and a large range of enzymes, each with a specific recognition sequence, is now commercially available. An example is the restriction enzyme *BamHI* (endonuclease I isolated from *Bacillus amyloliquefaciens H*) which cleaves DNA at the sequence GGATCC. In contrast to the restriction enzymes, DNA ligases have no specific sequence requirement but are capable of joining together compatible ends of DNA molecules. These features of the restriction enzymes and DNA ligases make it possible to perform *in vitro* cleavage and rejoining ('splicing') of DNA fragments as illustrated in Figure 2.1.

In order to clone a DNA molecule it must be introduced into a host cell in a form which allows it to be replicated. Replication requires the presence of a specific sequence (replication origin) which can be provided by circular 'minichromosomes' called plasmids. Therefore, to generate clonable DNA, a restriction fragment is joined to a plasmid DNA molecule which can replicate in a suitable host cell. A very important host–plasmid system is the bacterium *Escherichia coli* and its naturally occurring small (2–10 kbp) plasmids. Some of these plasmids have been modified so that they are particularly useful for cloning purposes, and they are hence called vectors. Apart from the replication origin, an important feature of the vector is the presence of a gene coding for a selectable marker, such as resistance to an antibiotic. When recombinant plasmids are introduced into *E. coli* (in a process called transformation) the resistance gene will allow the cells to

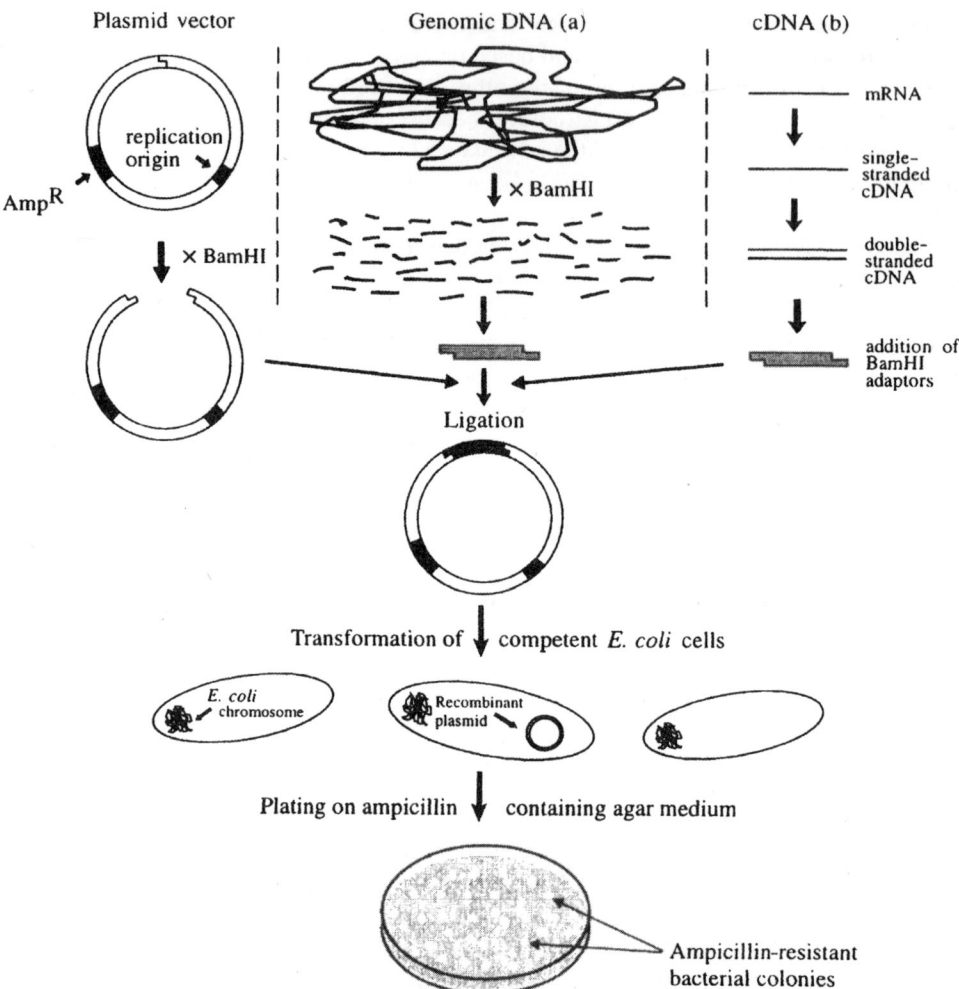

Plasmid vector   Genomic DNA (a)   cDNA (b)

Figure 2.1 Overview of the steps involved in the cloning of (a) chromosomal DNA and (b) cDNA fragments into a plasmid vector. The plasmid vector carries a replication origin and a gene which confers resistance to the antibiotic ampicillin (*Amp^R*). Outside these regions, a unique site for the restriction endonuclease (*BamHI*) is present. In order to clone DNA fragments into this site, vector and foreign DNA are cleaved with *BamHI*, mixed and joined together with the aid of the enzyme DNA ligase. The mixture of ligated DNA molecules is transformed into *E. coli* cells. Only cells which have received a plasmid will grow in the presence of ampicillin. Generation of double-stranded cDNA from eukaryotic mRNA requires the following steps. A chemically synthesized oligo-dT primer is annealed to the 3′ polyadenylated tails of the mRNA molecules. Single-stranded cDNA is then synthesized by the aid of reverse transcriptase. Following removal of the mRNA molecules by hydrolysis with sodium hydroxide to give a population of single-stranded cDNA molecules, these are used as templates in a new round of DNA synthesis catalysed by the enzyme DNA polymerase. Synthetic DNA adaptors for *BamHI* are subsequently ligated to the double-stranded cDNA molecules and the resulting fragments can be inserted into the *BamHI* site of the vector.

grow in the presence of the antibiotic, while cells without the plasmid will be eliminated. A recombinant plasmid is able to multiply until each cell contains several copies (often in the range from 20 to 200), and as the doubling time for *E. coli* under optimal conditions is 20–30 minutes, it is possible to multiply a single copy of the recombinant plasmid to more than 10^{10} molecules within 12 hours, thereby making further manipulation of a particular gene possible.

2.2.2 Cloning of cDNA

Cloning and expression of human genes, which are of special interest within pharmaceutical contexts, present certain special challenges. This is because most human genes (in contrast to bacterial genes) contain internal regions, called introns, that do not code for any part of the corresponding protein. Introns can range in size from less than 100 bp to several kbp, and mammalian genes can be very large due to the presence of several introns. In human and other eukaryotic cells, intron-containing RNA are processed to generate mature mRNA without introns through processes that bacteria cannot perform. When human chromosomal DNA (also called genomic DNA) is cloned in *E. coli* (Figure 2.1a), each individually cloned restriction fragment rarely contains a complete gene and in most cases intron sequences will be present. Therefore, while cloning of chromosomal DNA in *E. coli* is very useful for characterization of human gene structure, it is less useful as a starting point for expression of a protein by a particular gene.

Fortunately, a method has been devised that allows cloning of a DNA fragment containing only the coding regions of a human gene: by an enzymatic *in vitro* reaction using an enzyme called reverse transcriptase, it is possible to make a DNA copy of the mature mRNA. This DNA copy (called cDNA for copy DNA or complementary DNA) contains an uninterrupted coding region which can be cloned and expressed in *E. coli* (Figure 2.1b).

To generate cDNA coding for a protein of interest, mRNA is isolated from specific tissues or organs known to produce this protein. The isolated mRNA consist of a population of molecules representing all the genes expressed in the particular tissue, and cloning of the complete cDNA population results in a so-called tissue-specific cDNA library. From this library, clones that contain the desired cDNA must be identified and isolated before further analysis. This can sometimes be a formidable task, since only a small proportion of the cDNA molecules generated by reverse transcription corresponds to the protein of interest. A standard identification method is the DNA hybridization technique illustrated in Figure 2.2. This method requires some knowledge about the sequence of the desired cDNA, e.g. a partial amino acid sequence of the encoded protein. Based on a known amino acid sequence, a short DNA fragment with the corresponding coding capacity can be designed and synthesized in a radioactive form. This radioactive DNA probe is able to bind (hybridize) to the desired cDNA through base pairing, and thereby give rise to a signal from the desired clone. If no sequence information is available, other methods must be used which usually require that the protein encoded by the cDNA becomes expressed by the host cell. If a specific antibody against the desired protein is available, a screening procedure similar to the hybridization technique outlined in Figure 2.2 can be used, except that the antibody is used instead of a DNA probe. In such cases, the detection signal (radioactivity or light emission) is normally produced by a secondary antibody that recognizes the specific antibody.

In some cases the biological activity of a protein may be the only means by which to identify the cDNA clone encoding it. Thus, it becomes necessary to use host cells which

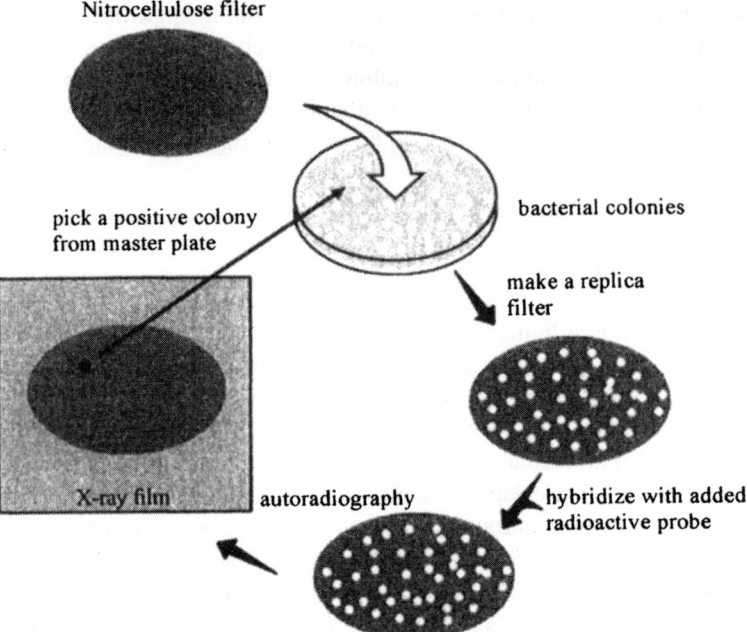

Nitrocellulose filter

pick a positive colony
from master plate

bacterial colonies

make a replica
filter

X-ray film   autoradiography

hybridize with added
radioactive probe

Figure 2.2 Overview of the steps involved in the screening of a plasmid library with a radioactive DNA probe. Plasmid libraries are typically screened by spreading several thousand plasmid-containing cells on each of a series of agar plates. After the bacterial colonies have grown to visible size, nitrocellulose filters are carefully laid onto the surface of the plates. Colonies from the plate adhere to the filter, creating a precise replica on the filter of the colonies on the plate. The filters are treated to solubilize the cells and immobilize their content of DNA to the filter surface. The filters are now incubated in a solution containing a radioactively labelled DNA probe complementary in sequence to a portion of the desired cloned gene. The filters are carefully washed to remove unbound probe, leaving behind only the probe molecules specifically bound to complementary sequences. The location of the bound probe is determined by exposing the filters to X-ray film (autoradiography). The position is represented by a spot of exposure on the film. By orienting the film with the original agar plate, the colony harbouring the desired clone can be identified and the gene isolated.

preserve this activity. An obvious choice is mammalian cells grown in culture, and many different mammalian host–vector systems now exist for such purposes. Other host cells, notably frog oocytes, have also proven very versatile in their ability to express complex functional proteins, such as mammalian membrane receptors or ion channels.

2.2.3 *PCR cloning and DNA database mining*

Since its invention in the early 1980s, the polymerase chain reaction (PCR) has become increasingly important as a cloning tool (Innis *et al.*, 1990). The PCR technique is based on enzymatic *in vitro* amplification of a specific DNA sequence. Figure 2.3 illustrates how a sequence, often called the target or template DNA, can be amplified from a population of single-stranded cDNA.

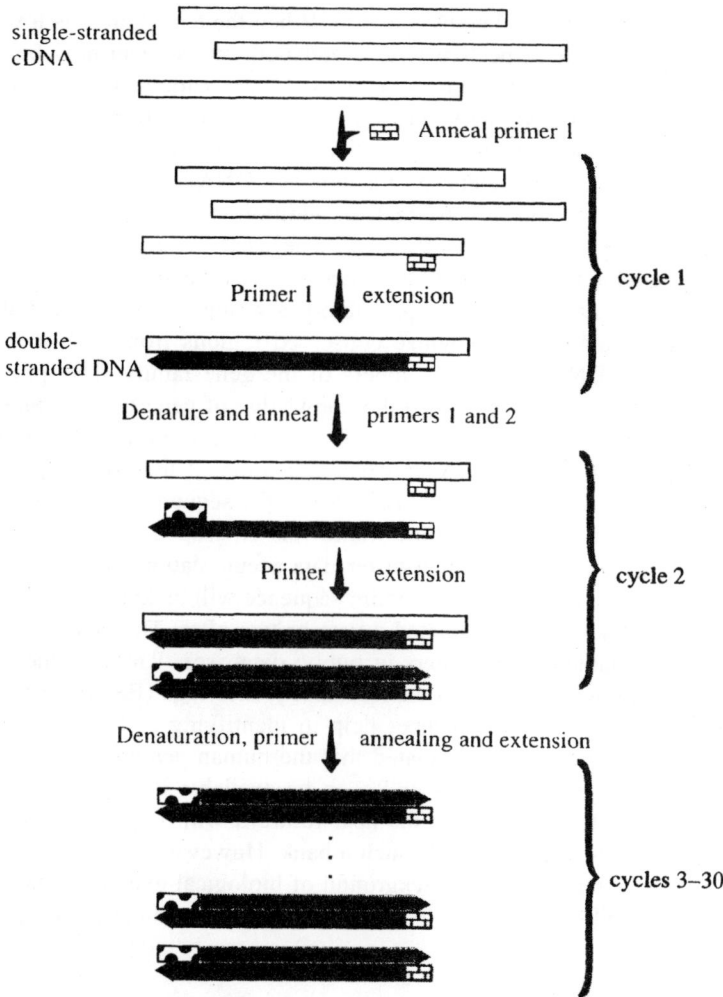

Figure 2.3 The polymerase chain reaction using single-stranded cDNA as template. The reaction mixture is heated and primer 1 is annealed to the mixture of single-stranded cDNA molecules. Added *Taq* polymerase synthesizes a new strand of DNA complementary to the template and extends to the end of the molecule. The reaction mixture is heated again and the original and newly synthesized DNA strands separate. Binding sites for primers 1 and 2 now become available and complementary strands are being synthesized using the original strand and its complementary strand as templates. Repeated rounds of denaturation, annealing and primer extension will produce double-stranded DNA molecules identical to the region defined by the two primers in the original target sequence.

Repeated rounds of DNA synthesis are catalysed by a thermostable DNA polymerase in a suitable buffer containing the cDNA template, a specific set of two short synthetic DNA primers and a mixture of the four deoxynucleotides. The primer set is designed so that one primer can bind near one end of the cDNA and the other can bind near the other end of the complementary strand. When the first primer anneals to its target it creates a starting point for the DNA polymerase and a complementary copy of the template is generated. By heating the reaction mixture, the two complementary strands are separated, and targets for both primers are thereby exposed. When the temperature is lowered, both primers can anneal to their respective targets and both strands can now be copied by DNA polymerase. Each consecutive cycle consists of the same series of strand separation, primer annealing and DNA polymerization reactions, and after about 20 cycles the target sequence may become amplified more than a million-fold.

Some knowledge about the sequence of the template DNA is needed in order to design primers which are able to bind to the target. In addition to using protein sequence data as described previously, other types of data can be used for primer design. Thus, the sequences now available for a vast number of genes have made it clear that genes coding for particular functions have evolved as gene families which exhibit substantial sequence homology. Primers designed to recognize conserved regions from such a gene family may readily amplify cDNA for new members of the gene family. This procedure has proved very successful and has expanded the knowledge of gene families rapidly.

A dedicated effort to sequence the entire human genome of around 3×10^9 bp, including introns and other non-coding DNA stretches, was initiated around 1990, as part of the so-called Human Genome Project (HUGO). Initially, the sequence information accumulated very slowly, and a 15–20 year time frame was envisaged for its completion, but improved technology has enabled much faster data accumulation (Rowen *et al.*, 1997). Thus, it is now expected that 90% of the entire sequence will be known in the year 2000, and the remaining 10% will be completed a year or two after. The generated sequences have been made available to researchers in public databases. Partial sequences from a number of cDNA libraries – so-called expressed sequence tags (ESTs) – are also available in databases, and the EST sequences help in identifying coding from non-coding genome sequences. It is currently estimated that the human genome contains 50 000 to 100 000 genes, and all of these genes will soon be available in sequence databanks as well as in DNA clone banks. Thus, in the near future it will be possible to acquire the DNA encoding any human protein from such a bank. However, the selection of interesting genes will always require a vast background of biological expertise, and the experimental use of DNA will require knowledge of the basic recombinant DNA procedures.

2.3   Expression of recombinant proteins

Production of recombinant proteins and peptides to be used as pharmaceuticals requires efficient expression systems that generate proteins with the correct structure and function. In this section, some requirements for expression will be outlined together with a discussion of the relative merits and drawbacks of different host–vector expression systems (Hodgson, 1993).

2.3.1   *Transcription, translation and protein modifications*

The first step in gene expression is transcription into RNA. Transcription is initiated with the binding of the enzyme RNA polymerase to a specific DNA sequence at the beginning of the coding region, the so-called promoter. Many different promoters and other elements that influence transcription efficiency have been characterized in both prokaryotes and eukaryotes. Promoters are most often species-specific, and in higher organisms also tissue-specific. Transcription termination is also determined by specific DNA sequences, located downstream of the coding region. In prokaryotes, the RNA molecule produced by

Table 2.1 Examples of post-translational protein modifications

Modification	Occurrence
Cleavage of peptide bonds (maturation) Formation of disulphide bonds Phosphorylation of tyrosine, serine or threonine	Prokaryotes and eukaryotes
N-linked glycosylation of asparagine O-linked glycosylation of threonine or serine	Eukaryotes
Hydroxylation of proline or lysine Myristylation of N-terminal glycine Prenylation of cysteine	
γ-carboxylation of glutamic acid Amidation of C-terminal glycine Sulphation of tyrosine	Higher eukaryotes

transcription of protein-coding genes is used without any major modifications as mRNA to direct protein synthesis. However, in eukaryotes, complex enzymatic processing steps which remove intron regions from the primary transcription product are required for the generation of mature mRNA.

The next step in gene expression is translation of the mRNA into protein. The basic translation mechanisms are the same in prokaryotes and eukaryotes. Thus, the codon for translation initiation is with rare exceptions universal and leads to translation products containing methionine at their N-terminus. The N-terminal methionine is, however, often removed already during protein elongation by a ribosome-associated methionyl amino peptidase. Most proteins destined for secretion from cells are synthesized with a so-called signal peptide of 20–30 amino acids at their N-terminus. The signal sequence mediates contact between the membrane and the nascent protein, and it is usually cleaved off during the translocation over the membrane before the protein is fully synthesized. Compared to prokaryotes, eukaryotes have a very complex pathway for secretion of proteins, and many essential modifications of proteins occur in the secretory pathway. Therefore, maturation of the primary translation product to a protein with correct secondary and tertiary structure cannot always be obtained if *E. coli* is used for expression of a eukaryotic protein. To overcome this problem, several efficient eukaryotic expression systems have been developed in addition to the 'classical' *E. coli* systems. Examples of post-translational protein modifications in prokaryotes and eukaryotes are listed in Table 2.1 (Creighton, 1993).

2.3.2   *Choice of expression system*

E. coli

Since 1977, when a human protein (somatostatin) was expressed in *E. coli* in a functional form for the first time (Itakura *et al.*, 1977), many hundreds of eukaryotic genes have been expressed in this organism but very often not in a functional form. Because *E. coli* is incapable of performing many of the post-translational modifications that occur in mammalian cells, it is most suitable for production of proteins with no or limited modifications. However, its ease of handling and its safety as a production organism speak in favour of maintaining *E. coli* as an host even for proteins which are not completely matured in this organism. In order to obtain functional proteins, various ingenious methods have been devised that allow the final maturation to occur *in vitro* and recent progress has also improved the *in vitro* processing capacity tremendously (Georgiou and Valax, 1996). A number of approved recombinant pharmaceuticals are produced in *E. coli*, e.g. growth hormone, interferon-α and interleukin 2.

Intensive work has been performed aimed at optimizing the signals involved in expression in *E. coli*. Promoters from highly transcribed bacterial genes, or synthetic promoters, are used for high production levels. Regulated promoters that can be turned on or off by altering the growth conditions (temperature, added chemicals, etc.) are often used for production of recombinant proteins that are harmful to the cell. Examples of such promoters are the *lac* promoter from the β-galactocidase operon, or the bacteriophage λ promoters P_L and P_R. The advantage of these promoters is that synthesis of recombinant proteins can be delayed until the cell density has reached a high level where a short protein production phase results in generation of large amounts of product.

Yeast

Saccharomyces cerevisiae (baker's yeast) is an eukaryotic microorganism that performs many post-translational modifications which also occur in humans. This makes it useful for production of peptides and proteins of medium complexity (Romanos *et al.*, 1992). The long tradition of using yeast for food production contributes significantly to its acceptability as a producer of pharmaceutical proteins.

S. cerevisiae contains a natural plasmid (2 μm DNA) which can be used as vector when equipped with a selectable marker. Instead of using a gene conferring resistance to an antibiotic, a yeast gene which complements a mutated gene in the host genome is used for plasmid selection. This principle (which can also be used in *E. coli* and other hosts) is preferable to antibiotic selection for large-scale work.

It is often an advantage to produce recombinant proteins from yeast in a secreted form, because secreted proteins may be subjected to desired post-translational modifications in the secretory pathway, and because they become easier to purify. However, undesirable changes may occur during passage through the secretory pathway, e.g. incorrect N- and O-linked glycosylation. Since the glycosyl side-chains made in yeast and humans are different, yeast-produced glycoproteins may be immunogenic in man. Thus, glycoproteins are generally not suitable for production in yeast. Insulin is an example of an approved recombinant protein produced in *S. cerevisiae*.

Mammalian cells

Mammalian systems are the obvious choice for production of glycoproteins or proteins with other complex post-translational modifications. Some of the modifications (e.g. sulphation, amidation and γ-carboxylation) are often essential for the biological activity of the proteins. For example, the activity of a number of blood coagulation factors (factors II, VII, IX and X) is completely dependent on γ-carboxylation of glutamic acid, a modification which occurs only in cells of higher eukaryotes.

Many different mammalian host–vector systems have become available for expression of recombinant proteins (Birch and Froud, 1994). Examples of immortalized mammalian cell lines, which can be grown in culture and therefore are suitable as host cells, are CHO (Chinese hamster ovary), BHK (baby hamster kidney), COS (African green monkey kidney) and HeLa (human epitheloid carcinoma) cells. Mammalian viruses which replicate autonomously in cells can be used as vectors for recombinant DNA, but it is also possible to perpetuate introduced DNA in mammalian cells without an autonomously replicating vector, because transfected DNA molecules can integrate into mammalian chromosomes. Viruses which have been modified to be particularly useful as vectors include SV40 (simian virus), Epstein-Barr virus and retroviruses. The first two are self-replicating, while the third mediates cellular uptake and integration into the genome with significantly increased efficiency compared to DNA constructs without retroviral elements. To allow easy isolation of cells in which recombinant DNA has been taken up, a selectable marker gene such as resistance to the cytotoxic drug neomycin is introduced into the vector. The promoters used for gene expression in mammalian cells are often strong viral promoters such as the SV40 and CMV (cytomegalovirus) promoters.

It is important to realize that post-translational modifications may not be identical to those occurring in authentic protein, even when mammalian cells are used for production. In the body, proteins are produced in organs composed of specialized cells, and it has become increasingly clear that differences exist in the processing capabilities of different tissues. Also, immortalized cell lines may have diverged from their progenitors not only in their growth characteristics but also in other ways. Proteins with aberrant post-translational modification may be recognized as 'non-self' by the immune system and therefore evoke an immune response that will eliminate the protein. Therefore, careful characterization of recombinant proteins is equally important whether production occurs in mammalian cells or in microorganisms.

From an economic point of view, mammalian cell culture systems are less attractive than microbial systems. The low growth rate, moderate expression levels, complex medium requirements and the need for advanced cultivation equipment are the main reasons for the relatively high cost of producing biologicals from mammalian cell lines. However, production costs are usually an issue only if large amounts of the protein are necessary for therapeutic purposes. The coagulation factors FVIIa and FVIII are examples of marketed recombinant proteins produced in mammalian cell cultures.

An alternative to the expression of recombinant proteins in mammalian cell cultures is the use of so-called transgenic animals (Velander *et al.*, 1997). Transgenic animals can be generated by injection of a foreign gene into fertilized eggs by *in vitro* micromanipulation. The injected foreign gene integrates into the chromosome of the egg, normally at random locations. The eggs are then implanted into the oviduct of a foster mother which after a normal period of pregnancy will give birth to transgenic progeny with the foreign gene incorporated permanently into the genome of all cells. When using transgenic animals as producers of recombinant proteins, it is important that expression of the transgene is regulated, because the compound is usually biologically active and therefore should be produced and stored in organs or compartments where it does not affect the animal. The most attractive transgenic production systems employ mammary gland specific regulatory elements that target accumulation of the protein in the milk of the animal. Transgenic sheep have been made which express substantial levels of recombinant proteins in the milk, e.g. tissue plasminogen activator, α_1-antitrypsin and coagulation factor IX. None of the proteins produced by transgenic animals have yet been marketed as pharmaceuticals.

Insect cells

Insect cell expression systems have become very popular, especially for small-scale production of mammalian recombinant proteins. High expression levels of active proteins can usually be obtained much faster than in mammalian cells, which is a key attraction for the systems.

Most vectors are based on a lytic insect virus belonging to the baculovirus family, and the foreign cDNA is inserted in the viral genome without interfering with the lytic life cycle of the virus. Protein production thus occurs during a lytic infection of insect cells with recombinant viruses, and usually less than a week is needed from infection to maximal product yield (Griffiths and Page, 1997). The fact that baculoviruses are non-infectious to vertebrates and their promoters inactive in mammalian cells gives insect systems a potential advantage over other systems when expressing oncogenes or other genes that are harmful to mammals. Thus, baculovirus has been used for expression of an HIV protein (gp160) which is at present in phase III clinical trial as a recombinant AIDS vaccine.

2.4   Protein design

Recombinant DNA technology makes it feasible not only to produce natural proteins, but also to design and produce new types of protein molecules. New proteins can be broadly classed in two categories: (1) natural protein variants harbouring replacements, insertions and deletions of small numbers of amino acids, and (2) chimeric proteins consisting of domains originating from different proteins. These approaches have been very useful in structure–function studies of many proteins, e.g. in the determination of which part of the protein is responsible for catalytic function, tissue targeting or stability. The techniques play an important role in the rational design of protein drugs with improved character-istics such as prolonged stability *in vivo* (Blundell, 1994). A third category comprising peptides and proteins designed from scratch has recently become a booming activity as the result of the introduction of the so-called epitope-display technology.

2.4.1   *Protein variants*

Protein variants can be generated by a genetic engineering method called site-directed mutagenesis. Using this method, it is possible to change, delete or insert one or a few nucleotides within a coding region (Figure 2.4). Expression of a mutated gene will result in a protein variant with a specific amino acid alteration, whose function is then tested in various ways. Engineering small changes in proteins can have a variety of objectives, such as changing the solubility or stability properties or the affinity for substrates or receptors.

For example, extensive work has been performed to design insulin molecules with improved characteristics for the replacement therapy necessary in diabetes treatment (Bristow, 1993). The insulin molecule contains four glutamic acid residues, and by changing one or more of these to the neutral amino acid glutamine the isoelectric point is shifted towards a higher pH, thus resulting in an insulin variant with a lower solubility at the physiological pH (around 7.3). Such insulin variants constitute improved alternatives to existing slow-release formulations made by complexing native insulin with protamine or zinc ions. Similarly, other amino acid substitutions can lead to new quick-acting insulin preparations. As anticipated, many variants have proved undesirable at various stages in the test programmes, e.g. because of altered affinity for the receptors for insulin and the related hormone insulin-like growth factor. Another major problem for the clinical applica-tion of the analogues is the risk of adverse reactions due to immunogenicity or altered physiological functions of degradation products. These risks are very serious, especially

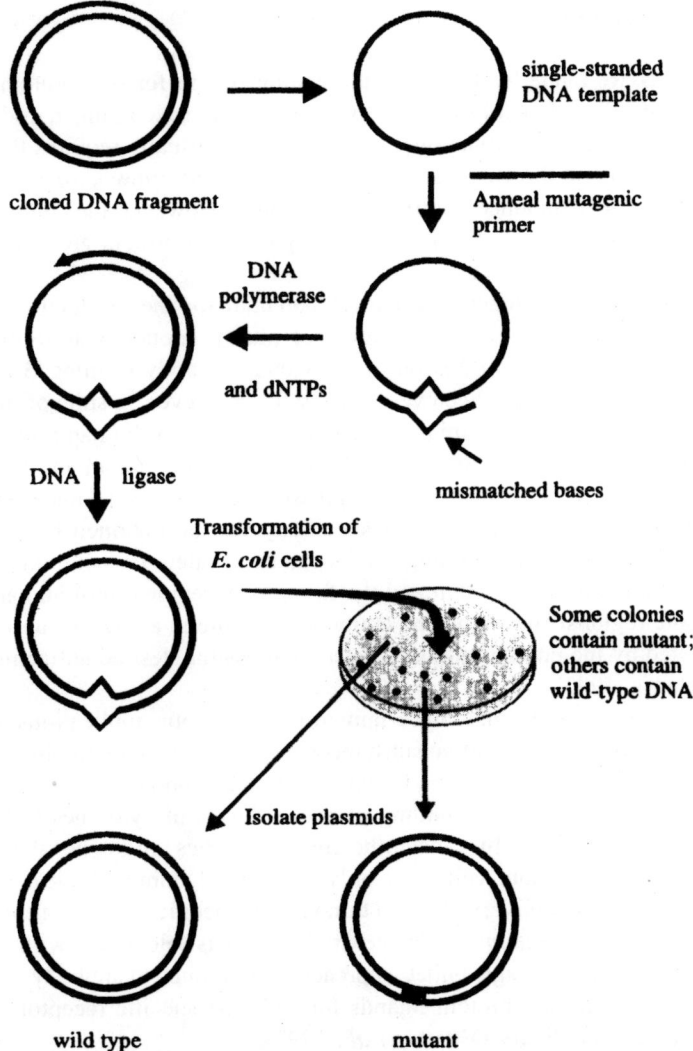

Figure 2.4 Oligonucleotide-directed mutagenesis by enzymatic primer extension. A 'mutagenic' oligonucleotide encoding the desired mutation embedded in wild-type flanking sequences is annealed to single-stranded DNA template. The sequence of the oligonucleotide is complementary to the template except for the nucleotides that define the mutation. The mutagenic oligonucleotide serves as a primer for DNA synthesis by DNA polymerase. Once the entire template has been copied, the ends of the newly synthesized strand are linked by DNA ligase. The heteroduplex DNA is transformed into *E. coli*. Replication results in segregation into separate wild-type and mutant plasmids and further repair mechanisms generate colonies containing either one or the other plasmid. Plasmid DNA is isolated from the resulting colonies and is screened to identify mutants.

for a pharmaceutical such as insulin which is administered daily for years, and extensive test programmes for adverse reactions are therefore necessary. However, several insulin analogues are now in phase III clinical trial, and one fast-acting analogue (Insulin Lispro, so named because the alteration involves the amino acids Lys and Pro) has already been marketed.

2.4.2 *Protein chimeras*

Chimeric proteins are made by fusion of the coding region for one protein (or protein domain) with that of another followed by expression of the recombined coding regions. The successful generation of a functional chimeric protein usually requires that structure–function relationships of the two starting proteins are well known, so that the desired active domains will be maintained within the new protein. However, the chimeric approach can also be used as a means to discover which regions of a protein are important for its activity.

The chimeric protein approach has a great potential for the production of so-called humanized monoclonal antibodies. Production of murine monoclonal antibodies based on the immortalization and proliferation of individual antibody-forming B-cells, the so-called hybridoma technique, has been very successful. However, in spite of major efforts, it has been difficult to establish similar production methods for human monoclonal antibodies. Instead, with access to cloned antibody genes of both human and murine origin, a production based on recombinant DNA expression is possible. In mice, expansion of antibody-producing cells with a required specificity can be obtained by immunization with any antigen, and the corresponding cDNA can be isolated. By combining the coding regions corresponding to the constant and the framework regions from human antibodies with the variable regions from a murine antibody, a chimeric antibody can be generated which is tolerated by the human immune system and has the desired antigenic specificity (Winter and Harris, 1993).

Such humanized antibodies may have numerous therapeutic applications, either alone or coupled to other biologically active substances. For example, they might be used in the elimination of toxic substances such as bacterially derived endotoxins, or as regulators of the humoral immune response by binding of excess levels of cytokines (interleukine 1, tumour necrosis factor, etc.). In cancer therapy, antibodies might be able to mediate selective destruction of tumour cells, not only via normal immunological mechanisms, but also via targeted toxin delivery. Thus, a tumour-cell specific antibody fused to parts of a cellular toxin, such as *Pseudomonas* exotoxin A, interacts selectively with and destroys tumour cells. This so-called 'magic bullet' approach is not limited to antibody–toxin fusions; chimeras between toxins and protein ligands for cell-type specific receptors are another possibility for targeted delivery (Vitetta *et al.*, 1993).

2.4.3 *Epitope display libraries*

Vast libraries of peptides without any predetermined homology to known proteins or peptides can be created through cloning complex mixtures of combinatorially synthesised oligonucleotides into specialized expression 'display' vectors. The filamentous phage display system, whereby the expressed peptides are displayed as fusion to phage coat proteins, has been effective in the discovery of many new ligands for various target proteins (Cortese, 1996). Affinity purification of the population of peptide phage-particles on the target protein is used to recover peptides with binding activity. Repeated rounds of binding selection and repropagation of bound eluted phage will enrich for tighter binders, and the identity of the selected peptide is simply revealed by sequencing of the appropriate segment of the DNA of each captured phage (Figure 2.5). An exceptionally elegant demonstration of the power of this technology was the identification of a small peptide that binds to and activates the receptor for erythropoietin (Wrighton *et al.*, 1996). Surprisingly,

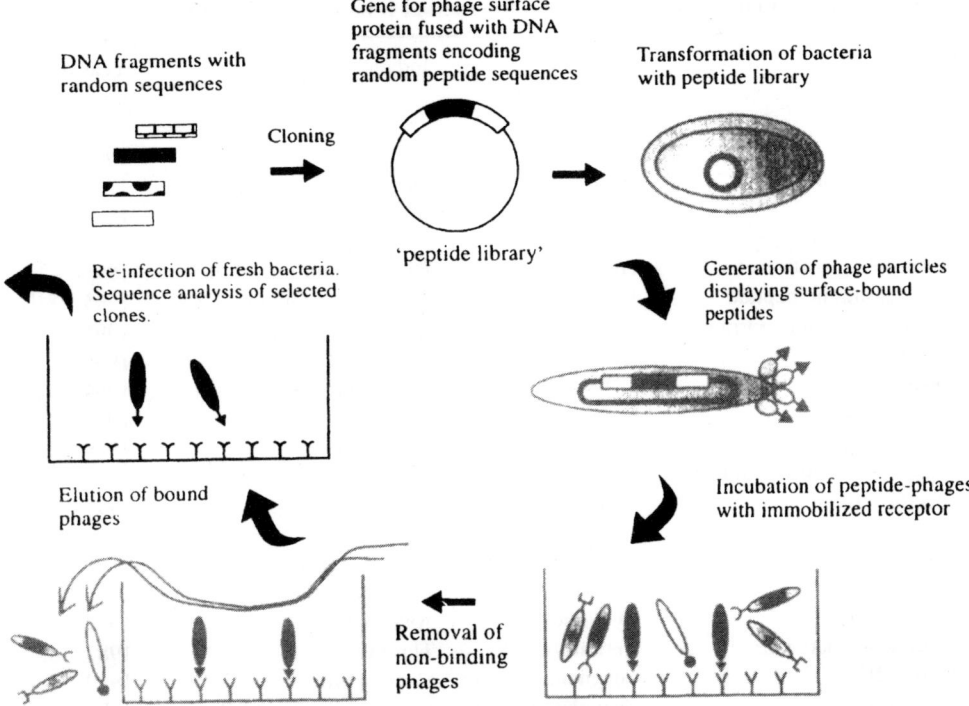

DNA fragments with random sequences

Cloning

Gene for phage surface protein fused with DNA fragments encoding random peptide sequences

Transformation of bacteria with peptide library

'peptide library'

Re-infection of fresh bacteria. Sequence analysis of selected clones.

Generation of phage particles displaying surface-bound peptides

Elution of bound phages

Incubation of peptide-phages with immobilized receptor

Removal of non-binding phages

Figure 2.5   Phage-display of peptide epitopes. Cloning of random DNA sequences in frame with the coding region of a phage coat protein (gIIIp) results in the formation of phage particles each displaying one unique peptide epitope on the phage surface. Phage particles displaying an epitope recognizing a specific target (e.g. a receptor or an enzyme) can be isolated by repeated rounds of affinity purification and repropagation in *E. coli* of the eluted bound phages.

the peptide showed no sequence or structural homology to the cytokine it mimics. Peptides with improved binding characteristics can now be generated from the selected peptide phage-particle by random or directed mutagenesis of the different amino acid residues of the peptide by site-directed mutagenisis procedures. The size of the phage-displayed epitope is not limited to that of a small peptide. Thus, a flush of publications during the past few years have described the generation of phage libraries displaying antibody fragments (single chain Fv and Fab), cDNA derived proteins and proteins of random sequence fused to scaffold proteins designed to offer a endless diversity of molecular shapes (Dougall *et al.*, 1994).

2.5   Recombinant protein therapeutics – status and future trends

Table 2.2 lists the recombinant proteins which, to our knowledge, have been marketed as drugs at the time of writing. Only a few of these drugs were previously produced by non-recombinant means, and only one (insulin) was available in ample amounts. As can be seen, they represent a broad variety of biological substances from hormones and cytokines through enzymes and blood coagulation regulators to vaccines.

Table 2.2 Recombinant protein drugs on the market (PharmaProjects, 1997)

Product	Category	Indication	Year of introduction
Insulin	Hormone	Diabetes	1982
Growth hormone	Hormone	Growth defects	1985
Glucagon	Hormone	Insulin-induced hypoglycaemia	1993
Insulin-like growth factor I	Hormone	Diabetes and growth defects	1995
Atrial natriuretic peptide	Hormone	Acute heart failure	1995
Follicle stimulating hormone	Hormone	Infertility	1996
Interferon-α	Cytokine	Cancer therapy	1986
Interferon-β	Cytokine	Cancer therapy	1994
Interferon-γ	Cytokine	Cancer therapy	1990
Interleukin 2	Cytokine	Cancer therapy	1989
Granulocyte colony-stimulating factor (G-CSF)	Cytokine	Leucopenia	1991
Granulocyte–macrophage colony-stimulating factor (GM-CSF)	Cytokine	Leucopenia	1991
Erythropoetin	Cytokine	Anaemia	1989
Plasminogen activator	Antithrombotic	Thrombosis	1987
Factor VIII	Coagulation factor	Haemophilia A	1989
Factor VIIa	Coagulation factor	Haemophilia A and B	1996
Factor IX	Coagulation factor	Haemophilia B	1997
α_1-antitrypsin	Enzyme inhibitor	α_1-antitrypsin deficiency	1992
Hirudin	Enzyme inhibitor		
DNase I	Enzyme	Cystic fibrosis	1994
Glucocerebrosidase	Enzyme	Gaucher's disease	1994
Hepatitis B subunit vaccine	Vaccine	Hepatitis prevention	1988
Hepatitis A vaccine	Vaccine		
Pertussis vaccine	Vaccine	Pertussis prevention	
Cholera vaccine	Vaccine	Cholera prevention	

In addition to the compound classes listed in the Table 2.2, several recombinant antibodies have also appeared on the market. So far these have been used for diagnostic purposes rather than as therapeutics, but therapeutic antibodies are expected to appear soon, and will probably, together with recombinant vaccines, form the fastest growing compound class. According to various sources, there are at present around 250 new proteins in advanced clinical trials. A sizable proportion of these represent improved versions of already known proteins, where specific modifications have been introduced to improve, for example, their pharmacokinetic properties. However, around 100 represent truly novel pharmaceutical substances with no precedent in medical therapy. It is estimated that around 2000 gene technology-based drugs, comprising both proteins and non-proteins, are in early stages of development. The information coming out of the human genome and EST projects is a very important resource for the selection of new potential protein drugs, and so-called DNA database mining has become a standard procedure in drug discovery both in major pharmaceutical companies and in small biotech companies.

Protein therapeutics are traditionally viewed as less convenient drugs than low-molecular therapeutics, from both the production and the user points of view. For the producer, the complexity of proteins poses challenges for purification, stability and administration. For the user, administration of protein therapeutics is more cumbersome because, unlike low-molecular weight drugs, proteins cannot yet be administered orally and usually have to be injected. Thus, at the same time as many resources are spent on discovering new therapeutic proteins, pharmaceutical companies are also trying to find low-molecular weight analogues that can supplant the known protein therapeutics. At a more basic research level, gene therapy is being explored as another substitute for protein therapeutics. In the long run, gene therapy may be able to replace protein therapeutics, but currently there are considerable technical obstacles (Sandhu *et al.*, 1997). Thus, in the near future protein therapeutics are expected to retain and increase their importance. With improvements in formulation and administration routes they may never become obsolete.

References

BIRCH, J.R. and FROUD, S.J., 1994, Mammalian cell culture systems for recombinant protein production, *Biologicals*, **22**(2), 127–133.

BLUNDELL, T.L., 1994, Problems and solutions in protein engineering – towards rational design, *Tibtech*, **12**, 145–148.

BRISTOW, A.F., 1993, Recombinant-DNA-derived insulin analogues as potentially useful therapeutic agents, *Tibtech*, **11**, 301–305.

CORTESE, R. (ed.), 1996, Combinatorial Libraries: Synthesis, Screening and Application Potential. Berlin, New York: Walter de Gruyter.

CREIGHTON, T.E., 1993, *Proteins: Structures and Molecular Properties*, 2nd edn. New York: W.H. Freeman & Co.

DOUGALL, W.C., PETERSON, N.C. and GREENE, M.I., 1994, Antibody-structure-based design of pharmacological agents, *Tibtech*, **12**, 372–379.

GEORGIOU, G. and VALAX, P., 1996, Expression of correctly folded proteins in *Escherichia coli*, *Current Opinion in Biotechnology*, **7**(2), 190–197.

GRIFFITHS, C.M. and PAGE, M.J., 1997, Production of heterologous proteins using the baculovirus/insect expression system, *Methods in Molecular Biology*, **75**, 427–440.

HODGSON, J. 1993, Expression systems: a user's guide, *Biotechnology*, **11**, 887–893.

INNIS, M.A., GELFAND, D.H., SNINSKY, J.J. and WHITE, T.J., 1990, *PCR Protocols. A Guide to Methods and Applications*. London: Academic Press.

ITAKURA, K., HIROSE, T., CREA, R., RIGGS, A.D., HEYNEKER, H.L., BOLIVER, F. and BOYER, H.-W., 1977, Expression in *E. coli* of a chemically synthesized gene for the hormone somatostatin, *Science*, **198**, 1056.

PharmaProjects, 1997, Richmond: PJB Publications.

ROMANOS, M.A., SCORER, C.A. and CLARE, J.J., 1992, Foreign gene expression in yeast: a re-view, *Yeast*, **8**, 423–488.

ROWEN, L., MAHAIRAS, G. and HOOD, L., 1997, Sequencing the human genome, *Science*, **278**, 605–607.

SAMBROOK, J., FRITSCH, E.F. and MANIATIS, T., 1989, *Molecular Cloning. A Laboratory Manual*, 2nd edn. Cold Spring Harbor, NY: Cold Spring Harbor Laboratory Press.

SANDHU, J.S., KEATING, A. and HOZUMI, N., 1997, Human gene-therapy, *Critical Reviews in Biotechnology*, **17**(4), 307–326.

VELANDER, W.H., LUBON, H. and DROHAN, W.N., 1997, Transgenic livestock as drug factories, *Scientific American*, **276**, 70–74.

VITETTA, E.S., THORPE, P.E. and UHR, J.W., 1993, Immunotoxins: magic bullets or misguided missiles? *Immunology Today*, **14**(6), 252–259.

WINTER, G. and HARRIS, W.J., 1993, Humanized antibodies, *Immunology Today*, **14**(6), 243–246.

WRIGHTON, N.C., FARRELL, F.X., CHANG, R., KASHYAP, A.K., BARBONE, F.P., MULCAHY, L.S., JOHNSON, D.L., BARRETT, R.W., JOLIFFE, L.K. and DOWER, W.J., 1996, Small peptides as potent mimetics of the protein hormone erythropoietin, *Science*, **273**, 458–463.

3

Protein Purification

LARS HOVGAARD[1], LARS SKRIVER[2] AND SVEN FROKJAER[1]

[1]*Department of Pharmaceutics, The Royal Danish School of Pharmacy, Copenhagen, Denmark*
[2]*L&K Biosciences, Vedbaek, Denmark*

3.1   Introduction

Therapeutic proteins are available from a number of different sources, e.g. animal tissue, body fluids (blood), plants, microorganisms, and cell culture systems. However, most commercially available therapeutic proteins are produced by large-scale fermentation using either microorganisms or mammalian cells as sources; e.g. recombinant human insulin, human growth hormone, erythropoeitin (EPO), and interferons.

Large-scale fermentation of such recombinant organisms can produce substantial quantities of proteins. The majority of recombinant proteins from microbial sources are produced by a limited number of microorganisms which are classified as GRAS ('generally recognized as safe'). GRAS microorganisms include bacteria such as *Bacillus subtilis* and *Escherichia coli*, and also yeast and other fungi.

Many proteins obtained by fermentation are secreted by the microorganisms into the culture medium. Such extracellular proteins are normally much simpler to purify in the subsequent downstream processing than intracellular proteins, as there is no requirement to disrupt the cells in order to harvest the protein. Thus for extracellular secreted proteins the amount of foreign protein and cellular components that needs to be separated from the product of interest is much less. Mammalian cell cultures such as Chinese hamster ovary (CHO) cells and baby hamster kidney (BHK) cells represent another very important source of therapeutic proteins. Blood coagulation factors and monoclonal antibodies are among the examples produced in such cell culture systems. Compared to microbial sources, the mammalian cell culture systems are generally a more complex protein source. Besides basal nutritional requirements, various supplements have to be added to the culture media. Such supplements may include vitamins, growth hormone, mineral salts, amino acids, antibiotics, and serum. The added mixture of proteins and other biomolecules makes the subsequent purification protocol more complex. Addition of serum also increases the risk of contaminating the culture medium, and potentially the final product, with blood-borne pathogens, hence much effort is devoted to developing serum-free cell culture conditions. Mammalian cells are capable of conducting important post-translational metabolic reactions such as glycosylation of proteins and routinely provide protein products that are properly folded and functional. This is therefore the preferred cell system for industrial production of large-size proteins.

The purpose of this chapter is to give a brief overview of downstream processing of protein products. Several useful books covering protein separation and purification from different points of view are available (Jakoby, 1971, 1984; Jakoby and Wilchek, 1974; Deutscher, 1990; Hejnaes *et al.*, 1998). Another important source of information is booklets from manufacturers of separation equipment and media.

3.2 Fractionation strategies

Detailed information about the characteristics of the protein and preferably also about the properties of the most important impurities is a prerequisite not only for making a successful purification scheme, but also for achieving the optimal quality of the final drug product. Important data involve approximate molecular weight, isoelectric point, solubility, presence of labile groups, and other physicochemical data. One should also define criteria with regard to the stability of the protein to be purified. The parameters affecting the chemical and physical stability of proteins are temperature, pH, organic solvents, heavy metals, oxygen, mechanical stress, and a potential risk of proteolytic degradation. Based on such information, it is possible to select fractionation techniques and conditions that effectively separate the protein of interest from other proteins and biomolecules. A typical overall purification process is outlined in Figure 3.1. Detailed information on the various fractionation techniques is given in Chapter 4.

3.2.1 *Initial fractionation step*

The purpose of the initial fractionation (capture) is to obtain a solution of the protein suitable for further purification by chromatography. If the protein is intracellular, the initial step involves a cell disruption by homogenization followed by a removal of any remaining cells and cell debris, often by centrifugation or filtration. Ideally, one should use a highly selective step as early as possible in the purification process. This may reduce the number of subsequent steps that must be used to achieve the desired level of purity. In the case of extracellular proteins, a concentration and clarification step is often employed as the initial step. This can be achieved by unit operations such as precipitation and ultrafiltration. The capture is assigned to remove most of the foreign (host cell) proteins and to concentrate the product. The capture is preferably performed by chromatographic methods such as selective absorption affinity matrix or by ion exchange chromatography.

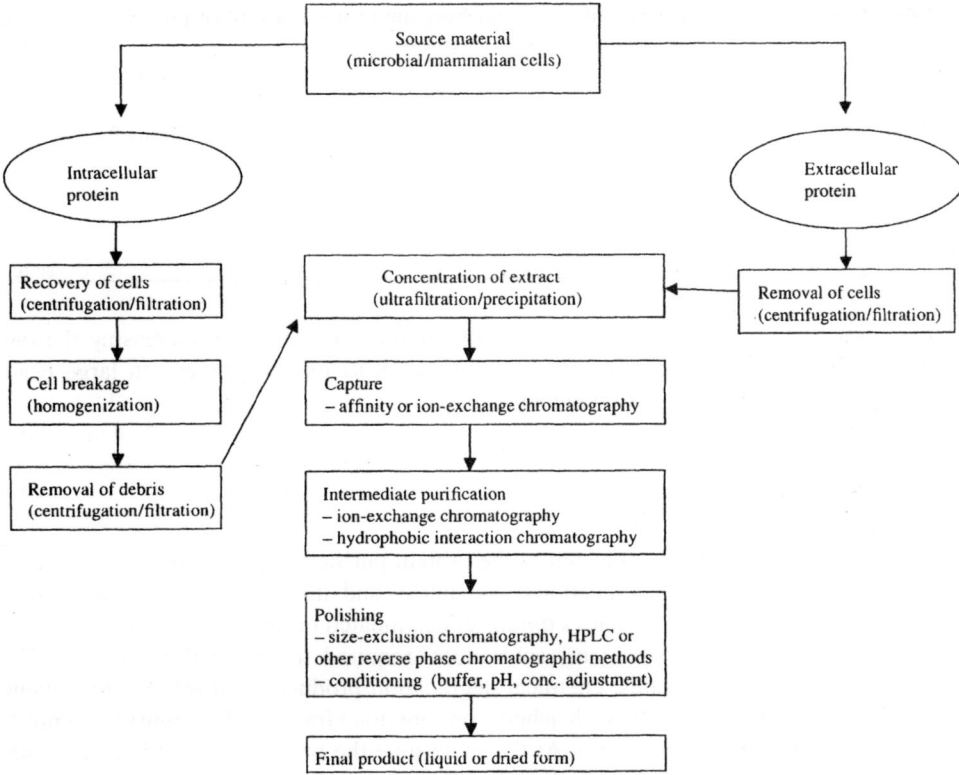

Figure 3.1 Outline of protein production from microbial and mammalian sources.

These initial steps should also result in the removal of most of the bulk foreign proteins including proteases and membrane fragments that might bind the protein of interest.

Precipitation of a protein in an extract may be achieved by changing the mother liquor solution by adding salts (e.g. ammonium sulphate, sodium sulphate), organic solvents (e.g. ethanol, acetone), polymers (e.g. polyethylene glycol), or by changing the pH or the temperature of the solution.

3.2.2 *Intermediate purification step*

In order to achieve a high protein purity by removal of host cell proteins, nucleic acids, lipids, salts and other small molecules, various chromatographic steps can be applied. The number and types of such chromatographic steps needed at this stage in the process depend primarily on the resolution achieved by each step and the economic constraints determined by the balance between the overall purity and total product recovery.

Although many different chromatographic techniques are also available on an industrial scale, the most common techniques are ion exchange, affinity chromatography, hydrophobic interaction chromatography and size-exclusion chromatography (see Chapter 4).

The order in which the different chromatographic steps are applied in a protein purification protocol is determined by factors such as capacity, process time, and cost. Monoclonal antibody-based affinity chromatography may be attractive because of its high selectivity. However, the rather viscous fermentation broth will limit the hydrodynamic properties of

Table 3.1   Examples of factors that may adversely affect the stability of proteins

Chemical	Physical	Biological
Detergents	Extreme pH	Proteolytic activity
Urea	Elevated temperatures	
Organic solvents	Light	
Oxidants	Freezing and thawing	
Heavy metals	High shear forces	

such a system and will reduce the column life significantly, thereby increasing the cost substantially. In such cases more robust media, such as ion exchangers on large beads with high purity, might be advantageous.

3.2.3   *Final polishing step*

The purpose of the last polishing step in the protein purification protocol is to produce a product which fulfils the specifications (e.g. purity and limits for specific degradation products, product derivatives such as oxidized, deamidated or degraded forms of product, and contaminants such as pyrogenic substances) required for its final processing. This means that possible aggregates, chemical degradation products, and other contaminants including ligands which may have leached from previous fraction steps must be removed to a certain limit by this procedure. At the same time the protein product is conditioned for its use in the final pharmaceutical preparation.

Size-exclusion chromatography is often used in this final step. It may effectively remove dimeric or higher molecular weight aggregates and low molecular weight degradation products and at the same time it can be used to change the purified protein into a new buffer system suitable for further processing or storage. As a production process has strong 'artistic' elements and can be difficult to protect by patents, detailed information on protein purification schemes used in purification of therapeutic proteins is generally considered as strictly confidential for obvious commercial reasons.

3.2.4   *The finished product*

Optimization of stability of the protein is of importance not only during downstream processing but also during storage of the purified product. Loss of activity may be caused by a number of factors, as outlined in Table 3.1. This should be minimized by setting strict limits on exposure, residual content, and product handling. Furthermore, it should be kept in mind that denatured proteins are often more susceptible to chemical degradation.

In some instances, the addition of specific stabilization agents may enhance the stability of the finished product. Such additives include substances such as substrates, co-factors, or other compounds which may interact with the protein of interest and thereby stabilize the native conformation. A variety of non-specific additives may also be incorporated into the finished product, e.g. glycerol, polyethylene glycol, sugars, and salt such as ammonium sulphate or sodium chloride, as they may have a significant stabilizing influence on some – but definitely not all – proteins. Bulk enzymes such as proteases are often marketed as solutions containing 20% sodium chloride as stabilizer.

Many processes leading to loss in biological activity of proteins in solution may be minimized by removal of water. Drying may be achieved by various techniques such as spray-drying, vacuum-drying, and drum-drying. However, these methods are relatively harsh due to high-temperature exposure and/or high mechanical stress, and are therefore unsuitable for labile proteins. Labile proteins are best dried by lyophilization, as discussed in more detail in Chapter 9.

3.3 Protein stability in downstream processing

After the large-scale manufacture of given desired proteins in expression organisms, the downstream processing poses a major challenge for the retention of protein structure and thus activity. It is of the utmost importance to design or tailor-make the downstream process for the protein, as the possible unfolding of proteins caused by the expression and initial purification conditions may not be fully reversible. The unfortunate result of the unfolding is often a loss in biological activity, as this typically depends inherently on the folding of the native state protein. In the protein, hydrophobic forces and hydrogen bonding are predominantly responsible for the overall stability. If viewed in an isolated fashion these effects are weak and can easily be broken. However, they act synergistically and are responsible for the immense activity that is seen in proteins. In the processing the proteins may undergo unfolding and, due to possible chemical modification, intermediate formation or aggregation, the native state may not be regained after purification. Therefore, it is extremely important to maintain native structure throughout the processing, thus ensuring that the biological activity levels are comparable to the level desired for the use of the protein in the final product.

The major concerns when designing a purification scheme for a protein are the unknowns. Very often the protein chemist has very little information about chemical and physical stability. Moreover, the specific effects of solvent pH, ionic strength, redox potential, additives, co-solvents, etc. are generally unknown. During the processing the protein is thus at risk of unintentional unfolding and destabilization, resulting in loss of activity and yield.

This chapter outlines the types of instability encountered for proteins as well as the physical and chemical factors affecting instability in the downstream processing. Further details are given in the References; see also Chapters 5 and 6.

3.3.1 *Protein conformation stability*

The conformational stability of a protein is important in relation to its activity and use. The unfolding of globular proteins in aqueous solutions is easily induced. Only 20–60 kJ/mol separates the native/active state from the unfolded/inactive state of a protein under physiological conditions (Pace, 1975; Privalov, 1979). Proteins of very different natures exhibit similar thresholds for unfolding.

Native state

In the globular protein specific features and folding patterns govern the native conformation. These structures are α-helix, β-sheet and random-coil structures. The specific folding optimizes stability for a given protein due to a complex combination of weak bonds. Some of these are hydrophobic and often referred to as van der Waals' forces. They include the burial of non-polar amino acid residues in the protein interior, α-helical dipoles and caps and weak polar aromatic interactions (Alber, 1989). Moreover, hydrophilic and electrostatic interactions are also very abundant in the folded protein structure. Among the types are the complexes between oppositely charged amino acids (Sali *et al.*, 1991; Serrano *et al.*, 1992), and hydrogen bonds between hydrogen atoms on hetero-atoms and neighbouring hetero-atoms (Stickle *et al.*, 1992). The latter play a major role in the α-helical formations in the native state of proteins (Dill, 1990). Other more specific factors affecting protein native state stability involve certain amino acids, e.g. Phe, Trp and Tyr, as they are all hydrophobic (Burley and Petsko, 1985).

Thus it is apparent that the native state is determined by a complex of different types of interaction, and that all of this is strongly dependent on the environmental situation (Seckler and Jaenicke, 1992; Oobatake and Ooi, 1993).

Intermediate state

In the unfolding process the protein passes through several transition conformations. One has been termed the molten globule state, and refers to a state of the protein in which it retains much of its native folding but has lost its tertiary structure (Ptitsyn and Uversky, 1994). It is believed that the protein occasionally assumes this state in the downstream processing and that the state is easily unfolded to an inactive form. But if this is prevented the native state can be assumed again (Privalov, 1979).

Unfolded state

The unfolded state of a protein defines a random coil structure whose specific nature varies from individual molecule to molecule. This inherently implies that all biological activity is absent. It is important to note that the chemical structure of the unfolded protein in almost all cases is identical to that of the native state (Jaenicke, 1987; Creighton, 1990). Thus the downstream bioactivity testing is an important measure for the folding state of the recovered protein.

3.3.2 *Protein instability*

The instability of proteins can be attributed to two distinctly different phenomena, physical instability and chemical instability. The former is more directly related to the conformational stability discussed above, although chemical instability may also be an integral part of a complex degradation path of a protein.

Physical events

The physical instability of a protein involves the unfolding and subsequent undesirable interaction of several protein molecules. It does not involve chemical alterations such as covalent bond breaking or formation. Very often this kind of instability leads to the

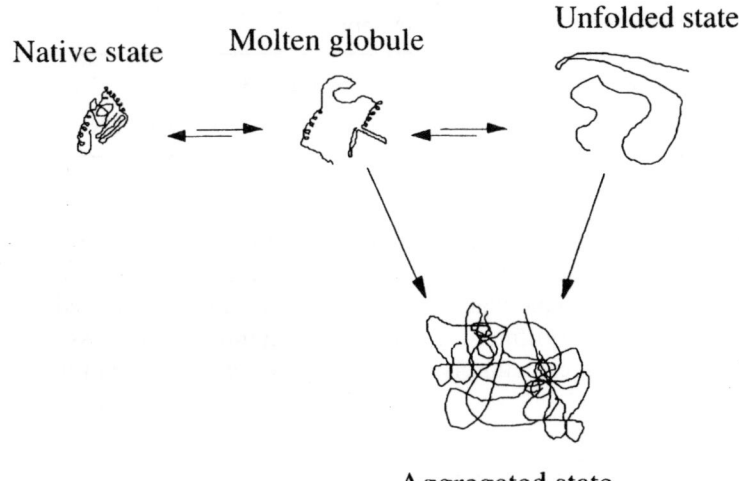

Figure 3.2 Unfolding and aggregation of proteins.

formation of higher order complexes, aggregates or precipitates. If the physical instability leads to irreversible aggregation, the product's biological activity is greatly reduced (Brange *et al.*, 1997). The phenomenon is difficult to predict, and the result often disastrous.

Denaturation When a protein loses its tertiary structure it is often said to denature. All intramolecular hydrogen bonds and hydrophobic interactions are disrupted. The organized native structure is lost, and all components are exposed (Jakoby, 1984; Tanford, 1968). The complexity of the phenomenon, of course, depends on the size and folding of the native protein (Privalov, 1979). Very often the denaturation is an abrupt process occurring in a narrow range of environmental factors, i.e. temperature, pH, ionic strength, redox potential or denaturant concentration. It is important to note that the denaturation reactions in many cases are reversible, but that the renaturation must be performed under tightly controlled conditions. In many cases the result of the renaturation is an insoluble gel or protein aggregate which is totally without activity (Matthiesen *et al.*, 1996; Hejnaes *et al.*, 1992).

Aggregation When denatured, proteins are prone to aggregate due to favourable inter-molecular hydrophobic interactions. This reaction may be very fast and may lead to an insoluble polymerized product (Figure 3.2). The aggregation is assumed to be governed by the initial dimerization process resulting in a second-order reaction (Kiefhaber *et al.*, 1991). However, in the case of insulin it has been suggested that even partially unfolded conformation may lead to aggregates (Brange *et al.*, 1997; Speed *et al.*, 1996).

Precipitation Controlled mild precipitation is being used in the downstream purification of proteins to limit or prevent irreversible aggregation in solution. Different techniques are used. Iso-precipitation involves the phenomenon of poor solubility of proteins around the iso-electric point, merely a pH effect. Salting-in and salting-out are two opposed

Table 3.2   The Hoffmeister series

Cations	$NH_4^+ > K^+ > Na^+ > Li^+ > Mg^{++} > Ca^{++} > Gdn^{++}$
Anions	$SO_4^{--} > HPO_4^{--} > CH_3COO^- > Cl^- > NO_3^- > SCN^-$

phenomena. The former denotes the situation where proteins' solubility is increased by low salt concentrations. The latter is a phenomenon of high-concentration salt precipitation of proteins. One technique that is very well known and widely used is addition of ammonium sulphate. The Hoffmeister series of salting-in effectiveness of cations and anions can be useful if precipitation is needed (Hejnaes *et al.*, 1998) (Table 3.2).

Chemical events

The chemical instability of proteins can involve the cleavage or formation of covalent bonds within the protein primary structure. This results in new chemical entities which often impose a severe separation problem for the chemist, as these derivatives are almost identical. Very specialized separation techniques are thus needed downstream. The reader is encouraged to refer to literature on the specific chemical processes, as they will be mentioned only briefly in this chapter.

Proteolytic degradation   In the living cell, a fine balance between the degrading enzymes and the synthesized protein is maintained by the subcellular compartmentation. However, all this is destroyed when the cell is lysed and disrupted as part of the protein isolation procedure. The cellular enzymes are mixed with the protein of interest and poses a potential fatal track. Therefore, the initial steps of a downstream purification are aimed at minimizing this proteolysis. The proteolysis may be of little chemical significance and thus difficult to detect. Moreover, proteolytic removal of just a few amino acids terminally may affect the biological activity. Therefore, high-power analytical tools such as mass spectroscopy, spectroscopy, capillary electrophoresis, peptide mapping, and isoelectric focusing are used to determine the degree of proteolysis (Chapters 4 and 9). In order to minimize the enzyme activities, environmental factors such as pH, protein stabilizing compounds and enzyme inhibitor cocktails can be used with caution (Hejnaes *et al.*, 1998).

Non-enzymatic degradation   In general the peptide bond is stable at alkaline pH. In dilute acid, however, where the carboxyl group of aspartic acid is not dissociated, peptide bonds of aspartic acid are cleaved very quickly. Even under normally stable conditions Asp-Pro is a major destabilizing link in a protein (Piszkiewicz *et al.*, 1970). N-terminal degradation has been reported for human growth hormone (Battersby *et al.*, 1994).

Deamidation   The deamidation process involves the nucleophilic attack of nitrogen in the peptide chain on the carbonyl residue in the amide side-chain of asparaginine, resulting in the loss of one amino group. The chemical result is an Asp residue or an iso-Asp residue. This reaction is documented in a long list of proteins: insulin (Berson and Yalow, 1966; Brange *et al.*, 1992), human growth hormone (Lewis *et al.*, 1970), cytochrome-c (Flatmark, 1966), etc. Moreover, the deamidation has also been shown to occur at the glutamine amide bond. The extent of deamidation can in general be minimized by the use of buffers with low ionic strength, low temperatures and neutral to slightly acidic pH (Scotchler and Robinson, 1974).

β-elimination and racemization   A very frequently reported degradation reaction is the β-elimination, which involves the abstraction of a β-hydrogen from cystein, serine or threonine residue in the protein backbone. A resulting carbanion can either form the unsaturated derivative or cause racemization (D-, L-amino). The abstraction of the β-hydrogen is proportional to the alkalinity, hence pH should not be high (Hejnaes *et al.*, 1998). Moreover, the reaction is affected at high temperatures and by divalent metals (Sen *et al.*, 1977; Lee *et al.*, 1977; Nashef *et al.*, 1977).

Oxidation   Several oxidation reactions in proteins have been reported (Brot *et al.*, 1982). In the alkaline or neutral medium the residues of the amino acids cystein, histidin, methionine, tryptophane and tyrosine are especially prone to oxidation. In acidic conditions, however, methionine is sensitive. Often the oxidation reactions cause a great loss in biological activity and even immunogenicity (Morley *et al.*, 1965). As the amino acid cystein serves protein stability in a unique way through intramolecular disulphide bonds, its stability is especially well characterized (Gilbert, 1990; Martin and Viswanatha, 1975). The disulphide bond is degraded via a variety of chemical mechanisms. It has been shown to be very sensitive to nucleophils in neutral and alkaline media and to the presence of electrophils in strongly acidic media. However, divalent metals, e.g. Cu^{++} in the presence of OH^- and O_2, are also a major reason for oxidation. Among the proteins that exhibit this cystein instability are insulin, chymotrypsin and ribonuclease (Martin and Viswanatha, 1975; Gilbert, 1990).

Stability of arginine

The last type of instability reaction to be discussed in this chapter is the conversion of arginine to ornithine. The reaction involves the alkaline hydrolysis of the guanidinium group in arginine and is generally smaller than for Cys, Lys, Tht, and Set.

3.3.3   *Essential process-related parameters*

Given that the crude protein product is present in an immensely complex mixture of fragmented cells and subcellular components as well as in a wealth of cellular proteins and enzymes, it is a serious challenge for the purification scientist to pick the protein of interest. Moreover, to pick and purify this protein without losing the bioactivity is a task for experts. Hence this section will only outline the effects of parameters that are used in this separation and isolation process. Table 3.3 gives some of the effects and examples of common parameters varied in the downstream processing of proteins. As mentioned earlier, little is known about the properties of a protein at the time of isolation. Hence the pitfalls of the process are numerous. It is important to keep in mind that the described parameters can all happen unexpectedly and at the same time. Therefore, the chemist faces a tremendous challenge. Process strategy and design are prerequisites for the following easy formulation of the drug product. This involves careful assessment of choice of matrices and solutes in combination with physical factors involved in the downstream purification processes.

Table 3.3 Parameters in the downstream processing of proteins

Parameter	Effect	Example	Reference
pH	Isoelectric point stabilization	pI-precipitation	Pace (1990)
	Protonation/ deprotonation	Carboxyl (pKa 3–4.7)	Hejnaes et al. (1998)
		Imidazolium (pKa 8–8.5)	
		Sulphydryl (pKa 8–8.5)	
		Amino (pKa 7.6–10.6)	
		Q-Hydroxy (pKa 9.4–10.4)	
	Other	Deamidation (alkaline and acidic conditions)	Lee et al. (1977), Florence (1980), Porter et al. (1993)
		Racemization	
		Hydrolysis	
Temperature	High-temperature denaturation	Entropy exceeds enthalpy	Privalov et al. (1986)
	Low-temperature denaturation	Hydrogen bond breaking	
		Breakage of hydrophobic interaction	
Redox potential	Reducing	Cleavage of disulphide bonds	Brot et al. (1982)
	Oxidizing	Methionine oxidation	
Co-solvents	Sugars, some amino acids, some salts, polyols	Stabilizing through increase in surface tension, *no* protein binding	Hejnaes et al. (1998)
	Weakly interacting salts	Stabilizing through increase in surface tension, binding to charged protein residues	
	Steric exclusion and repulsion from charged groups	PEG and 2-methyl-2,4-pentanediol	
Protein concentration	Aggregation reactions	Insulin fibrillation	Brange et al. (1997)
Pressure	Denaturation	570 MPa, 30°C, secondary structure disruption	Takeda et al. (1995)

References

ALBER, T., 1989, Mutational effects on protein stability, *Annu. Rev. Biochem.*, **58**, 765–798.
BATTERSBY, J.E., HANCOCK, W.S., CANOVA-DAVIS, E., OESWEIN, J. and O'CONNOR, B., 1994, Diketopiperazine formation and N-terminal degradation in recombinant human growth hormone, *Int. J. Pept. Protein Res.*, **44**, 215–222.

BERSON, S.A. and YALOW, R.S., 1966, Deamidation of insulin during storage in frozen state, *Diabetes*, **15**, 875–879.

BRANGE, J., LANGKJAER, L., HAVELUND, S. and VOLUND, A., 1992, Chemical stability of insulin. 1. Hydrolytic degradation during storage of pharmaceutical preparations, *Pharm. Res.*, **9**, 715–726.

BRANGE, J., ANDERSEN, L., LAURSEN, E.D., MEYN, G. and RASMUSSEN, E., 1997, Toward understanding insulin fibrillation, *J. Pharm. Sci.*, **86**, 517–525.

BROT, N., WERTH, J., KOSTER, D. and WEISSBACH, H., 1982, Reduction of N-acetyl methionine sulfoxide: a simple assay for peptide methionine sulfoxide reductase, *Anal. Biochem.*, **122**, 291–294.

BURLEY, S.K. and PETSKO, G.A., 1985, Aromatic–aromatic interaction: a mechanism of protein structure stabilization, *Science*, **229**, 23–28.

CREIGHTON, T.E., 1990, Protein folding, *Biochem. J.*, **270**, 1–16.

DEUTSCHER, M.P., 1990, *Methods in Enzymology (Vol. 182), Guide to Protein Purification*, San Diego: Academic Press.

DILL, K.A., 1990, Dominant forces in protein folding, *Biochemistry*, **29**, 7133–7155.

FLATMARK, T., 1966, On the heterogeneity of beef heart cytochrome c. 3. A kinetic study of the non-enzymic deamidation of the main subfractions (CyI-Cy 3), *Acta Chem. Scand.*, **20**, 1487–1496.

FLORENCE, T.M., 1980, Degradation of protein disulphide bonds in dilute alkali, *Biochem. J.*, **189**, 507–520.

GILBERT, H.F., 1990, Molecular and cellular aspects of thiol–disulfide exchange, *Adv. Enzymol. Relat. Areas Mol. Biol.*, **63**, 69–172.

HEJNAES, K.R., BAYNE, S., NORSKOV, L., SORENSEN, H.H., THOMSEN, J., SCHAFFER, L., WOLLMER, A. and SKRIVER, L., 1992, Development of an optimized refolding process for recombinant Ala-Glu-IGF-1, *Protein Eng.*, **5**, 797–806.

HEJNAES, K., MATTHIESEN, F. and SKRIVER, L., 1998, Protein stability in downstream processing. Subraminian, G. (ed.), *Bioseparation and Bioprocessing*, Vol. 3, New York: Wiley and Sons.

JAENICKE, R., 1987, Folding and association of proteins, *Prog. Biophys. Mol. Biol.*, **49**, 117–237.

JAKOBY, W.B., 1971, *Methods in Enzymology (Vol. 22), Enzyme Purification and Related Techniques*, New York, San Francisco, London: Academic Press.
    1984, *Methods in Enzymology (Vol. 104), Enzyme Purification and Related Techniques, Part C*, Orlando: Academic Press.

JAKOBY, W.B. and WILCHEK, M., 1974, *Methods in Enzymology (Vol. 34), Affinity Techniques of Enzyme Purification, Part B*, New York, London: Academic Press.

KIEFHABER, T., RUDOLPH, R., KOHLER, H.H. and BUCHNER, J., 1991, Protein aggregation *in vitro* and *in vivo*: a quantitative model of the kinetic competition between folding and aggregation, *Biotechnology (NY)*, **9**, 825–829.

LEE, H.S., OSUGA, D.T., NASHEF, A.S., AHMED, A.I., WHITAKER, J.R. and FEENEY, R.E., 1977, Effects of alkali on glycoproteins. b-Elimination and nucleophilic addition reactions of substituted threonyl residues of antifreeze glycoprotein, *J. Agric. Food Chem.*, **25**, 1153–1158.

LEWIS, U.J., CHEEVER, E.V. and HOPKINS, W.C., 1970, Kinetic study of the deamidation of growth hormone and prolactin, *Biochim. Biophys. Acta*, **214**, 498–508.

MARTIN, B.M. and VISWANATHA, T., 1975, Selective reduction of a disulfide bond in chymotrypsin A-alpha, *Biochem. Biophys. Res. Commun.*, **63**, 247–254.

MATTHIESEN, F., HEJNAES, K.R. and SKRIVER, L., 1996, Stabilization of recombinantly expressed proteins, *Ann. NY Acad. Sci.*, **782**, 413–421.

MORLEY, J.S., TRACY, H.J. and GREGORY, R.A., 1965, Structure–function relationships in the active C-terminal tetrapeptide sequence of gastrin, *Nature*, **207**, 1356–1359.

NASHEF, A.S., OSUGA, D.T., LEE, H.S., AHMED, A.I., WHITAKER, J.R. and FEENEY, R.E., 1977, Effects of alkali on proteins. Disulfides and their products, *J. Agric. Food Chem.*, **25**, 245–251.

OOBATAKE, M. and OOI, T., 1993, Hydration and heat stability effects on protein unfolding, *Prog. Biophys. Mol. Biol.*, **59**, 237–284.

PACE, C.N., 1975, The stability of globular proteins, *CRC Crit. Rev. Biochem.*, **3**, 1–43.

1990, Conformational stability of globular proteins, *Trends Biochem. Sci.*, **15**, 14–17.

PISZKIEWICZ, D., LANDON, M. and SMITH, E.L., 1970, Anomalous cleavage of aspartyl–proline peptide bonds during amino acid sequence determinations, *Biochem. Biophys. Res. Commun.*, **40**, 1173–1178.

PORTER, W.R., STAACK, H., BRANDT, K. and MANNING, M.C., 1993, Thermal stability of low molecular weight urokinase during heat treatment. I. Effects of protein concentration, pH and ionic strength, *Thromb. Res.*, **71**, 265–279.

PRIVALOV, P.L., 1979, Stability of proteins: small globular proteins, *Adv. Protein Chem.*, **33**, 167–241.

PRIVALOV, P.L., GRIKO, YU. V., VENYAMINOV, S. YU. and KUTYSHENKO, V.P., 1986, Cold denaturation of myoglobin, *J. Mol. Biol.*, **190**, 487–498.

PTITSYN, O.B. and UVERSKY, V.N., 1994, The molten globule is a third thermodynamical state of protein molecules, *FEBS Lett.*, **341**, 15–18.

SALI, D., BYCROFT, M. and FERSHT, A.R., 1991, Surface electrostatic interactions contribute little of stability of barnase, *J. Mol. Biol.*, **220**, 779–788.

SCOTCHLER, J.W. and ROBINSON, A.B., 1974, Deamidation of glutaminyl residues: dependence on pH, temperature, and ionic strength, *Anal. Biochem.*, **59**, 319–322.

SECKLER, R. and JAENICKE, R., 1992, Protein folding and protein refolding, *FASEB J.*, **6**, 2545–2552.

SEN, L.C., GONZALEZ-FLORES, E., FEENEY, R.E. and WHITAKER, J.R., 1977, Reactions of phosphoproteins in alkaline solutions, *J. Agric. Food Chem.*, **25**, 632–638.

SERRANO, L., KELLIS, J.T. JR, CANN, P., MATOUSCHEK, A. and FERSHT, A.R., 1992, The folding of an enzyme. II. Substructure of barnase and the contribution of different interactions to protein stability, *J. Mol. Biol.*, **224**, 783–804.

SPEED, M.A., WANG, D.I. and KING, J., 1996, Specific aggregation of partially folded polypeptide chains: the molecular basis of inclusion body composition, *Nat. Biotechnol.*, **14**, 1283–1287.

STICKLE, D.F., PRESTA, L.G., DILL, K.A. and ROSE, G.D., 1992, Hydrogen bonding in globular proteins, *J. Mol. Biol.*, **226**, 1143–1159.

TAKEDA, N., KATO, M. and TANIGUCHI, Y., 1995, Pressure- and thermally-induced reversible changes in the secondary structure of ribonuclease A studied by FT-IR spectroscopy, *Biochemistry*, **34**, 5980–5987.

TANFORD, C., 1968, Protein denaturation, *Adv. Protein Chem.*, **23**, 121–282.

4

Peptide and Protein Characterization

MIROSLAV BAUDYŠ AND SUNG WAN KIM

Department of Pharmaceutics and Pharmaceutical Chemistry, University of Utah, Salt Lake City, Utah, USA

4.1 Introduction

Due to tremendous progress in recombinant DNA techniques and technology, a large number of biosynthetic, pharmaceutically relevant polypeptides and proteins became available in the past decade and have been successfully approved for use as protein drugs. New manufacturing methods also enabled production of pharmaceutical proteins on a large scale, making their use as drugs economically feasible (Blohm *et al.*, 1988; Copsey and Delnatte, 1988). Analytical chemistry has played a key role not only in the design and validation of these methods, but also in the regulatory acceptance of protein pharmaceuticals, even though the reliability and efficacy of a protein drug can only be established through appropriately designed human clinical trials. Generally, polypeptide drugs must fulfil the same standards on purity, potency and safety as apply for all other conventional pharmaceuticals.

Proteins are monodisperse linear macromolecules with well-defined regular three-dimensional structures consisting of primary, secondary, tertiary and quaternary structure (Creighton, 1993). Because of the complexity of a protein molecule, a number of different subtle chemical alterations can occur during purification, formulation, handling and storage, such as side-chain deamidation, polypeptide chain cleavage, oxidation and reactions involving disulphides and cysteines (Ahern and Klibanov, 1988; Manning *et al.*, 1989; Kosen, 1992). Very small structural changes can have a significant impact on protein activity either directly (e.g. modification of the active site) or indirectly through their influence on the physicochemical properties (e.g. conformational changes). Suitable analytical methods must therefore have high maximum resolution to afford separation between very slightly altered macromolecules, such as those produced by methionine oxidation to methionine sulphoxide. For example, recombinant human growth hormone can undergo oxidation on one of the two different methionine residues (Teshima and Canova-Davis, 1992).

Generally, achieving homogeneity of a protein preparation is of primary importance. To demonstrate purity of a protein drug, the application of multiple analytical methods is always required (Holthuis and Driebergen, 1995; Benedek and Swadesh, 1991). Applied analytical methods are expected to detect and quantify reliably not only the active protein but also accompanying impurities, which can be sometimes even more important. The sources of impurities in biosynthetic protein preparations can be roughly divided into three categories: host-related (expression system used), process-related (involves specific purification and separation methods applied), and product-related (Garnick *et al.*, 1988). Product-related contaminants are in most cases related to decomposition processes that occur during the purification process, formulation and subsequent storage of a formulated protein drug in a dosage form. In addition, microheterogeneity of a protein product can originate during protein biosynthesis in heterologous cells and thus can be host-related (Kagawa *et al.*, 1988). For example, glycoproteins as a rule exhibit high microheterogeneity which can be exclusively attributed to structural differences of the glycan moiety or moieties (Lis and Sharon, 1993; Kobata, 1992). Other post-translational modifications such as acylation, formulation, phosphorylation and sulphation (Creighton, 1993; Wold, 1981) can further increase the complexity of a protein pharmaceutical. Of course, the larger the protein molecule and the higher the microheterogeneity, the more difficult and expensive it is to separate individual forms during a large-scale purification process. Especially when attempting to keep the costs down as much as possible, a purified protein drug for clinical use will be heterogeneous (Geigert and Ghirst, 1996), which will complicate formulation and stability studies. This again points to the requirement for using analytical methods with maximum resolving power and high reproducibility (Jones, 1994).

Another problem in the analysis of formulated proteins is the presence of excipients which are added to increase the stability of the active protein component (Hanson and Edmond-Rouan, 1992) in the formulation. The most commonly used stabilizers in protein formulations include surfactants, inactive carrier proteins (serum albumin) and sugars (Hovgaard *et al.*, 1992; Arakawa and Timasheff, 1982). The presence of excipients may severely interfere with the assay(s) originally developed to quantify and monitor the active ingredient during production and storage, and will require modification or development of new analytical assay(s) (Advant *et al.*, 1997). This is particularly true for formulations with low concentrations of active protein component, where an additional concentration step might be necessary. The validation of this step would be extremely critical since a number of these excipients can complicate and interfere with the final assay. The same considerations apply for protein delivery systems which are based on polymer or inactive protein carriers (Cleland and Langer, 1994a; Serres *et al.*, 1996).

This chapter provides an overview of the analytical methods currently available and used for characterization of pharmaceutical proteins. A thorough understanding of the underlying principles of each method and the influence of the operating parameters selected by an investigator is crucial to avoid artefacts and jumping to false conclusions. Generally, most of the efforts in developing analytical assay methods should concentrate on maximizing resolving capability and achieving optimal assay conditions. The most frequently used assay methods are listed in Table 4.1 along with the protein's structural parameters/features and alterations/changes that these methods are most sensitive to. The chapter cannot cover all frequently used methods in depth, but rather presents an overview of approaches to characterizing proteins from both chemical and physical viewpoints. Biological assays determining potency are not addressed because of low precision and the inability to resolve different forms of the biopharmaceutical.

4.2 Chromatography

Biochemical methods based on various types of high performance liquid chromatography (HPLC) have been essential for structural and physicochemical characterization as well as separation and purification of proteins for more than two decades, and have been extensively reviewed (Oliver, 1989; Gooding and Regnier, 1990; Mant and Hodges, 1996). Consequently, the availability of HPLC methods was also essential for progress in the field of pharmaceutical proteins (Benedek and Swadesh, 1991). Proteins differ in their size, shape, net charge and hydrophobicity, and these differences can be utilized to separate them. Thus, the major modes of HPLC employed in protein separation take advantage of differences in protein size (size-exclusion chromatography, SEC), net charge (ion-exchange chromatography, IEC) or hydrophobicity (reversed phase HPLC, RP HPLC, and, to a lesser extent, hydrophobic interaction chromatography, HIC). Once the separation mode is selected, other 'degrees of freedom' to achieve better resolution are provided by mobile phase conditions, such as composition, pH, ionic strength and temperature, which may be manipulated to maximize the separation capability of a particular column (stationary phase).

4.2.1 Reversed phase chromatography

RP HPLC has become the most popular chromatographic technique for the separation and analysis of peptides and proteins, as well as their fragments prepared by chemical or enzymatic cleavage (Aguilar and Hearn, 1996). This can be attributed to a number of factors, among the most important being excellent resolution and reproducibility as well as ease of use. This method was initially developed for the separation of small molecules but quickly gained popularity for separation of amino acid derivatives and peptides (Hancock, 1984). The experimental system comprises an alkyl chain modified silica-based sorbent, to which peptides and proteins can bind through their hydrophobic regions, and are then eluted from with gradients of increasing concentration of an organic solvent such as acetonitrile or propanol. Mobile phase usually also contains an ion-pairing agent, e.g. trifluoroacetic acid, phosphoric acid. Generally, the more hydrophobic the polypeptide,

Table 4.1 Common methods used for chemical and physical characterization of pharmaceutical proteins and monitoring of structural alterations that occur during purification, formulation and storage (see text for more details)

Method	Will determine/detect	Comments
*Chromatographic methods**		
Reversed phase HPLC (RP HPLC)	Purity, microheterogeneity/neutral or charge alteration	Frequently used in peptide mapping or diagonal mapping technique (e.g. disulphides identification); easily detects Met, Trp oxidation, Asn, Gln deamidation
Ion-exchange chromatography (IEC) and chromatofocusing	Charge microheterogeneity/charge alteration	Easily detects Asn, Gln deamidation, differences in sialic acid content in glycoproteins
Size-exclusion chromatography (SEC)	Molecular weight/association, aggregation	Combination of detectors allows determination of glycoprotein molecular weight; aggregates must be soluble
Electrophoretic methods		
Polyacrylamide gel electrophoresis (PAGE)	Purity, charge heterogeneity/charge alteration	Separates according to mass to charge ratio; detects Asn, Gln deamidation, differences in sialic acid content
Sodium dodecyl sulphate PAGE (SDS PAGE)	Molecular weight/covalent aggregates through disulphides	Identifies multichain proteins and quaternary structure; for glycoproteins, results must be interpreted cautiously
Isoelectric focusing (IEF)	Isoelectric point (pI)/charge heterogeneity, alteration	Frequently, presence of neutral surfactants is required to maintain solubility around pI
Two-dimensional PAGE (2D PAGE)	Protein spots are distributed in the X-Y plane according to pI and molecular weight	Increases resolution power substantially by combining isoelectric focusing and SDS electrophoresis
Capillary electrophoresis (CE)	Charge heterogeneity, structural microheterogeneity/charge, structural alterations	Separates according to mass-to-charge ratio; frequently used for peptide mapping and diagonal techniques
Chemical methods		
N- and C-terminal sequencing	N- and C-terminal primary structure/proteolytic cleavage and processing	Detects sites of post-translational modification
Limited enzymatic or chemical cleavage	Batch-to-batch consistency of the primary structure including post-translational modifications	In combination with chromatographic or electrophoretic peptide mapping, or mass spectrometry

Mass spectrometry methods

Method	Information detected	Comments
Fast atom bombardment mass spectrometry (FAB MS)	Molecular weight and sequence[†]/any mass change (1 Da)	Upper mass limit 5000–7000 Da
Matrix-assisted laser desorption/ionization mass spectrometry (MALDI MS)	Molecular weight/any mass change (1 Da)	Molecular weight range unlimited; suitable for complex peptide and protein mixtures
Electrospray ionization mass spectrometry (ESI MS)	Molecular weight and sequence[†]/any mass change (1 Da)	Molecular weight range up to 100 000 Da; suitable for coupling with HPLC or CE

Spectroscopic optical methods

Method	Information detected	Comments
UV absorption spectroscopy	Protein concentration/altered tertiary structure	Knowledge of molar absorption coefficient; tertiary structure changes detected through UV difference spectroscopy
Fluorescence spectroscopy	Changes in tertiary structure/altered chromophore environment	Only fluorescent amino acids Tyr, Trp will respond
Circular dichroism (CD) spectroscopy	Secondary structure content/altered secondary structure	Far-UV CD (170–250 nm) – possible aromatic amino acid interference; near-UV CD (250–320 nm) – only aromatic residues Phe, Tyr, Trp sensitive to change
Infrared spectroscopy (IR)	Secondary structure content/altered secondary structure	IR spectrum can be acquired in dry state; Fourier transform technique is exclusively used (FTIR)
Static light scattering (SLS)	Molecular weight/protein association, aggregation	Currently used in combination with SEC (flow-through SLS detector)
Quasielastic laser light scattering (QELS)	Diffusion coefficient (hydrodynamic radius)/protein association, aggregation	Cannot quantify aggregated fraction

Hydrodynamic mass transport methods

Method	Information detected	Comments
Viscosity	Molecular weight/protein association, aggregation	Measures effect on the flow properties of the bulk solvent
Analytical ultracentrifugation	Molecular weight/protein association, aggregation	Sedimentation velocity technique requires diffusion coefficient; equilibrium method can also yield thermodynamic parameters of association

* Most current methods exist in column and capillary format
† For sequence data, tandem MS instrumentation is required

the higher is the organic acid modifier concentration needed to release it from the binding to the stationary phase. As a consequence, RP HPLC is very suitable for hydrophilic proteins and peptides, but recovery of very hydrophobic proteins, e.g. membrane proteins, can be variable and poor (Kato *et al.*, 1987). Protein molecules can easily undergo conformational change and unfolding, especially in the presence of an organic modifier and/or increased temperature, which can increase retention time due to the exposure of hydrophobic residues which would be buried in the native state. This phenomenon can lead to non-ideal chromatographic behaviour, as observed for a series of globular proteins (Lin and Karger, 1990).

RP HPLC is not only used for the resolution of structural variants of recombinant and native proteins (Ohgami *et al.*, 1989), but is also frequently applied in a technique known as peptide mapping. In this technique, a particular protein is fragmented into a characteristic set of peptides either using a selected proteolytic enzyme with high specificity, such as trypsin, or by chemical means (BrCN), and the mixture is resolved in a characteristic profile or 'map' on the RP column. This 'fingerprinting' method has been routinely applied to proteins as large as tissue plasminogen activator (Chloupek *et al.*, 1989). Usually, not all the peaks are completely resolved by 'one-dimensional' chromatography and rechromatography is often required in a different solvent and/or column system. The most commonly used detection technique for eluted peptides and proteins is UV detection in the near UV range (~280 nm for Tyr and Trp) and far UV range (200–230 nm) for peptide bonds. This requires use of a solvent system transparent to the wavelength used. Tryptophan residues can be specifically monitored by fluorescence detection. More recently, mass spectrometry detection techniques have been introduced which now allow liquid chromatography–mass spectrometry (LC-MS) to be routinely performed (Siuzdak, 1996). Such two-dimensional techniques provide valuable extra information on the mass composition of eluted peaks, generating a three-dimensional map (retention time, mass and relative content). Thus, LS-MS techniques enable identification of nearly every component of a complex peptide mixture (Ling *et al.*, 1991). However, the prerequisite for their introduction was the rapid development in the field of capillary RP HPLC (Novotny, 1996), which enabled separation and characterization of picomolar amounts of peptides and proteins (Stults *et al.*, 1996). Finally, RP HPLC is very frequently used in a diagonal arrangement which allows identification of post-translational modifications such as phosphorylation (Aitken and Learmonth, 1997a) or disulphide bonds (Aitken and Learmonth, 1997b). In principle, the peptide mixture is separated, individual fractions collected, and each fraction undergoes a selected chemical modification reaction specific for a particular chemical group (post-translational modification). RP HPLC after derivatization should produce pure peptide(s) even from a very complex mixture, since the elution position of other peptides will be unaffected.

4.2.2   *Hydrophobic interaction chromatography*

Hydrophobic interaction chromatography (HIC) is based, as is RP HPLC, on the interaction of the hydrophobic areas of the protein with a hydrophobic ligand. HIC, first reported as salting-out chromatography (Porath, 1960), operates under relatively mild conditions as compared to RP HPLC due to less hydrophobic supports and a lower ligand density. The separation principle relies on surface hydrophobicity of the folded proteins since no organic modifier is usually used. Proteins are adsorbed at a high salt concentration and elution and separation are achieved by a descending gradient of salt concentration. HIC provides, as a rule, better recoveries of intact proteins than RP HPLC for the reasons given above. More recently, the introduction of new matrix supports based on perfusive and non-porous particles has reduced the separation time of HIC from hours to minutes (Wu and Karger, 1996).

4.2.3 Ion-exchange chromatography

Ion-exchange chromatography (IEC) has been a widely used technique for the purification and analysis of proteins and peptides for more than 40 years (Choudhary and Horvath, 1996; Henry, 1989). In IEC, the stationary phase (matrix) interacts with oppositely charged proteins via electrostatic interactions, provided the ionic strength is sufficiently low. Since proteins are ampholytes, pH of the mobile phase is of primary importance to ensure that dissolved proteins are sufficiently either below or above their isoelectric points (pI), depending on IEC matrix used. In addition to the protein net charge, the distribution of charged residues on its surface is believed to influence the binding and subsequent separation (Holthuis and Driebergen, 1995). The adsorbed proteins are usually eluted by a gradual increase of ionic strength via a salt gradient that diminishes ionic interactions. Other types of interaction (usually unwanted) can also occur, depending on the matrix structure, and change the elution profile (Stahlberg *et al.*, 1992). The mild separation conditions ensure that the proteins are eluted in the native state and only surface charged residues affect separation. This implies that the net charge, and hence chromatographic behaviour, will change upon denaturation. A number of proteins, e.g. insulin (Blundell *et al.*, 1972), can specifically associate into dimers, tetramers, etc. which will complicate the elution profile. By using denaturants such as 7 M urea, this association can be abolished and uniformity of the protein net charge achieved, and thus ideal single peak behaviour can be restored (Baudyš *et al.*, 1995). Depending on the surface charge of the matrix, two different forms of IEC are practised: cation exchange chromatography and anion exchange chromatography. Three main types of porous matrix materials, i.e. polysaccharide-, silica- and polymer-based, are mostly used. Several new types of IEC supports have been introduced which enable even faster protein separation and characterization by reducing stationary phase mass transfer resistance: non-porous pellicular sorbents (Choudhary and Horvath, 1996), gigaporous sorbents (Lloyd and Warner, 1990; Regnier, 1991), gel-in-the-cage stationary phase (Boschetti, 1994) and continuous bed stationary phase (Liao *et al.*, 1998).

Chromatofocusing is a particular branch of IEC (Sluyterman and Elgersma, 1978). As in isoelectric focusing, the basis for separation is a pH gradient, which in this case, however, is generated in the column by an ampholytic mobile phase. Initially, the sample is loaded onto a positively charged ion exchanger equilibrated at high pH. The column is then eluted with an ampholyte buffer mixture at low pH. As it passes through the column, it forms a pH gradient due to titration of the fixed groups on the support. The gradient gradually moves down the column and proteins are separated according to their pI. The resolution is enhanced by a focusing effect (Sluyterman and Wijdenes, 1978).

4.2.4 Size-exclusion chromatography

Size-exclusion chromatography (SEC), also termed gel permeation chromatography or gel filtration, separates proteins based on their hydrodynamic properties, i.e. their shape and size (Welling and Welling-Wester, 1989). The larger molecules do not enter the pores of the stationary phase and are excluded (and eluted) in the void volume V_o, which is the volume outside the particles. The volume inside the stationary phase V_i is differentially accessible to smaller molecules depending on size of pores. Consequently, the elution volume V_e of a separated protein is between V_o and $V_o + V_i = V_t$ (total accessible volume). This strictly requires that no interactions occur between the support and a solute. Thus, SEC is of relatively low resolution, mostly given by V_i/V_o ratio, and only a limited number of peaks can be obtained (Barth and Boyes, 1992). It is, however, a simple and fast method for estimating molecular weight of globular proteins based on V_e, given that the SEC column was calibrated with appropriate protein standards of known molecular weight. This method is not reliable for proteins of non-globular shape or glycoproteins, or if any interaction with the matrix occurs. Some of these limitations can be overcome by running the protein in a denaturant buffer, which renders this method a measure of polypeptide length, in much the same way as SDS electrophoresis (Mann and Fish, 1972). Wen et al. (1996) described an SEC technique with on-line light scattering, UV absorbance and refractive index detectors, which enables direct determination of molecular weights of proteins and glycoproteins, or to determine stoichiometry of protein complexes. The two main advantages are that the molecular weight measurement is independent of V_e and that the polypeptide molecular weight for glycoproteins can be directly measured regardless of the carbohydrate moiety.

4.3   Electrophoresis

The weak acidic and basic groups of proteins make them polyelectrolytes with amphoteric features. This property enables separation and characterization of proteins and peptides in an externally applied electrical field according to their charge, size and shape (Vesterberg, 1993). This phenomenon is called electrophoresis and is dependent on the pH and composition of the medium. The detailed theoretical understanding of electrophoresis is incomplete (Mosher et al., 1992). On the other hand, there is an overwhelming amount of practical knowledge about the electrophoresis of proteins, so one can easily select the appropriate electrophoretic technique for any particular application. The separation is governed by differences in electrophoretic mobility, which is proportional to the ratio of charge and frictional coefficient of a given macromolecule (Compton and O'Grady, 1991). The most popular electrophoretic techniques for protein characterization by far are different forms of gel electrophoresis. It is a relatively inexpensive method that gives results which are easy to interpret.

4.3.1   Gel electrophoresis

The first analytical high-resolution polyacrylamide gel electrophoresis (PAGE) system for native proteins was introduced by Ornstein (1964) and Davis (1964). In this discontinuous arrangement, the buffer composition differs for the electrolyte chambers, stacking gel and resolving gel. The difference in mobility of leading (Cl⁻) and terminal (glycinate) ions causes the sample proteins to be initially concentrated into a narrow zone in the stacking gel. Once the proteins enter the resolving gel, they are slowed down and separated by the sieving effect of the gel. A more detailed theoretical description is given by Kleparnik and Bocek (1991). Thus, proteins with the same mass to charge ratio will be separated if they are of different sizes. Conversely, if the sample proteins are of the same size, the separation will be governed by the differences in charge. Charge heterogeneity can arise from deamidation during storage of a pharmaceutical protein such as insulin (Brange *et al.*, 1992), or by post-translational processing during production.

The most common discontinuous system based on that of Ornstein and Davis, however, employs the denaturant sodium dodecyl sulphate (SDS) (Laemmli, 1970). Reduced proteins bind a relatively constant amount of SDS (1.4 g) per gram of protein (Reynolds and Tanford, 1970). Thus, the mass to charge ratio becomes constant and fractionation in the gel is controlled strictly by molecular sieving of the gel matrix. With a fixed polyacrylamide concentration and thus pore size, only proteins within a limited range of molecular weight can be separated. To increase resolution power and broaden the protein molecular weight separation range, gradient gels have been introduced (Hames, 1990). They consist of gradients of polyacrylamide concentration, and so the pores become smaller as the proteins move through the gel. The molecules migrate to the limits of porosity and essentially stop moving (band sharpening effect). Some researchers prefer continuous systems such as the Weber and Osborn (1969) continuous SDS PAGE, which uses sodium phosphate as a buffer. SDS PAGE is commonly used to characterize molecular weight and purity of pharmaceutical proteins, as well as to evaluate protein stability over time during formulation and storage. It enables easy detection of proteolytic degradation and formation of covalent aggregates through intermolecular disulphide bridges (Nguyen and Ward, 1993). This analysis is performed in the presence or absence of a reducing agent such as mercaptoethanol. Various glycoproteins and lipoproteins bind SDS irregularly and the resultant SDS–polypeptide complexes migrate anomalously with respect to their molecular weight (Garfin, 1995).

Isoelectric focusing (IEF) is an electrophoretic technique for separating protein molecules in pH gradients (Righetti, 1983). The charge of a protein, as an amphoteric molecule, is determined by the pH of its surroundings. At a particular pH, which is termed the isoelectric point p*I*, the protein net charge is zero and its mobility in an electric field is zero. IEF is carried out in large pore polyacrylamide or agarose gels (Garfin, 1990) and a linear, stationary pH gradient that increases steadily from the anode to the cathode is established with synthetic carrier ampholytes (Just, 1983). Thus, the protein will migrate in the electric field to the point where the pH is the same as the p*I* and migration will stop. Because of focusing effect, IEF is a high-resolution method that can separate proteins differing in p*I* by less than 0.05 pH units. IEF is frequently used to assess the heterogeneity of glycoproteins (Bažil *et al.*, 1986). To maintain proteins' solubility around their p*I*, the presence of urea (3–8 M) or neutral surfactants (1–2%) is often required. There are a number of additional problems associated with conventional IEF which cannot be discussed here because of lack of space, but the result is that a new method has been developed in which the pH gradient is covalently immobilized on the gel matrix: the IPG method (Righetti *et al.*, 1996). By selecting a suitably narrow pH gradient, the resolution limit of Δp*I* of 0.001 was demonstrated (Cossu and Righetti, 1987).

Detection of separated protein bands in gels is mostly accomplished by staining them with dyes or metals (Allen *et al.*, 1984; Syrovy and Hodny, 1991). Other detection methods can also be used, such as detection of enzyme activity using suitable substrates, specific chemical reactions (glycoprotein staining) or immunodetection. Coomassie Brilliant Blue R-250 is the most commonly used stain for protein detection, and has a sensitivity limit of 0.1–1.0 μg of protein per band (Wilson, 1983). This staining solution concomitantly performs another function, which is protein fixation in gels. In the case of silver staining, protein bands have to be fixed first. The sensitivity limit is at least one order lower than that for dye staining (Merril, 1990). Alternatively, single-step staining of SDS-PAGE gels can be achieved by using copper salts (Lee *et al.*, 1987). This method does not require fixation, and protein bands can be quantitatively eluted after chelating the copper. Quantitation of the amount of protein present determined by densitometry is usually problematic since it is known that the band intensities are not simply related to the amount of protein loaded (Syrovy and Hodny, 1991). Thus, for purity or stability estimates of a purified protein pharmaceutical, the results of PAGE must be taken cautiously and interpreted in parallel with results provided by chromatographic methods.

Instead of direct staining, proteins can be electrophoretically transferred onto the surface of an inert membrane, such as nitrocellulose or polyvinylidene difluoride (PVDF) membranes, in a process called electroblotting (Baldo and Tovey, 1989). This procedure produces an exact replica retaining the resolution of the gel. The protein pattern on the replica can be probed with primary antibodies specific for the protein studied by immunoblotting. Secondary antibodies specific for the primary antibody type and conjugated with a suitable reporter group, the most popular being an enzyme such as horseradish peroxidase, are then used to bind to primary antibodies. Analysed protein is finally detected with a specific enzyme substrate that yields a coloured, insoluble product. Alternatively, after the blotting is complete using fibre glass or PVDF membrane, proteins are visualized by a general protein stain, the protein spot of interest is excised and its N-terminal sequence is determined (Mutsadaira, 1987).

4.3.2 Two-dimensional gel electrophoresis

To increase further the resolution of PAGE for separation of complex protein mixtures, two-dimensional techniques have been introduced. The most common technique is the O'Farrell method (1975), reviewed by Dunbar *et al.* (1990). In this procedure, the sample is first separated by IEF in gels cast in glass tubes in the presence of urea and neutral surfactants. The gel rods are extruded from the glass tubes, briefly equilibrated in SDS sample buffer, and placed horizontally on the top of a slab SDS PAGE gel, which is run using the Laemmli system. After separation in the second, perpendicular dimension, the gel is stained to detect protein spots which are distributed in the X-Y plane of the gel according to their pI and molecular weight. Recently, the reproducibility of 2D PAGE has been substantially improved by the replacement of classical IEF using carrier ampholytes with the IPG technique (Dunn and Corbett, 1996). Still, the utility of 2D PAGE for studying pharmaceutical proteins remains limited simply because of the complexity of the method and the fact that the resolution power does not provide any advantage as compared to one-dimensional techniques.

4.3.3 Capillary electrophoresis

The commercial introduction of capillary electrophoresis (CE) instrumentation a decade ago enabled automation of electrophoretic analysis (Landers, 1993). The separation is carried out in narrow diameter (25–100 µm) fused silica capillaries, allowing the application of high electrical fields without thermal problems (efficient heat dissipation), and hence high-speed separation and resolution is possible (Jorgenson and Lukacs, 1983). The use of narrow diameter capillaries also eliminates the need for an anticonvective medium, such as gel. Another important factor to consider in CE is electroosmosis (EO). EO flow (EOF) is a bulk flow of liquid inside the capillary induced by the applied electric field and caused by a permanent inner surface charge of the capillary. The negatively charged silica surface attracts cations, with their hydration shells forming an electrical double layer. In the electric field, the positive charge starts to move towards the anode, dragging the solvent along (Van de Goor, 1995). EOF usually dominates over electrophoretic mobility, and thus not only positively charged, but also neutral and negatively charged peptides are swept through the detector to the anode. EOF can be controlled by coating the capillary with non-interactive coatings (Chien and Burgi, 1992). Most CE applications use absorbance detection at far UV (~200 nm) to compensate for the very short optical path, which is on average 200 times shorter than the conventional 1 cm path length flow cells used for HPLC. Laser-induced fluorescence is also employed. The need to obtain additional structural information after separation has led to the coupling of CE with mass spectrometry (MS), as in the case of HPLC. The most common CE MS interfacing methods are based on either electrospray ionization (ESI) or fast atom bombardment (FAB) (Rickard and Towns, 1996).

The separation of proteins and peptides by CE is based on the mass to charge ratio and so can be viewed as an orthogonal and complementary method to RP HPLC, which predominantly separates according to hydrophobicity. Since coelution of the same proteins or peptides in both methods is highly unlikely, their complementary use can be taken as a proof of identity and purity of the polypeptide under investigation. For optimal resolution using CE, maximum charge differences between species are necessary. In other words, selection of the optimal pH and appropriate buffer are the most important parameters governing the resolution efficiency, as shown for recombinant CD4 variants (Teshima and Wu, 1996). CE is frequently used for peptide mapping, and so again complements chromatographic methods (Cobb and Novotny, 1989).

The examples described so far used a zone CE (CZE) mode, in which zones are migrating independently in a background electrolyte. Hjerten and Zhu (1985) introduced a capillary isoelectric focusing mode. This required the elimination of EO by using coated capillaries, since IEF is a stationary method. To detect focused zones, however, the gradient must be moved through the detector. This can be achieved by hydraulic or electrophoretic mobilization (Garfin, 1995). As with gel IEF, the presence of neutral or zwitterionic detergents is often unavoidable in order to maintain the solubility of focused proteins, as demonstrated by the heterogeneity characterization of the glycoforms of recombinant tissue plasminogen activator (Yim, 1991). CE is also used to determine protein molecular weight by adopting SDS PAGE principles. However, a number of technical obstacles were encountered using crosslinked gels (Karger *et al.*, 1996). Only after the introduction of replaceable linear polymers such as dextrans, which act as a sieving matrix due to the polymer chain entanglement (Hjerten *et al.*, 1989), did this method start to gain popularity. It also enables accurate determination of molecular weight for glycoproteins using a Ferguson plot (Karger *et al.*, 1996).

4.4 Structural characterization

Chromatographic and electrophoretic methods, regardless of how powerful they are as analytical tools, provide only limited amounts of information. Structural data (covalent structure), if available, represent a higher qualitative level of information for peptide and protein characterization. However, without the application of separation methods the task of obtaining structural data would be nearly impossible, and thus chromatographic and electrophoretic methods play a very important role in structural data acquisition.

4.4.1 *Primary structure*

Every protein chemist probably performs amino acid analysis first when characterizing a protein structurally (Irvine, 1997; Smith, 1997). Next, N-terminal sequence is determined by a cyclic chemical procedure known as Edman degradation (Edman and Begg, 1967). In each cycle, the N-terminal amino acid is removed and identified using RP HPLC while the amino terminal of the next amino acid is exposed. An automated version allows for a routine determination of a sequence of up to 30 amino acid residues (Hewick *et al.*, 1981). A similar automated chemical procedure has recently been introduced for the C-terminal sequence determination, although with much lower efficiency (~3 residues) (Bailey and Miller, 1997). This type of information is valuable since it identifies the sites of hydrolytic cleavage in a pharmaceutical protein that occurred during storage, allowing the design of a formulation with minimum degradation. Further, N- and C-terminal sequencing confirms the integrity of the polypeptide chain in the final product evaluation. It will also detect any significant protein impurities in the final product. Determination of the complete sequence is nevertheless usually required for any recombinant protein product in the validation process. Batch-to-batch consistency of the primary structure, including post-translational modifications of the product, is most conveniently examined by the peptide mapping techniques discussed in section 4.2 above. A well-developed fingerprint method has the capability of detecting changes below the 5% level (Jones, 1994). Individual peaks can be easily collected and sequenced, and thus the internal sequences of the investigated protein obtained. In this manner, the amount of the structural information obtained is dramatically increased, but the procedure is time-consuming and expensive. As mentioned previously, recent developments of mass spectrometry (MS) interfaces (especially ESI and FAB) for LC and CE enable one to obtain direct, on-line, accurate information on the peptide mass composition of the eluted peaks (peptide mass vs retention time distribution). By combining chromatographic, sequencing, and MS methods, it is now possible to characterize and identify negligible changes in a protein molecule that involve one amino acid side-chain only, such as deamidation of an asparagine or glutamine residue, or oxidation of methionine to methionine sulphoxide (Ahern and Manning, 1992; Cleland and Langer, 1994b). These modification reactions occur frequently during storage of a formulated protein pharmaceutical (Cleland *et al.*, 1993).

A growing number of protein pharmaceuticals, such as tissue plasminogen activator, are glycoproteins. Thus, structural characterization of the carbohydrate moiety or moieties of these glycoproteins, including carbohydrate type, branching and heterogeneity, is becoming an integral step in the development process (Guzzetta and Hancock, 1996). The attachment of the carbohydrate moieties to a protein and their reprocessing occur in the endoplasmic reticulum and the Golgi apparatus, and are under enzymatic control, which results in substantial heterogeneity. The most commonly known example for N-linked glycans is the heterogeneity in the number of sialic acids per molecule, which is responsible for the charge heterogeneity (Teshima and Wu, 1996). On the other hand, the number of possible glycan structures is limited due to the high stereospecificity of the transglycosidases involved in the synthesis (Varki, 1993). For structural characterization, the glycopeptides are first identified using comparative peptide mapping before and after treatment with enzymes that specifically remove the glycan moieties from the glycoprotein

(Reinhold *et al.*, 1996; Küster *et al.*, 1997), or by the LC/ES/MS method using a contour plot technique (Hancock *et al.*, 1996). These procedures also determine the sites of attachment of the glycans to the polypeptide chain. The glycan structure determination requires the use of sophisticated methods such as NMR spectroscopy, used to determine linkage positions and anomeric configurations, matrix-assisted laser desorption/ionization (MALDI) MS in combination with normal phase HPLC and exoglycosidase sequencing (Küster *et al.*, 1997), or ES MS in combination with collision-induced dissociation and subsequent detection of the fragments (ES-MS/CID/MS), which also provides linkage and branching information (Reinhold *et al.*, 1996).

4.4.2 *Mass spectrometry*

In the past two decades, remarkable progress has been made in using mass spectrometry (MS) for structural analysis of macromolecules and, at present, MS must be considered an established technique in protein chemistry. This tremendous progress was enabled by the introduction of new ionization methods that permitted the production of gas phase peptide and protein ions without prior chemical modification and with minimum fragmentation (Caprioli and Suter, 1995). MS analysis basically comprises three steps: (1) formation of gas phase ions in vacuum (ionization step); (2) ion separation according to their mass to charge ratio (*m/z*); and (3) detection of separated ions and determination of *m/z*.

The first method, fast atom bombardment (FAB) MS, was introduced in the early 1980s (Barber *et al.*, 1981; Dell and Morris, 1982). The sample must be dissolved in an appropriate non-volatile liquid matrix, usually glycerol or nitrobenzyl alcohol. Desorption of peptide ions is induced by bombardment with a high-energy beam of Ar or Xe atoms, or Cs^+ ions. The detection limit of most peptides in a standard FAB MS experiment is at the picomolar level. The upper mass limit that is considered routine is about 5000 to 7000 Da, depending on the nature of polypeptide. Continuous flow (CF) FAB MS was developed to allow the introduction of samples into the spectrometer directly in aqueous solutions on-line (Caprioli, 1990), and is more than 10 times as sensitive as conventional FAB MS. Recently, a massive cluster impact (MCI) ionization technique using glycerol–NH_4^+ clusters has demonstrated an increased sample mass range through multiple charging of higher molecular weight polypeptides using quadrupole and magnetic sector instruments (Siuzdak, 1996); this technique may become a practical alternative to electrospray ionization (ESI) MS. The most prominent disadvantages of current FAB MS techniques are a severe drop in sensitivity at high mass, relatively low sensitivity compared to ESI MS or MALDI MS, high background matrix peaks in the 200–500 Da range and the requirement of solubility in a matrix.

MALDI MS, first described by Karas and Hillenkamp in 1988, permits the analysis of high molecular weight proteins with high sensitivity. The molecular weight range detected is practically unlimited since a time-of-flight (ToF) analyser is used for ion detection, and spectra of as little as 1 fmol of proteins have been reported (Karas *et al.*, 1989). MALDI uses a non-volatile solid matrix, such as nicotinic acid, benzoic acid derivatives, or cinnamic acid derivatives, which absorbs the laser UV radiation and results in the evaporation of the matrix and sample embedded within the matrix. The matrix protects the sample from radiation damage and facilitates the ionization process (Beavis, 1992). In contrast to FAB, very little suppression is observed in mixture analysis (Beavis and Chait, 1990), and MALDI MS is thus suitable for the analysis of complex peptide and protein mixtures. A great advantage of MALDI over other MS ionization techniques is its high tolerance to the presence of salts (Beavis and Chait, 1996). The main draw-backs of MALDI MS are lower resolution compared to other MS methods, and a high matrix background, which can constitute a problem for peptides with a mass of less than 1000 Da. The background peaks are highly dependent on the type of matrix used.

The two techniques described above are based on the desorption/ionization principle, whereas ESI MS is based on an entirely different ionization technique. ESI is a method used to produce gaseous ionized molecules from an aqueous solution by desolvation of liquid droplets in a high electric field at atmospheric pressure. The ions are then chan-nelled into the high vacuum of the analyser, which is usually a quadrupole type analyser (Roepstorff, 1995). Fenn *et al.* (1989) were first to use this technique for analysis of proteins. ESI is very effective and leads to very highly charged states. Consequently, ESI MS does not require instruments with a high mass range, and the molecular weight of proteins can be determined with high accuracy. Proteins up to 100 000 Da can be ana-lysed. The sensitivity limit is between those for FAB and MALDI MS, somewhere around 100 fmols. It is the most accurate method for protein studies among the MS techniques available, and the mode of ion formation makes it ideal for coupling with liquid phase dependent separation techniques such as HPLC or CE. The most promin-ent ESI MS disadvantage is a low tolerance for salts and mixtures, so the purity of the sample is important. The analysis of mixtures can also be confusing due to multiple charging.

Three MS methods are commonly used today for protein molecular weight determina-tion – ESI MS, MALDI MS and plasma desorption (PD) MS. Each method has specific advantages and limitations as discussed above. PD MS is perhaps the method of choice for routine analysis of polypeptides below 5000 Da because of its simplicity (Roepstorff, 1995). However, if a very limited quantity of sample is available, MALDI or ESI MS will be preferred. MALDI MS will also be chosen for analysing glycoproteins, where the extreme broadness in their peaks reflects the heterogeneity of their carbohydrate moiety or moieties.

MS techniques are currently being used to obtain mass spectrometric peptide maps, i.e. a list of molecular weights of peptides derived by cleavage of the protein with a specific protease. The experimental MS data are then compared with a database of peptide mass values calculated by applying the enzyme cleavage rules to a protein sequence database, such as SwissProt. This procedure is sufficiently tolerant to frequent database errors and is very sensitive. Therefore, it can be used to identify a protein separated by 2D-PAGE after blotting the protein of interest onto a PVDF membrane and subjecting it to *in situ* proteolytic digestion (Henzel *et al.*, 1993). MALDI MS is the method best suited for such purposes. MS peptide maps, usually in combination with HPLC or CE, are often used to identify different types of post-translational modifications, such as glycosylation, crosslinking through S–S bridges, acylation, or phosphorylation (Martin *et al.*, 1994; Smith and Zhou, 1990). It is not possible to address these techniques in detail here.

MS spectrometry is also making its way towards obtaining total sequence information of some peptides. A few pmols of a peptide is usually sufficient. Its molecular weight should not exceed 2500 Da. The ionization methods currently used are soft in order to create an abundance of molecular ions $(M+H)^{n+}$ and, consequently, the extent of fragmenta-tion of the peptide molecule is too low to allow structure determination. Thus, more complex procedures and instrumentation are required. This involves tandem MS (MS/MS) methodology that performs the ionization, the peptide ion backbone cleavage using collision induced dissociation (CID), and, finally, identification of the fragments. The two different types of tandem MS instruments commonly employed for peptide sequencing are the triple quadrupole spectrometer, and four sector (double focusing) magnetic/ electrostatic mass spectrometer (Caprioli and Suter, 1995). The first type utilizes the second quadrupole as a collision cell to generate fragment ions. The first quadrupole is used to select a particular molecular ion which collides in the second quadrupole with argon atoms. The formed fragments are analysed with the third quadrupole. This arrangement is used with ESI and is called low-energy CID MS/MS (Hunt *et al.*, 1991). The low-energy CID causes only fragmentation of the peptide backbone. The second type, four sector MS/MS, with a collision cell in the middle, provides high-energy CID which also causes fragmentation in side-chains. This technique, using FAB ionization, has been pioneered by Biemann (1990) and enables one to distinguish the isomeric amino acid residues Ile and Leu, as well as Gln and Lys, which have identical molecular mass.

Alternatively, a wet chemistry procedure, such as Edman degradation, can be combined with MALDI MS in a technique called protein ladder sequencing (Beavis and Chait, 1996). First, the cyclic Edman degradation is performed in the presence of a small amount of N-terminal amino blocking reagent which generates a set of truncated peptides, each differing from the next by a single amino acid residue. Second, MALDI MS is applied to analyse the mass composition of the complete fragment set and the sequence is read out. Finally, ESI MS has been applied to measurements of non-covalent interactions in protein–ligand and protein–protein complexes (Katta and Chait, 1991), and in protein folding studies using deuterium exchange (Miranker *et al.*, 1993).

4.5 Secondary and tertiary structure

Chemical changes that occur during purification and storage of a protein pharmaceutical are reflected in the changes in its covalent structure, which include post-translational modifications. Of equal importance for therapeutic proteins is their physical stability. Physical stability of a protein is directly correlated with resistance to changes in its secondary, tertiary and quaternary (when applicable) structures due to changes in the environment that surrounds the protein molecule. In other words, the integrity of the native conformation of the polypeptide chain must be preserved to ensure the protein's biological and pharmacological activity. Altered conformations can lead to altered (usually decreased) activity, and quite often to increased immunogenicity of the product, as in the case of insulin delivery with insulin pumps (Jeandidier *et al.*, 1995). This means that techniques that provide some kind of structural information must be used. Different physical methods will provide different levels of structural information with regard to the protein hydrodynamic properties, physicochemical behaviour (i.e. association, binding), three-dimensional (3D) structure, basic thermodynamic properties, etc. Obviously, the assessment of structural changes in the protein molecule is a very complex task and the data acquired by a single technique usually report on only some aspects of the structural alteration. Thus, the rational strategy is to apply a concerted approach in which the protein structure is evaluated by several different physical methods. This will usually yield a sufficient amount of structural information to provide a more complete picture of the physical (and chemical) state of the pharmaceutical protein studied.

The most commonly used physical methods to assess protein secondary and tertiary structure are spectroscopic techniques. In general, spectroscopy is the study of interaction of electromagnetic radiation with molecules, including protein molecules (Miura and Thomas, 1995). These interactions include absorption, emission and scattering of photons, and when quantitatively measured they provide structural information on the protein molecule studied. Depending on the radiation wavelength, the spectroscopy methods are classified as ultraviolet, visible, infrared, microwave or radiowave spectroscopy. X-ray and neutron diffraction methods, which involve inelastic scattering of photons or neutrons, are unique among physical methods in that they allow the determination of protein 3D structure at atomic resolution. However, they require preparation of protein crystals and are very complex (Blundell and Johnson, 1976; Stout and Jensen, 1989). An alternative method currently available for determining 3D structure of proteins is nuclear magnetic resonance (NMR), which involves radiowave absorption, and in which molecules are examined in solution (no requirement for crystals) under near-physiologic conditions (Evans, 1995; Bangerter, 1995). These 3D methods are outside the scope of the chapter and will not be discussed further.

Basic spectroscopic methods are capable of measuring only an average integral signal change. Thus, these methods cannot resolve contributions from individual amino acid residues in the polypeptide chain, and generally provide indirect information on the structure and conformation. The same is true for a situation when more than one protein conformer population is present in solution or solid state. In specific cases, such as when only a single tryptophan residue is available or a single optically active reporter group can be chemically introduced into a protein molecule in a site-directed manner (i.e. specific modification of a single cysteine with fluorescent moiety), more specific information about changes in the protein structure can be obtained.

4.5.1 Absorption and fluorescence spectroscopy

Transitions between different electronic energy levels of a peptide or protein molecule are accompanied by absorption of UV or visible light which can be measured by UV/visible spectroscopy. These transitions originate in specific amino acid residues and can be divided into three distinctive wavelength intervals (Wetlaufer, 1962). UV light absorption with $\lambda > 250$ nm can be attributed to the aromatic side-chains of Phe, Tyr and Trp, and cystine disulphide groups, and gives rise to a protein absorption maximum around 280 nm. Since UV absorption obeys the Beer-Lambert law, the concentration of a purified protein can be conveniently determined if the molar extinction coefficient ε_λ at 280 nm is known. The determination of the extinction (absorption) coefficient (Pace *et al.*, 1995) requires an accurate measurement of the protein concentration, for instance by quantitative amino acid analysis. If the coefficient is not available, simple photometric techniques, such as the Lowry *et al.* (1951) or Bradford (1976) method, can be used with the help of a defined protein standard, which is usually serum albumin.

UV light absorption having λ from 250 to 210 nm again comes mostly from aromatic side-chains, but substantial absorption also occurs from histidine, cystine, cysteine and methionine. Finally, for λ smaller than 210 nm, significant absorption results from electronic transitions in peptide bonds, in addition to other contributions named above (Fasman, 1992). The UV spectrum of a protein is by no means identical to the UV spectrum calculated by superposition of UV spectra of component amino acids, because the chromophore environment directly affects the wavelength of the absorption maximum and the corresponding absorption magnitude (λ_{max} and ε_{max}) (Pace *et al.*, 1995). In general, the hydrophobic environment of the protein interior induces a shift of the chromophore absorption band to a longer wavelength (the so called red shift). The effects of protein structure on absorption spectra can be studied by difference spectroscopy and utilized, for instance, to follow folding/unfolding of a protein molecule (Mach *et al.*, 1993).

A change in the chromophore environment in a protein molecule is, however, much easier to investigate using fluorescence spectroscopy (Cantor and Schimmel, 1980). This technique basically measures magnitude and wavelength of emitted light due to electronic transitions from the excited state S_1 to the ground state S_0 of a molecule. The energy of the transition to the ground state is lower than that for transition to the excited state (absorption) because a portion of the absorbed energy is converted into waste heat, and due to solvent relaxation (Jiskoot *et al.*, 1995). Thus fluorescence emission occurs at a longer wavelength than absorption. Proteins exhibit fluorescence in the 300–400 nm range due to the presence of the aromatic amino acids tryptophan and tyrosine. The fluorescence efficiency (quantum yield) increases significantly as the hydrophobicity of the fluorescent chromophore's (fluorophore) environment increases and is accompanied by a shift of the fluorescent band to a shorter wavelength (a blue shift). Thus, a single buried tryptophan residue in intact bovine growth hormone has an emission maximum of around 330 nm, while upon complete unfolding, this maximum is red-shifted to about 350 nm (Havel *et al.*, 1988).

4.5.2 Circular dichroism spectroscopy

Changes in the secondary structure of a protein are perhaps best monitored by circular dichroism (CD) spectroscopy (Bloemendal and Johnson, 1995). In principal, CD spectroscopy measures the dependence of $\Delta\varepsilon$, which is the difference of the absorption coefficients for the left- and right-handed polarized light, on the wavelength of incident light. Since CD is an absorption difference, it is observed only in absorption bands which originate from chromophores with inherent structural asymmetry or whose environment is asymmetric (chiral) (Miura and Thomas, 1995). Due to the predominance of peptide bonds relative to other chromophores in proteins, the UV CD spectra of proteins for the wavelength interval of 170 nm to 250 nm (far-UV CD spectrum) are dominated by the contributions of peptide bonds. In other words, a far-UV CD spectrum of a protein directly reflects the specific dihedral angles of its polypeptide chain. Accordingly, far-UV CD spectra are highly sensitive probes of polypeptide main-chain secondary structure. Far-UV CD provides a fast method for determining the overall secondary structure content of aqueous proteins, such as α-helix, antiparallel and parallel β-sheet, β-turns and irregular structures (Johnson, 1988). The prediction is based on the comparison of the known secondary structures of proteins, usually determined by X-ray crystallography, and their CD spectra, which results in a definition of the spectral contributions of secondary structure elements in globular proteins (Manning, 1989). However, the CD predictive method should be applied cautiously, not only for proteins with strongly distorted (dihedral angles) secondary structure elements (Hadden *et al.*, 1994), but also for proteins with significant numbers of aromatic amino acids (Manning and Woody, 1989). As mentioned above, far-UV CD is a valuable, fast and easy-to-use method for monitoring protein structure changes as a function of external conditions, for instance during protein refolding (Van der Vies *et al.*, 1992).

Aromatic amino acids Phe, Tyr and Trp give rise to the specific CD absorption bands in the near-UV region with wavelength from 250 nm to 320 nm, which is referred to as near-UV CD spectroscopy (Strickland, 1974). Disulphide groups (–S–S–) also contribute

to near-UV CD spectra and the shape and intensity of this band, in particular, is con-
formationally sensitive. Quantitative treatment, as for any type of UV spectrum, is nearly
impossible (Scharnagl and Schneider, 1991), hence near-UV CD is mostly used for a fast
qualitative comparison of structures of homologous proteins, site-directed mutants, or for
monitoring tertiary structure changes of a protein under different conditions. A highly
mobile CD active side-chain will usually give a weak CD signal, while a highly rigid
conformation is interpreted as a cause of intense near-UV CD bands (Bailey *et al.*, 1982).
Near-UV CD spectroscopy can be used to study protein association if aromatic side-chain(s)
are part of the interacting interface. Thus, concentration dependency of association of
insulin monomer into dimer can be conveniently followed by changes in near-UV CD
spectrum (Goldman and Carpenter, 1974). Non-associating monomeric insulin analogues
do not show any CD spectral changes with regard to their concentration (Brange *et al.*,
1988). Similarly, dimerization of human relaxin was investigated by near-UV CD spectroscopy (Shire *et al.*, 1991).

4.5.3 *Infrared spectroscopy*

Infrared (vibrational) spectroscopy can also provide insight into protein secondary structure.
Basically, it measures the energy absorbed during the transition of a protein molecule
from one vibrational mode to another (Krimm and Bandekar, 1986). The complexity of a
protein IR spectrum arises from numerous atoms or groups of atoms of a protein molecule
contributing to a particular vibration. This effect, however, quickly fades away with dis-
tance, hence it is possible in principle to separate contributions from groups of atoms
within the polypeptide chain, i.e. the amide linkage of a protein. At present, infrared
spectroscopic studies are carried out by use of Fourier transform infrared (FTIR) instruments. The vibrational modes associated with trans-peptide groups of proteins (amide
bands) are very sensitive to the conformation of the polypeptide backbone (Arrondo
et al., 1993). In theory, nine different backbone amide bands are considered informative
concerning protein secondary structure (Cooper and Knutson, 1995). In practice, only amide
I band (1620–1700 cm^{-1}) and amide III band (1200–1330 cm^{-1}) are currently being used
for estimation of the secondary structure content using local absorption maxima (Susi and
Byler, 1986; Surewicz *et al.*, 1993). However, to resolve overlapping local absorption
maxima of the naturally broad infrared bands, mathematical resolution enhancement
(deconvolution) techniques must be applied (Mantsch *et al.*, 1986). Alternatively, derivative procedures can be used (Kauppinen *et al.*, 1981). Another problem is the presence of
a huge H_2O band which interferes with amide I band and must be subtracted. It requires
the use of sophisticated computerized techniques, and even then the process is rather
subjective. When these factors are considered together, quantitative analysis of secondary
structure content by FTIR remains problematic. However, it provides a complement to
CD methods, and the CD and FTIR methods can be combined to improve estimation of
the secondary structure content (Sarver and Krueger, 1991).

Obviously, the presence of water is not a serious problem if the FTIR spectrum is
acquired in the solid state, which is used to evaluate aspects of protein structure in the
'dry state' (Arakawa *et al.*, 1993). This technique has recently become popular for
characterizing the effects of additives and excipients on structural changes of protein
pharmaceuticals during freezing or lyophilization (Pikal, 1994; Prestrelski *et al.*, 1994).
The conformation of many proteins can change substantially during freeze-drying. If,
however, the native conformation and, thus, bioactivity after storage and reconstitution
are recovered, this should not constitute a concern for a particular formulation.

4.5.4 *Other methods*

Hydrodynamic methods, which represent a large group of independent methods, allow macromolecular size, shape and interactions to be determined and can be divided into three broad categories (Stafford and Schuster, 1995). In principle, these methods analyse the transport of proteins through fluids. The first category includes various types of chromatographic and electrophoretic techniques which were discussed in detail earlier. The second category includes various types of viscosity measurements in which the effects of proteins on the flow properties of the bulk solvent are measured. By extrapolating to zero concentration, intrinsic viscosity can be obtained and molecular weight of a protein determined (Yang, 1961; Eisenberg, 1976).

Finally, the third category includes non-destructive analytical centrifugation methods, particularly sedimentation velocity and equilibrium techniques. Analytical centrifugation has been available as an analytical method for a long time (Schachman, 1959; Teller, 1973). Currently, the technique is enjoying a renaissance due to the recent development of simple, modern, computerized instruments. The first method uses relatively high speed (high *g* force) and measures the sedimentation velocity of a protein solute boundary by absorption or Rayleigh optical systems, which enables the calculation of a sedimentation coefficient. Moreover, if the diffusion coefficient is available, which can easily be obtained using photon correlation spectroscopy (Chu, 1974) and autocorrelation function processing (Koppel, 1972), the molecular weight of a protein can be calculated using the Svedberg equation.

The second ultracentrifugal method uses relatively low *g* force and the centrifuge is run until equilibrium between sedimentation and diffusion (chemical potential) forces is reached. There is practically no limit on the protein molecular weight being determined and it is an absolute method requiring no calibration or comparison with standards. Since the system is at equilibrium, it is possible to obtain thermodynamic information about the protein molecule and its interactions. For a non-interacting system composed of a single protein, the primary information obtained is the protein molecular weight and its purity. If the protein under study specifically associates (i.e. monomer–dimer equilibrium), the equilibrium is established not only between diffusion and sedimentation, but also between interacting species. By computing molecular weight averages across the cell for different concentrations and by analysis of the molecular weight distributions, assuming different models for association, one may obtain the stoichiometries and equilibrium constants of the interactions. If the equilibrium constants are measured as a function of temperature, the standard enthalpies and entropies of the interactions can also be determined. Such information is vital for complete understanding of the stability of biopharmaceuticals having quaternary structure. An example of a 'classic' pharmaceutical protein which associates is insulin forming dimers, tetramers and hexamers. Its stable formulation is actually hexameric (Brange *et al.*, 1988). Naturally, the association properties of insulin and its analogues with altered association characteristics were studied by ultracentrifugal analysis (Brems *et al.*, 1992).

The shape and size of a protein molecule can also be investigated by small-angle light scattering (Williams, 1986), small-angle X-ray scattering or small-angle neutron scattering (Chen and Bendedouch, 1986). The basis of these methods, similar to the wide-angle X-ray and neutron scattering crystallographic methods mentioned above, is the diffraction of a wave on a solid particle (molecule). Perhaps the most popular light scattering method for characterizing pharmaceutical proteins is the already noted quasielastic laser light scattering (QELS) technique (Chu, 1974), mainly because of its simplicity. It easily detects submicrometre size (~100 nm) aggregated protein particles because the intensity of scattered light changes with the sixth power of radius (Tanford, 1961), but consequently, the current methods have difficulty in quantitating the aggregated fraction, since the size of a protein molecule is usually 3–10 nm, and also because the autocorrelation function is composed of an unknown number of exponential terms. Classical static light scattering (Tanford, 1961) can be applied to determine the protein molecular weight, and the technique is now used in special flow-through detectors for chromatography as discussed in section 4.2.4.

4.6   Conclusion

This chapter has tried to cover the major methods commonly used in peptide and protein analysis and characterization, with emphasis on pharmaceutically relevant proteins. Of course, for a particular protein, the selection of the methods will depend first of all on the physical and chemical properties of the protein studied. Perhaps of the same importance is the form of the protein, i.e. its formulation design, which can incorporate different excipients and stabilizers, precluding the application of a number of analytical methods. Finally, the selection of the methods will be influenced by the purpose the analytical methods are being chosen for; either it is for protein content and purity monitoring during protein production, or for analysis of the final formulation of a biopharmaceutical protein with regard to its stability and degradation etc. In any case, a scientist working in any of these or related fields must have a deep knowledge and understanding of the underlying principles of these analytical methods, which will enable him or her to select the most appropriate set of these methods for a particular goal.

References

ADVANT, S.J., KOMAREK, D., ADAMS, G., ZHANG, Y. and SEETHARAM, R., 1997, Challenges in the analysis of formulated proteins in the presence of excipients, *Genetic Engineering News*, 1 June, 20.

AGUILAR, M.I. and HEARN, M.T.W., 1996, High-resolution reversed-phase high-performance liquid chromatography of peptides and proteins, in KARGER, B.L. and HANCOCK, W.S. (eds) *High Resolution Separation and Analysis of Biological Macromolecules, Part A: Fundamentals*, Methods in Enzymology, Vol. 270, pp. 3–24, San Diego: Academic Press.

AHERN, T.J. and KLIBANOV, A.M., 1988, Analysis of processes causing thermal inactivation of enzymes, *Methods of Biochemical Analysis*, **33**, 91–127.

AHERN, T.J. and MANNING, M.C. (eds), 1992, *Stability of Protein Pharmaceuticals, Part A: Chemical and Physical Pathways of Protein Degradation*, Pharmaceutical Biotechnology, Vol. 2, New York: Plenum Press.

AITKEN, A. and LEARMONTH, M., 1997a, Analysis of sites of protein phosphorylation, in SMITH, B.J. (ed.) *Protein Sequencing Protocols*, pp. 293–306, Totowa, NJ: Humana Press.

1997b, Quantitation and location of disulfide bonds in proteins, in SMITH, B.J. (ed.) *Protein Sequencing Protocols*, pp. 317–328, Totowa, NJ: Humana Press.

ALLEN, R.C., SARAVIS, C.A. and MAURER, H.R., 1984, *Gel Electrophoresis and Isoelectric Focusing of Proteins: Selected Techniques*, Berlin: Walter de Gruyter.

ARAKAWA, T. and TIMASHEFF, S.W., 1982, Stabilization of protein structure by sugars, *Biochemistry*, **21**, 6536–44.

ARAKAWA, T., PRESTRELSKI, S.J., KENNEY, W.C. and CARPENTER, J.F., 1993, Factors affecting short-term and long-term stabilities of proteins, *Advanced Drug Delivery Reviews*, **10**, 1–10.

ARRONDO, J.L.R., MUGA, A., CASTRESANA, J. and GONI, F.M., 1993, Quantitative studies of the structure of proteins in solution by Fourier transform infrared spectroscopy, *Progress in Biophysics and Molecular Biology*, **59**, 23–56.

BAILEY, J.C., MARTIN, S.R. and BAYLEY, P.M., 1982, A circular dichroism study of epidermolytic toxins A and B from *Staphylococcus aureus*, *Biochemical Journal*, **203**, 775–778.

BAILEY, J.M. and MILLER, C.G., 1997, Automated methods for C-terminal protein sequencing, in SMITH, B.J. (ed.) *Protein Sequencing Protocols*, pp. 259–269, Totowa, NJ: Humana Press.

BALDO, B.A. and TOVEY, E.R. (eds), 1989, *Protein Blotting: Methodology, Research and Diagnostic Applications*, Basel: Karger.

BANGERTER, B.W., 1995, Nuclear magnetic resonance, in GLASEL, J.A. and DEUTSCHER, M.P. (eds) *Introduction to Biophysical Methods for Protein and Nucleic Acid Research*, pp. 317–379, San Diego: Academic Press.

BARBER, M., BORDOLI, R.S., SEDGWICK, R.D. and TYLER, A.N., 1981, Fast atom bombardment of solids (F.A.B.): a new ion source for mass spectrometry, *Journal of Chemical Society, Chemical Communications*, 325–327.

BARTH, H.G. and BOYES, B.E., 1992, Size exclusion chromatography, *Analytical Chemistry*, **64**, 428R–442R.

BAUDYŠ, M., UCHIO, T., MIX, D., WILSON, D. and KIM, S.W., 1995, Physical stabilization of insulin by glycosylation, *Journal of Pharmaceutical Sciences*, **84**, 28–33.

BAŽIL, V., HOREJSI, V., BAUDYŠ, M., KRISTOFOVA, H., STROMINGER, J.L., KOSTKA, V. and HILGERT, I., 1986, Biochemical characterization of a soluble form of the 53-kDa monocyte surface antigen, *European Journal of Immunology*, **16**, 1583–1589.

BEAVIS, R.C., 1992, Matrix assisted ultraviolet laser desorption: evolution and principles, *Organic Mass Spectrometry*, **27**, 653–659.

BEAVIS, R.C. and CHAIT, B.T., 1990, Rapid sensitive analysis of protein mixtures by mass spectrometry, *Proceedings of National Academy of Sciences USA*, **87**, 6873–6877.

1996, Matrix assisted laser desorption ionization mass spectrometry of proteins, in KARGER, B.L. and HANCOCK, W.S. (eds) *High Resolution Separation and Analysis of Biological Macromolecules, Part A: Fundamentals*, Methods in Enzymology, Vol. 270, pp. 519–551, San Diego: Academic Press.

BENEDEK, K. and SWADESH, J.K., 1991, HPLC of proteins and peptides in the pharmaceutical industry, in FONG, G.W. and LAM, S.K. (eds) *HPLC in the Pharmaceutical Industry*, pp. 241–302, New York: Marcel Dekker.

BIEMANN, K., 1990, Sequencing of peptides by tandem mass spectrometry and high energy collision induced dissociation, in MCCLOSKEY, J.A. (ed.) *Mass Spectrometry*, Methods in Enzymology, Vol. 193, pp. 455–479, San Diego: Academic Press.

BLOEMENDAL, M. and JOHNSON, W.C., JR, 1995, Structural information on proteins from circular dichroism spectroscopy: possibilities and limitations, in HERRON, J.N., JISKOOT, W. and CROMMELIN, D.J.A. (eds) *Physical Methods to Characterize Pharmaceutical Proteins*, pp. 65–100, New York: Plenum Press.

BLOHM, D., BOLLSCHWEILER, C. and HILLEN, H., 1988, Pharmaceutical proteins, *Angewandte Chemie: International Edition in English*, **27**, 207–225.

BLUNDELL, T.L., DODSON, G.G., HODGKIN, D.M.C. and MERCOLA, D.A., 1972, Insulin: the structure in the crystal and its reflection in chemistry and biology, *Advances in Protein Chemistry*, **26**, 279–402.

BLUNDELL, T.L. and JOHNSON, L.N., 1976, *Protein Crystallography*, New York: Academic Press.

BOSCHETTI, E., 1994, Advanced sorbents for preparative protein separation purposes, *Journal of Chromatography A*, **658**, 207–236.

BRADFORD, M.M., 1976, A rapid and sensitive method for the quantitation of microgram quantities of protein utilizing the principle of protein–dye binding, *Analytical Biochemistry*, **72**, 248–254.

BRANGE, J., RIBEL, U., HANSEN, J.F., DODSON, G., HANSEN, M.T., HAVELUND, S., MELBERG, S.G., NORRIS, F., NORRIS, K., SNEL, L., SØRENSEN, A.R. and VOIGHT, H.O., 1988, Monomeric insulins obtained by protein engineering and their medical applications, *Nature*, **333**, 679–682.

BRANGE, J., LANGKJAER, L., HAVELUND, S. and VØLUND, A., 1992, Chemical stability of insulin. I. Hydrolytic degradation during storage of pharmaceutical preparations, *Pharmaceutical Research*, **9**, 715–726.

BREMS, D.N., ALTER, L.A., BECKAGE, M.J., CHANCE, R.E., DiMARCHI, R.D., GREEN, L.K., LONG, H.B., PEKAR, A.H., SHIELDS, J.E. and FRANK, B.H., 1992, Altering the association properties of insulin by amino acid replacement, *Protein Engineering*, **5**, 527–533.

CANTOR, C.R. and SCHIMMEL, P.R., 1980, *Biophysical Chemistry. Part II. Techniques for the Study of Biological Structure and Function*, San Francisco: Freeman.

CAPRIOLI, R.M. (ed.), 1990, *Continuous Flow Fast Atom Bombardment Mass Spectrometry*, New York: Wiley.

CAPRIOLI, R.M. and SUTER, M.J.F., 1995, Electrophoretic methods, in GLASEL, J.A. and DEUTSCHER, M.P. (eds) *Introduction to Biophysical Methods for Protein and Nucleic Acid Research*, pp. 147–204, San Diego: Academic Press.

CHEN, S.H. and BENDEDOUCH, D., 1986, Structure and interactions of proteins in solution studied by small angle neutron scattering, in HIRS, C. and TIMASHEFF, S. (eds) *Enzyme Structure, Part K*, Methods in Enzymology, Vol. 130, pp. 79–116, Orlando: Academic Press.

CHIEN, R.L. and BURGI, D.S., 1992, On column sample concentration using field amplification in CZE, *Analytical Chemistry*, **64**, 489A–496A.

CHLOUPEK, R.C., HARRIS, R.J., LEONARD, C.K., KECK, R.G., KEYT, B.A., SPELLMAN, M.W., JONES, A.J.S. and HANCOCK, W.S., 1989, Study of the primary structure of recombinant tissue plasminogen activator by reversed phase high performance liquid chromatographic tryptic mapping, *Journal of Chromatography*, **463**, 375–396.

CHOUDHARY, G. and HORVATH, C., 1996, Ion exchange chromatography, in KARGER, B.L. and HANCOCK, W.S. (eds) *High Resolution Separation and Analysis of Biological Macromolecules, Part A: Fundamentals*, Methods in Enzymology, Vol. 270, pp. 47–82, San Diego: Academic Press.

CHU, B., 1974, *Laser Light Scattering*, New York: Academic Press.

CLELAND, J.L. and LANGER, R., 1994a, Formulation and delivery of proteins and peptides: design and development, in CLELAND, J.L. and LANGER, R. (eds) *Formulation and Delivery of Proteins and Peptides*, ACS Symposium Series, Vol. 567, pp. 1–19, Washington, DC: American Chemical Society.

(eds), 1994b, *Formulation and Delivery of Proteins and Peptides*, ACS Symposium Series, Vol. 567, Washington, DC: American Chemical Society.

CLELAND, J.L., POWELL, M.F. and SHIRE, S.J., 1993, The development of stable protein formulations: a close look at protein aggregation, deamidation and oxidation, *Critical Reviews in Therapeutic Drug Carrier Systems*, **10**, 307–377.

COBB, K.A. and NOVOTNY, M.V., 1989, High sensitivity peptide mapping by capillary zone electrophoresis and microcolumn liquid chromatography using immobilized trypsin for protein digestion, *Analytical Chemistry*, **61**, 2226–2231.

COMPTON, B.J. and O'GRADY, E.A., 1991, Role of charge suppression and ionic strength in the free zone electrophoresis of proteins, *Analytical Chemistry*, **63**, 2597–2602.

COOPER, E.A. and KNUTSON, K., 1995, Fourier transform infrared spectroscopy investigations of protein structure, in HERRON, J.N., JISKOOT, W. and CROMMELIN, D.J.A. (eds) *Physical Methods to Characterize Pharmaceutical Proteins*, pp. 101–143, New York: Plenum Press.

COPSEY, D.N. and DELNATTE, S.Y.J., 1988, *Genetically Engineered Human Therapeutic Drugs*, New York: Stockton Press.

COSSU, G. and RIGHETTI, P.G., 1987, Resolution of G_γ and A_γ foetal haemoglobin tetramers in immobilized pH gradients, *Journal of Chromatography*, **398**, 211–216.

CREIGHTON, T.E., 1993, *Proteins*, 2nd edn, New York: W.H. Freeman and Co.

DAVIS, B.J., 1964, Disc electrophoresis. II. Method and application to human serum proteins, *Annals of New York Academy of Sciences*, **121**, 404–427.

DELL, A. and MORRIS, H.R., 1982, Fast atom bombardment high field magnetic mass spectrometry of 6000 dalton polypeptides, *Biochemical and Biophysical Research Communications*, **106**, 1456–1461.

DUNBAR, B.S., KIMURA, H. and TIMMONS, T.M., 1990, Protein analysis using high resolution two dimensional polyacrylamide gel electrophoresis, in DEUTSCHER, M.P. (ed.) *Guide to Protein Purification*, Methods in Enzymology, Vol. 182, pp. 441–459, San Diego: Academic Press.

DUNN, M.J. and CORBETT, J.M., 1996, Two dimensional gel electrophoresis, in KARGER, B.L. and HANCOCK, W.S. (eds) *High Resolution Separation and Analysis of Biological Macromolecules, Part B: Applications*, Methods in Enzymology, Vol. 271, pp. 177–203, San Diego: Academic Press.

EDMAN, P. and BEGG, G., 1967, A protein sequenator, *European Journal of Biochemistry*, **1**, 80–91.

EVANS, J.N.S., 1995, *Biomolecular NMR Spectroscopy*, New York: Oxford University Press.

EISENBERG, H., 1976, *Biological Macromolecules and Polyelectrolytes in Solution*, Oxford: Oxford University Press.

FASMAN, G.D. (ed.), 1992, *CRC Practical Handbook of Biochemistry and Molecular Biology*, Boca Raton: CRC Press.

FENN, J.B., MANN, M., MENG, C.K., WONG, S.F. and WHITEHOUSE, C.M., 1989, Electrospray ionization for mass spectrometry of large biomolecules, *Science*, **246**, 64–71.

GARFIN, D.E., 1990, Isoelectric focusing, in DEUTSCHER, M.P. (ed.) *Guide to Protein Purification*, Methods in Enzymology, Vol. 182, pp. 459–477, San Diego: Academic Press.

1995, Electrophoretic methods, in GLASEL, J.A. and DEUTSCHER, M.P. (eds) *Introduction to Biophysical Methods for Protein and Nucleic Acid Research*, pp. 53–109, San Diego: Academic Press.

GARNICK, R.L., SOLLI, N.J. and PAPA, P.A., 1988, The role of quality control in biotechnology: an analytical perspective, *Analytical Chemistry*, **60**, 2546–2557.

GEIGERT, J. and GHIRST, B.F.D., 1996, Development and shelf-life determination of recombinant human granulocyte-macrophage colony-stimulating factor (Leukine®, GM-CSF), in PEARLMAN, R. and WANG, Y.J. (eds) *Formulation, Characterization and Stability of Protein Drugs, Case Histories*, Pharmaceutical Biotechnology, Vol. 9, pp. 329–342, New York: Plenum Press.

GOLDMAN, J. and CARPENTER, F.H., 1974, Zinc binding, circular dichroism, and equilibrium sedimentation studies in insulin (bovine) and several of its derivatives, *Biochemistry*, **13**, 4566–4574.

GOODING, K.M. and REGNIER, F.M. (eds), 1990, *HPLC of Biological Macromolecules: Methods and Practical Applications*, New York: Marcel Dekker.

GUZZETTA, A.W. and HANCOCK, W.S., 1996, Analyzing reversed phase peptide maps of recombinant human glycoproteins using LC/ES/MS, in HANCOCK, W.S. (ed.) *New Methods in Peptide Mapping for the Characterization of Proteins*, pp. 181–217, Boca Raton: CRC Press.

HADDEN, J.M., BLOEMENDAL, M., HARIS, P.I., SRAI, S.K.S. and CHAPMAN, D., 1994, Fourier transform infrared spectroscopy and differential scanning calorimetry of transferrins: human serum transferrin, rabbit serum transferrin and human lactoferrin, *Biochimica et Biophysica Acta*, **1205**, 59–67.

HAMES, B.D., 1990, One-dimensional polyacrylamide gel electrophoresis, in RICKWOOD, D. and HAMES, B.D. (eds) *Gel Electrophoresis of Proteins: a Practical Approach*, pp. 1–147, Oxford: IRL Press.

HANCOCK, W.S. (ed.), 1984, *Handbook of HPLC for The Separation of Amino Acids, Peptides and Proteins*, Boca Raton: CRC Press.

HANCOCK, W.S., APFFEL, A., CHAKEL, J., SOUDERS, C., M'TIMKULU, T., PUNGOR, E. and GUZZETTA, A.W., 1996, Reversed phase peptide mapping of glycoproteins using liquid chromatography/electrospray ionization mass spectrometry, in KARGER, B.L. and HANCOCK, W.S. (eds) *High Resolution Separation and Analysis of Biological Macromolecules, Part B: Applications*, Methods in Enzymology, Vol. 271, pp. 403–427, San Diego: Academic Press.

HANSON, M.A. and EDMOND-ROUAN, S.K., 1992, Introduction to formulation of protein pharmaceuticals, in AHERN, T.J. and MANNING, M.C. (eds) *Stability of Protein Pharmaceuticals, Part B: In Vivo Pathways of Degradation and Strategies for Protein Stabilization*, pp. 209–233, New York: Plenum Press.

HAVEL, H.A., KAUFFMAN, E.W. and ELZINGA, P.A., 1988, Fluorescence quenching studies of bovine growth hormone in several conformational states, *Biochimica Biophysica Acta*, **955**, 154–163.

HENRY, M.P., 1989, Ion exchange chromatography of proteins and peptides, in OLIVER, R.W.A. (ed.) *HPLC of Macromolecules: a Practical Approach*, pp. 91–125, Oxford: IRL Press.

HENZEL, W.J., BILLECI, T.M., STULTS, J.T., WONG, S.C., GRIMLEY, C. and WATANABE, C., 1993, Identifying proteins from two dimensional gels by molecular mass searching of peptide fragments in protein sequence databases, *Proceedings of National Academy of Sciences USA*, **90**, 5011–5015.

HEWICK, R.M., HUNKAPILLER, M.W., HOOD, L.E. and DRYER, W.J., 1981, A gas–liquid solid phase peptide and protein sequenator, *Journal of Biological Chemistry*, **256**, 7990–7997.

HJERTEN, S. and ZHU, M., 1985, Adaptation of the equipment for high performance electrophoresis to isoelectric focusing, *Journal of Chromatography*, **346**, 265–270.

HJERTEN, S., VALTCHEVA, L., ELENBRING, K. and EAKER, D., 1989, High performance electrophoresis of acidic and basic low molecular weight compounds and of proteins in the presence of polymers and neutral surfactants, *Journal of Liquid Chromatography*, **12**, 2471–2499.

HOLTHUIS, J.J.M. and DRIEBERGEN, R.J., 1995, Chromatographic techniques for the characterization of proteins, in HERRON, J.N., JISKOOT, W. and CROMMELIN, D.J.A. (eds) *Physical Methods to Characterize Pharmaceutical Proteins*, pp. 243–299, New York: Plenum Press.

HOVGAARD, L., MACK, E.J. and KIM, S.W., 1992, Insulin stabilization and GI absorption, *Journal of Controlled Release*, **19**, 99–108.

HUNT, D.F., KRISHNAMURTHY, T., SHABANOWITZ, J., GRIFFIN, P.R., YATES, J.R., III, MARTINO, P.A., McCORMACK, A.L. and HAUER, C.R., 1991, Peptide sequence analysis by triple quadrupole and quadrupole Fourier transform mass spectrometry, in DESIDERIO, D.M. (ed.) *Mass Spectrometry of Peptides*, pp. 139–158, Boca Raton: CRC Press.

IRVINE, G.B., 1997, Amino acid analysis, in SMITH, B.J. (ed.) *Protein Sequencing Protocols*, pp. 131–138, Totowa, NJ: Humana Press.

JEANDIDIER, N., BOIVIN, S., SAPIN, R., ROSART-ORTEGA, F., ÜRING-LAMBERT, B., REVILLE, P. and PINGET, M., 1995, Immunogenicity of intraperitoneal insulin infusion using programmable implantable devices, *Diabetologia*, **38**, 577–584.

JISKOOT, W., HLADY, V., NALEWAY, J.J. and HERRON, J.N., 1995, Application of fluorescence spectroscopy for determining the structure and function of proteins, in HERRON, J.N., JISKOOT, W. and CROMMELIN, D.J.A. (eds) *Physical Methods to Characterize Pharmaceutical Proteins*, pp. 1–63, New York: Plenum Press.

JOHNSON, W.C., JR. 1988, Secondary structure of proteins through circular dichroism spectroscopy, *Annual Reviews of Biophysics and Biophysical Chemistry*, **17**, 145–166.

JONES, A.J.S., 1994, Analytical methods for the assessment of protein formulations and delivery systems, in CLELAND, J.L. and LANGER, R. (eds) *Formulation and Delivery of Proteins and Peptides*, ACS Symposium Series, Vol. 567, pp. 22–45, Washington, DC: American Chemical Society.

JORGENSON, J.W. and LUKACS, K.D., 1983, Capillary zone electrophoresis, *Science*, **222**, 266–272.

JUST, W.W., 1983, Synthesis of carrier ampholytes for isoelectric focusing, in HIRS, C. and TIMASHEFF, S. (eds) *Enzyme Structure*, Methods in Enzymology, Vol. 91, pp. 281–298, New York: Academic Press.

KAGAWA, Y., TAKASAKI, S., UTSUMI, J., HOSOI, K., SHIMIZU, H., KOCHIBE, N. and KOBATA, A., 1988, Comparative study of the asparagine-linked sugar chains of natural human interferon-β1 and recombinant interferon-β1 produced by three different mammalian cells, *Journal of Biological Chemistry*, **263**, 17508–17515.

KARAS, M. and HILLENKAMP, F., 1988, Laser desorption ionization of proteins with molecular masses exceeding 10 000 daltons, *Analytical Chemistry*, **60**, 2299–2301.

KARAS, M., INGENDOH, A., BAHR, U. and HILLENKAMP, F., 1989, Ultraviolet laser desorption/ ionization mass spectrometry of femtomolar amounts of large proteins, *Biomedical Environmental Mass Spectrometry*, **18**, 841–843.

KARGER, B.L., FORET, F. and BERKA, J., 1996, Capillary electrophoresis with polymer matrices: DNA and protein separation and analysis, in KARGER, B.L. and HANCOCK, W.S. (eds) *High Resolution Separation and Analysis of Biological Macromolecules, Part B: Applications*, Methods in Enzymology, Vol. 271, pp. 293–319, San Diego: Academic Press.

KATO, Y., KITAMURA, T., NAKAMURA, K., MITSUI, A., YAMASAKI, Y. and HASHIMOTO, T., 1987, High-performance liquid chromatography of membrane proteins, *Journal of Chromatography*, **391**, 395–407.

KATTA, V. and CHAIT, B.T., 1991, Observation of the heme–globin complex in native myoglobin by electrospray ionization mass spectrometry, *Journal of American Chemical Society*, **113**, 8534–8535.

KAUPPINEN, J.K., MOFFATT, D.J., MANTSCH, H.H. and CAMERON, D.G., 1981, Fourier self-deconvolution: a method for resolving intrinsically overlapped bands, *Applied Spectroscopy*, **35**, 271–276.

KLEPARNIK, K. and BOCEK, P., 1991, Theoretical background for clinical and biomedical applications of electromigration techniques, *Journal of Chromatography*, **569**, 43–62.

KOBATA, A., 1992, Structures and functions of the sugar chains of glycoproteins, *European Journal of Biochemistry*, **209**, 483–501.

KOPPEL, D.E., 1972, Analysis of macromolecular polydispersity in intensity correlation spectroscopy: the method of cummulants, *Journal of Chemical Physics*, **57**, 4814–4820.

KOSEN, P.A., 1992, Disulfide bonds in proteins, in AHERN, T.J. and MANNING, M.C. (eds) *Stability of Protein Pharmaceuticals, Part A: Chemical and Physical Pathways of Protein Degradation*, pp. 31–67, New York: Plenum Press.

KRIMM, S. and BANDEKAR, J., 1986, Vibrational spectroscopy and conformation of peptides, polypeptides and proteins, *Advances in Protein Chemistry*, **38**, 181–364.

KÜSTER, B., WHEELER, S.F., HUNTER, A.P., DWEK, R.A. and HARVEY, D.J., 1997, Sequencing of N-linked oligosaccharides directly from protein gels: in-gel deglycosylation followed by matrix assisted laser desorption/ionization mass spectrometry and normal phase high performance liquid chromatography, *Analytical Biochemistry*, **250**, 82–101.

LAEMMLI, U.K., 1970, Cleavage of structural proteins during the assembly of the head of bacteriophage T4, *Nature*, **227**, 680–685.

LANDERS, J.P. (ed.), 1993, *CRC Handbook of Capillary Electrophoresis*, Boca Raton: CRC Press.

LEE, C., LEVIN, A. and BRANTON, D., 1987, Copper staining: a five minute protein stain for sodium dodecyl sulfate polyacrylamide gels, *Analytical Biochemistry*, **166**, 308–312.

LIAO, J.L., LAM, W.K.G., TISCH, T.L. and FRANKLIN, S.G., 1998, Continuous bed chromatography, *Genetic Engineering News*, 1 January, 17.

LIN, S. and KARGER, B.L., 1990, Reversed phase chromatographic behavior of proteins in different unfolded states, *Journal of Chromatography*, **499**, 89–102.

LING, V., GUZZETTA, A.W., CANOVA-DAVIS, E., STULTS, J.T., HANCOCK, W.S., COVEY, T.R. and SHUSHAN, B.I., 1991, Characterization of the tryptic map of recombinant DNA derived tissue plasminogen activator by high performance liquid chromatography–electrospray ionization mass spectrometry, *Analytical Chemistry*, **63**, 2909–2915.

LIS, H. and SHARON, N., 1993, Protein glycosylation. Structural and functional aspects, *European Journal of Biochemistry*, **218**, 1–27.

LLOYD, L.L. and WARNER, F.P., 1990, Preparative high performance liquid chromatography on a unique high speed macroporous resin, *Journal of Chromatography*, **512**, 365–376.

LOWRY, O.H., ROSEBROUGH, N.J., FARR, A.L. and RANDALL, R.J., 1951, Protein measurement with the folin phenol reagent, *Journal of Biological Chemistry*, **193**, 265–275.

MACH, H., RYAN, J.A., BURKE, C.J., VOLKIN, D.B. and MIDDAUGH, C.R., 1993, Partially structured self-associating states of acidic fibroblast growth factor, *Biochemistry*, **32**, 7703–7711.

MANN, K.G. and FISH, W.W., 1972, Protein polypeptide chain molecular weights by gel chromatography in guanidinium chloride, in HIRS, C.H.W. and TIMASHEFF, S.N. (eds) *Enzyme Structure, Part C*, Methods in Enzymology, Vol. 26, pp. 28–42, New York: Academic Press.

MANNING, M.C., 1989, Underlying assumptions in the estimation of secondary structure content in proteins by circular dichroism spectroscopy – a critical review, *Journal of Pharmacology and Biomedical Analysis*, **7**, 1103–1119.

MANNING, M.C. and WOODY, R.W., 1989, Theoretical study of the contribution of aromatic side chains to the circular dichroism of basic bovine pancreatic trypsin inhibitor, *Biochemistry*, **28**, 8609–8613.

MANNING, M.C., PATEL, K. and BORCHARDT, R.T., 1989, Stability of protein pharmaceuticals, *Pharmaceutical Research*, **6**, 903–918.

MANT, C.T. and HODGES, R.S., 1996, Analysis of peptides by high-performance liquid chromatography, in KARGER, B.L. and HANCOCK, W.S. (eds) *High Resolution Separation and Analysis of Biological Macromolecules, Part B: Applications*, Methods in Enzymology, Vol. 271, pp. 3–50, San Diego: Academic Press.

MANTSCH, H.H., CASAL, H.L. and JONES, R.N., 1986, Resolution enhancement of infrared spectra of biological systems, in CLARK, R.J.H. and HESTER, R.E. (eds) *Spectroscopy of Biological Systems*, pp. 1–46, Chichester: Wiley.

MARTIN, S.A., VATH, J.E., YU, W. and SCOBLE, H., 1994, Co-translational and post-translational processing of proteins, in MATSUO, T., CAPRIOLI, R.M., GROSS, M.L. and SEYAMA, Y. (eds) *Biological Mass Spectrometry: Present and Future*, pp. 313–330, New York: Wiley.

MERRIL, C.R., 1990, Gel staining techniques, in DEUTSCHER, M.P. (ed.) *Guide to Protein Purification*, Methods in Enzymology, Vol. 182, pp. 477–488, San Diego: Academic Press.

MIRANKER, T.W., ROBINSON, C.V., RADFORD, S.E., APLIN, R.T. and DOBSON, C.M., 1993, Detection of transient protein folding populations by mass spectrometry, *Science*, **262**, 896–900.

MIURA, T. and THOMAS, G.J., 1995, Optical and vibrational spectroscopic methods, in GLASEL, J.A. and DEUTSCHER, M.P. (eds) *Introduction to Biophysical Methods for Protein and Nucleic Acid Research*, pp. 261–315, San Diego: Academic Press.

MOSHER, R.A., SAVILLE, D.A. and THORMANN, W., 1992, *The Dynamics of Electrophoresis*, Weinheim: VCH Press.

MUTSADAIRA, P., 1987, Sequence of picomole quantities of proteins electroblotted onto polyvinylidene difluoride membranes, *Journal of Biological Chemistry*, **262**, 10035–10038.

NGUYEN, T.H. and WARD, C., 1993, Stability characterization and formulation development of Alteplase, a recombinant tissue plasminogen activator, in WANG, Y.J. and PEARLMAN, R. (eds) *Stability and Characterization of Protein and Peptide Drugs, Case Histories*, Pharmaceutical Biotechnology, Vol. 5, pp. 91–134, New York: Plenum Press.

NOVOTNY, M.V., 1996, Microcolumn liquid chromatography in biochemical analysis, in KARGER, B.L. and HANCOCK, W.S. (eds) *High Resolution Separation and Analysis of Biological Macromolecules, Part A: Fundamentals*, Methods in Enzymology, Vol. 270, pp. 101–133, San Diego: Academic Press.

O'FARRELL, P., 1975, High resolution two dimensional electrophoresis of proteins, *Journal of Biological Chemistry*, **250**, 4007–4021.

OHGAMI, Y., NAGASE, M., NABESHIMA, S., FUKUI, M. and NAKAZAWA, H., 1989, Characterization of recombinant DNA-derived human granulocyte macrophage colony stimulating factor by fast atom bombardment mass spectrometry, *Journal of Biotechnology*, **7**, 219–230.

OLIVER, R.V.H. (ed.), 1989, *HPLC of Macromolecules: a Practical Approach*, Oxford: IRL Press.

ORNSTEIN, L., 1964, Disc electrophoresis. I. Background and theory, *Annals of New York Academy of Sciences*, **121**, 321–349.

PACE, C.N., VAJDOS, F., FEE, L., GRIMSLEY, G. and GRAY, T., 1995, How to measure and predict the molar absorption coefficient of a protein, *Protein Science*, **4**, 2411–2423.

PIKAL, M.J., 1994, Freeze drying of proteins: process, formulation and stability, in CLELAND, J.L. and LANGER, R. (eds) *Formulation and Delivery of Proteins and Peptides*, ACS Symposium Series, Vol. 567, pp. 120–147, Washington, DC: American Chemical Society.

PORATH, J., 1960, Gel filtration of proteins, peptides and amino acids, *Biochimica et Biophysica Acta*, **39**, 193–207.

PRESTRELSKI, S.J., ARAKAWA, T. and CARPENTER, J.F., 1994, Structure of proteins in lyophilized formulations using Fourier transform infrared spectroscopy, in CLELAND, J.L. and LANGER, R. (eds) *Formulation and Delivery of Proteins and Peptides*, ACS Symposium Series, Vol. 567, pp. 148–169, Washington, DC: American Chemical Society.

REGNIER, F.E., 1991, Perfusion chromatography, *Nature*, **350**, 634–635.

REINHOLD, V.N., REINHOLD, B.B. and CHAN, S., 1996, Carbohydrate sequence analysis by electrospray ionization mass spectrometry, in KARGER, B.L. and HANCOCK, W.S. (eds) *High Resolution Separation and Analysis of Biological Macromolecules, Part B: Applications*, Methods in Enzymology, Vol. 271, pp. 377–402, San Diego: Academic Press.

REYNOLDS, J.A. and TANFORD, C., 1970, Binding of dodecyl sulfate to proteins at high binding ratios. Possible implications for the state of proteins in biological membranes, *Proceedings of National Academy of Sciences USA*, **66**, 1002–1007.

RICKARD, E.C. and TOWNS, J.K., 1996, Applications of capillary zone electrophoresis to peptide mapping, in KARGER, B.L. and HANCOCK, W.S. (eds) *High Resolution Separation and Analysis of Biological Macromolecules, Part B: Applications*, Methods in Enzymology, Vol. 271, pp. 237–264, San Diego: Academic Press.

RIGHETTI, P.G., 1983, *Isoelectric Focusing: Theory, Methodology and Applications*, Amsterdam: Elsevier.

RIGHETTI, P.G., GELFI, C. and CHIARI, M., 1996, Isoelectric focusing in immobilized pH gradients, in KARGER, B.L. and HANCOCK, W.S. (eds) *High Resolution Separation and Analysis of Biological Macromolecules, Part A: Fundamentals*, Methods in Enzymology, Vol. 270, pp. 235–255, San Diego: Academic Press.

ROEPSTORFF, P., 1995, Mass spectrometry in protein structural analysis, in HERRON, J.N., JISKOOT, W. and CROMMELIN, D.J.A. (eds) *Physical Methods to Characterize Pharmaceutical Proteins*, pp. 145–177, New York: Plenum Press.

SARVER, R.W. and KRUEGER, W.C., 1991, An infrared and circular dichroism combined approach to the analysis of protein secondary structure, *Analytical Biochemistry*, **199**, 61–67.

SCHACHMAN, H.K., 1959, *Ultracentrifugation in Biochemistry*, New York: Academic Press.

SCHARNAGL, C. and SCHNEIDER, S., 1991, UV–visible absorption and circular dichroism spectra of the subunits of c-phycocyanin. II. A quantitative discussion of the chromophore–protein and chromophore–chromophore interactions in the β-subunit, *Journal of Photochemistry and Photobiology B*, **8**, 129–158.

SERRES, A., BAUDYŠ, M. and KIM, S.W., 1996, Temperature and pH-sensitive polymers for human calcitonin delivery, *Pharmaceutical Research*, **13**, 196–201.

SHIRE, S.J., HOLLADAY, L.A. and RINDERKNECHT, 1991, Self-association of human and porcine relaxin as assessed by analytical centrifugation and circular dichroism, *Biochemistry*, **30**, 7703–7711.

SIUZDAK, G., 1996, *Mass Spectrometry for Biotechnology*, London: Academic Press.

SLUYTERMAN, L.A.A. and ELGERSMA, O., 1978, Chromatofocusing: isoelectric focusing on ion exchange columns, I. General principles, *Journal of Chromatography*, **150**, 17–30.

SLUYTERMAN, L.A.A. and WIJDENES, J., 1978, Chromatofocusing: isoelectric focusing on ion exchange columns, II. Experimental verification, *Journal of Chromatography*, **150**, 31–44.

SMITH, A.J., 1997, Postcolumn amino acid analysis, in SMITH, B.J. (ed.) *Protein Sequencing Protocols*, pp. 139–146, Totowa, NJ: Humana Press.

SMITH, D.L. and ZHOU, Z., 1990, Strategies for locating disulfide bonds in proteins, in McCLOSKEY, J.A. (ed.) *Mass Spectrometry*, Methods in Enzymology, Vol. 193, pp. 374–389, San Diego: Academic Press.

STAFFORD, W.F. and SCHUSTER, T.M., 1995, Hydrodynamic methods, in GLASEL, J.A. and DEUTSCHER, M.P. (eds) *Introduction to Biophysical Methods for Protein and Nucleic Acid Research*, pp. 111–145, San Diego: Academic Press.

STAHLBERG, J., JOENSSON, B. and HORVATH, C., 1992, Combined effect of coulombic and van der Waals interactions in the chromatography of proteins, *Analytical Chemistry*, **64**, 3118–3124.

STOUT, G.H. and JENSEN, L.H., 1989, *X-ray Structure Determination: a Practical Guide*, 2nd edn, New York: Wiley.

STRICKLAND, E.H., 1974, Aromatic contribution to circular dichroism spectra of proteins, *CRC Critical Reviews of Biochemistry*, **2**, 113–175.

STULTS, J.T., GILLECE-CASTRO, B.L., HENZEL, W.J., BOURELL, J.H., O'CONNELL, K.L. and NUWAYSIR, L.M., 1996, Packed capillary HPLC–electrospray ionization mass spectrometry, in HANCOCK, W.S. (ed.) *New Methods in Peptide Mapping for the Characterization of Proteins*, pp. 119–141, Boca Raton: CRC Press.

SUREWICZ, W.K., MANTSCH, H.H. and CHAPMAN, D., 1993, Determination of protein secondary structure by Fourier transform infrared spectroscopy: a critical assessment, *Biochemistry*, **32**, 389–394.

SUSI, H. and BYLER, D.M., 1986, Resolution enhanced Fourier transform infrared spectroscopy of enzymes, in HIRS, C.H.W. and TIMASHEFF, S.N. (eds) *Enzyme Structure: Part K*, Methods in Enzymology, Vol. 130, pp. 290–311, Orlando: Academic Press.

SYROVY, I. and HODNY, Z., 1991, Staining and quantification for proteins separated by polyacrylamide gel electrophoresis, *Journal of Chromatography*, **569**, 175–196.

TANFORD, C., 1961, *Physical Chemistry of Macromolecules*, New York: Wiley.

TELLER, D.C., 1973, Characterization of proteins by sedimentation equilibrium in the analytical centrifuge, in HIRS, C. and TIMASHEFF, S. (eds) *Enzyme Structure, Part C*, Methods in Enzymology, Vol. 27, pp. 346–441, New York: Academic Press.

TESHIMA, G. and CANOVA-DAVIS, E., 1992, Separation of oxidized human growth hormone variants by reversed-phase high-performance liquid chromatography. Effect of mobile phase and organic modifier, *Journal of Chromatography*, **625**, 207–213.

TESHIMA, G. and WU, S.L., 1996, Capillary electrophoresis analysis of recombinant proteins, in KARGER, B.L. and HANCOCK, W.S. (eds) *High Resolution Separation and Analysis of Biological Macromolecules, Part B: Applications*, Methods in Enzymology, Vol. 271, pp. 264–293, San Diego: Academic Press.

VAN DE GOOR, T.A.A.M., 1995, Capillary electrophoresis of proteins, in HERRON, J.N., JISKOOT, W. and CROMMELIN, D.J.A. (eds) *Physical Methods to Characterize Pharmaceutical Proteins*, pp. 301–327, New York: Plenum Press.

VAN DER VIES, S.M., VIITANEN, P.V., GATENBY, A.A., LORIMER, G.H. and JAENICKE, R., 1992, Conformational states of ribulosebiphosphate carboxylase and their interaction with chaperonin 60, *Biochemistry*, **31**, 3635–3644.

VARKI, A., 1993, Biological roles of oligosaccharides: all theories are correct, *Glycobiology*, **3**, 97–130.

VESTERBERG, O., 1993, A short history of electrophoretical methods, *Electrophoresis*, **14**, 1243–1249.

WEBER, K. and OSBORN, M., 1969, The reliability of molecular weight determinations by dodecyl sulfate–polyacrylamide gel electrophoresis, *Journal of Biological Chemistry*, **244**, 4406–4412.

WELLING, G.J. and WELLING-WESTER, S., 1989, Size exclusion chromatography of proteins, in OLIVER, R.W.A. (ed.) *HPLC of Macromolecules: a Practical Approach*, pp. 77–89, Oxford: IRL Press.

WEN, J., ARAKAWA, T. and PHILO, J.S., 1996, Size exclusion chromatography with on-line light scattering, absorbance and refractive index detectors for studying proteins and their interactions, *Analytical Biochemistry*, **240**, 155–166.

WETLAUFER, D.B., 1962, Ultraviolet spectra of proteins and amino acids, *Advances in Protein Chemistry*, **17**, 303–390.

WILLIAMS, R.C., 1986, Measurement of linear polymer formation by small angle light scattering, in HIRS, C. and TIMASHEFF, S. (eds) *Enzyme Structure, Part K*, Methods in Enzymology, Vol. 130, pp. 35–47, Orlando: Academic Press.

WILSON, C.M., 1983, Staining of proteins on gels: comparison of dyes and procedures, in HIRS, C. and TIMASHEFF, S. (eds) *Enzyme Structure*, Methods in Enzymology, Vol. 91, pp. 236–247, New York: Academic Press.

WOLD, F., 1981, *In vivo* chemical modification of proteins (posttranslational modification), *Annual Reviews of Biochemistry*, **50**, 783–814.

WU, S.L. and KARGER, B.L., 1996, Hydrophobic interaction chromatography of proteins, in KARGER, B.L. and HANCOCK, W.S. (eds) *High Resolution Separation and Analysis of Biological Macromolecules, Part A: Fundamentals*, Methods in Enzymology, Vol. 270, pp. 27–47, San Diego: Academic Press.

YANG, J.T., 1961, The viscosity of macromolecules in relation to molecular conformation, *Advances in Protein Chemistry*, **16**, 323–400.

YIM, K.W., 1991, Fractionation of the human recombinant tissue plasminogen activator (rtPA) glycoforms by high performance capillary zone electrophoresis and capillary isoelectric focusing, *Journal of Chromatography*, **559**, 401–410.

5

Chemical Pathways of Peptide and Protein Degradation

CHIMANLALL GOOLCHARRAN, MEHRNAZ KHOSSRAVI AND
RONALD T. BORCHARDT

Department of Pharmaceutical Chemistry, The University of Kansas, Lawrence, Kansas, USA

5.1 Introduction

With the advent of recombinant DNA technology and modern synthetic chemistry, peptides and proteins now comprise a significant portion of drugs under clinical development. However, the instability of these macromolecules presents unique challenges to the pharmaceutical scientists who are responsible for developing stable and efficacious formulations or delivery systems for these molecules (Manning *et al.*, 1989; Ahern and Manning, 1992a, 1992b; Wang and Pearlman, 1993; Cleland *et al.*, 1993; Pearlman and Wang, 1996; Sanders and Hendren, 1997; Schöneich *et al.*, 1997).

Unlike small organic molecules, peptides and proteins possess not only a primary structure (sequence of amino acids), but also, in the case of proteins, higher order structures (e.g. secondary, tertiary, and quaternary), which are required for biological activity. Instability of small synthetic peptides involves primarily chemical pathways of degradation. With proteins, the degradation pathways also include chemical transformations,

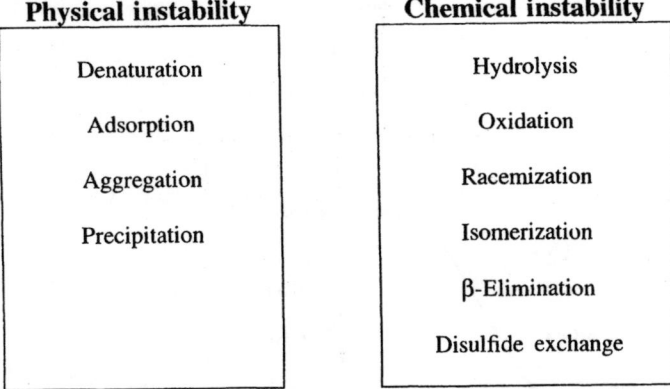

Figure 5.1   Major pathways of peptide and protein degradation.

which are similar to those observed in peptides, as well as physical instability (Manning *et al.*, 1989; Ahern and Manning, 1992a; Cleland *et al.*, 1993; Wang and Pearlman, 1993; Pearlman and Wang, 1996). Chemical instability can be defined as any process involving modification of the protein or peptide by bond formation (e.g. oxidation) or bond cleavage (e.g. deamidation), yielding a new chemical entity (Figure 5.1). In contrast, physical instability does not necessarily involve covalent modifications of the protein. Instead, it generally refers to changes in the higher order structure (secondary and above). This physical instability results in denaturation which can lead to adsorption to surfaces, aggregation and precipitation (Figure 5.1).

The primary objective of this chapter is to address the most prevalent chemical pathways for degradation of peptides and proteins observed during pharmaceutical production, purification, formulation and storage. For additional details about chemical, as well as physical, pathways of peptide and protein degradation, the reader is directed to several reviews (Manning *et al.*, 1989; Cleland *et al.*, 1993; Schöneich *et al.*, 1997) and books (Ahern and Manning, 1992a, 1992b; Wang and Pearlman, 1993; Pearlman and Wang, 1996; Sanders and Hendren, 1997).

5.2   Hydrolytic pathways

The most commonly observed hydrolytic reactions in peptides and proteins involve the side-chain amide groups of asparagine (Asn) and glutamine (Gln), and the peptide bond on the C-terminal side of an aspartic acid (Asp) or a proline (Pro) residue when it is in the penultimate position from the N-terminal end. These hydrolytic reactions are described in detail below.

5.2.1   *Deamidation of Asn and Gln residues*

Deamidation of side-chain amides of Asn and Gln residues in proteins and peptides involves intermolecular attack of water (Figure 5.2, e.g. Asp formation), intramolecular attack by the nitrogen of the N-terminal amine (Figure 5.3, e.g. pyroglutamate formation) or intramolecular attack by the nitrogen of the peptide bond on the C-terminal side (Figure 5.2, e.g. cyclicimide formation) of these amino acid residues (Schöneich *et al.*,

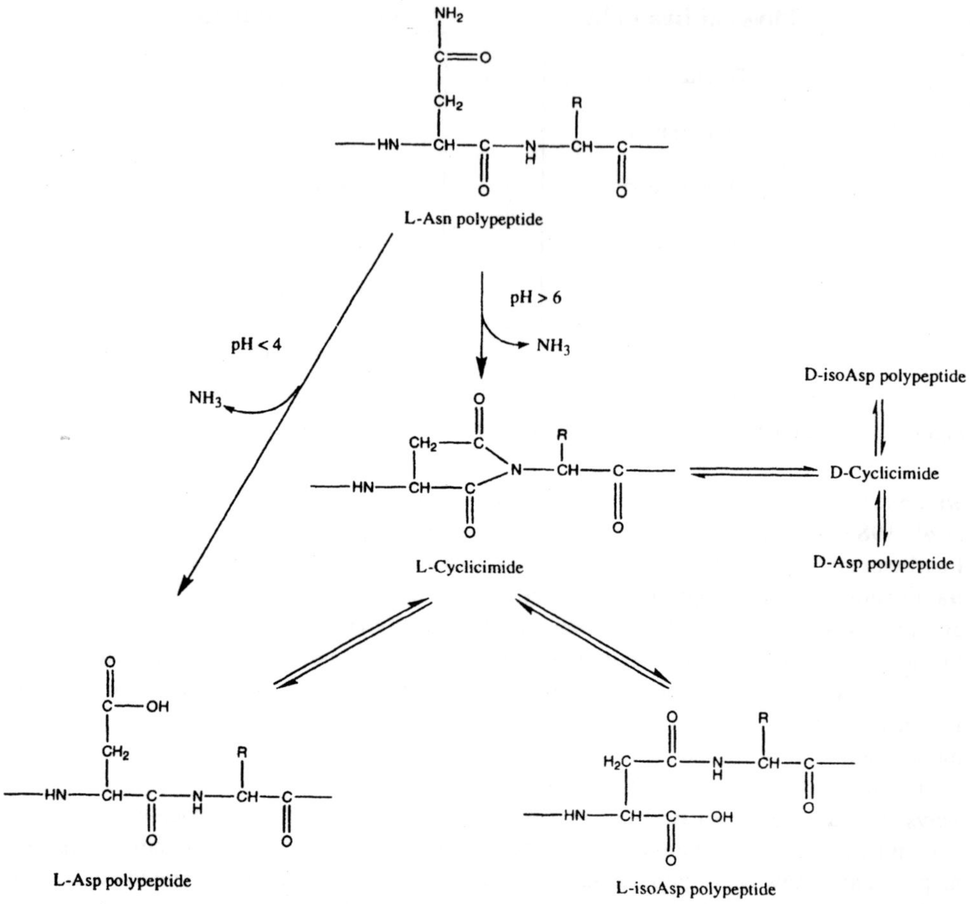

Figure 5.2 Pathways for the deamidation of Asn residues.

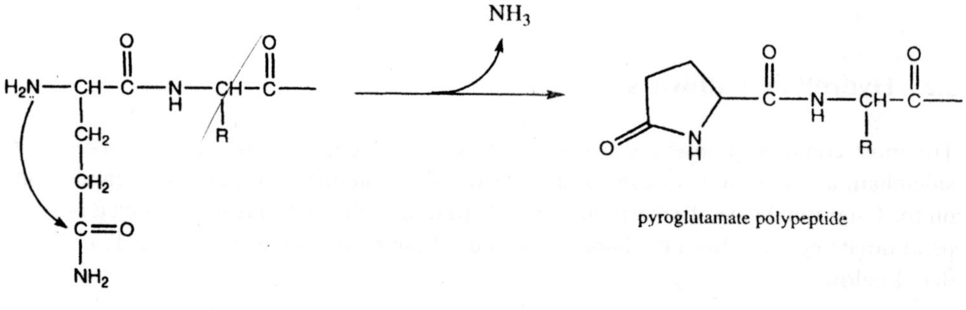

Figure 5.3 Formation of pyroglutamate from N-terminal Gln residue.

1997). Under acidic conditions (pH < 4.0), the predominant reaction with an Asn residue involves direct attack of water on the side-chain amide carbonyl carbon resulting in the formation of an Asp residue (Figure 5.2) (Patel and Borchardt, 1990a, 1990b). In contrast, at neutral or basic conditions (pH > 6.0), the predominant reaction with an Asn residue involves cyclicimide formation followed by attack of water on one of the two carbonyl carbon atoms to yield Asp and isoaspartic acid (isoAsp) residues (Figure 5.2) (Stephenson and Clarke, 1989; Patel and Borchardt, 1990a). It should be noted that this cyclicimide intermediate can undergo racemization yielding L- and D-Asp- and L- and D-isoAsp-polypeptides as degradation products (Geiger and Clarke, 1987). In general, Asn residues deamidate much more rapidly than Gln residues (Manning *et al.*, 1989; Cleland *et al.*, 1993; Schöneich *et al.*, 1997). However, Gln residues located at the N-terminal end of a polypeptide chain can rapidly deamidate to form a pyroglutamate residue (Figure 5.3).

Many factors, both exogenous (i.e. pH, buffer concentration, buffer species) and endogenous (i.e. primary sequence and secondary and higher ordered structures) can influence the rate of deamidation of Asn residues and the distribution of the degradation products (iso-Asp, Asp). The effects of exogenous factors on the kinetics of deamidation have been extensively studied by our laboratory using an Asn-hexapeptide, Val-Tyr-Pro-Asn-Gly-Ala, residues 22–27 of porcine adrenocorticotropic hormone (ACTH). The most significant exogenous factor determining the rate of deamidation of Asn residues is pH. Using the Asn-hexapeptide described above, we have shown that the degradation is pH-dependent and follows first-order kinetics (Patel and Borchardt, 1990a). Under acidic conditions (pH 1–2), this Asn-hexapeptide undergoes deamidation via direct attack of water on the Asn side-chain carbonyl carbon atom, followed by expulsion of ammonia and formation of an Asp-hexapeptide (Figure 5.2). Interestingly, the formation of a cyclicimide as a minor product was also observed under acidic conditions, although its formation was much slower than the direct hydrolysis reaction. The cyclicimide intermediate was shown experimentally to be highly stable in acidic medium, and under these conditions it does not break down to form the Asp- and isoAsp-hexapeptides. Therefore, the degradation of this Asn-hexapeptide under acidic conditions does not lead to the formation of an isoAsp degradation product. Under basic and neutral conditions (pH 5–12), the Asn-hexapeptide degrades via the formation of a cyclicimide which then undergoes further hydrolysis to produce a mixture of Asp- and isoAsp-hexapeptides in a ratio of approximately 1 to 4, respectively (Figure 5.2). The rate constants of deamidation at basic pH values were greater than the rate constants at acidic pH. The maximal stability of this Asn-hexapeptide was observed at pH 3–4 at 37°C. At all pH values, the ratio of Asp to isoAsp degradation products was independent of buffer and buffer concentrations.

Other exogenous factors reported to influence the rate of deamidation of Asn residues include temperature, buffer species, buffer concentration, ionic strength and formulation excipients (e.g. lactose, mannitol) (Oliyai and Borchardt, 1994b; Oliyai *et al.*, 1994a; Schöneich *et al.*, 1997).

The effect of endogenous factors on the rate of deamidation has been determined by several investigators. Using the model hexapeptides, Val-Tyr-X-Asn-Z-Ala, various laboratories have shown that the major endogenous factor that influences the rate of deamidation is the structural characteristics (i.e. size) of the residue on the C-terminal side (Z) of the Asn residue. For example, determination of the rates of deamidation of the series Val-Tyr-Pro-Asn-Z-Ala, in 0.1 M phosphate buffer (pH 7.4) at 37°C, generated half-lives as follows: Z = Gly, 1.5 days; Z = Ser, 6.8 days; Z = Ala, 20.2 days; Z = Leu, 70 days; and Z = Val, 106 days. As expected, as the steric bulk of the residue on the C-terminal side of Asn residue increased, the rate of deamidation decreased. At pH 1.0, the C-terminal residue (Z) was shown to have no effect on the rate of deamidation (Stephenson and Clarke, 1989; Patel and Borchardt, 1990b). Modifying the residue on the N-terminal side of an Asn residue had little or no effect on the rate of the reaction at acidic, neutral or basic pH values (Patel and Borchardt, 1990b). These observations reinforce the idea that at acidic pH, deamidation occurs by the direct hydrolysis of the side-chain amide of the

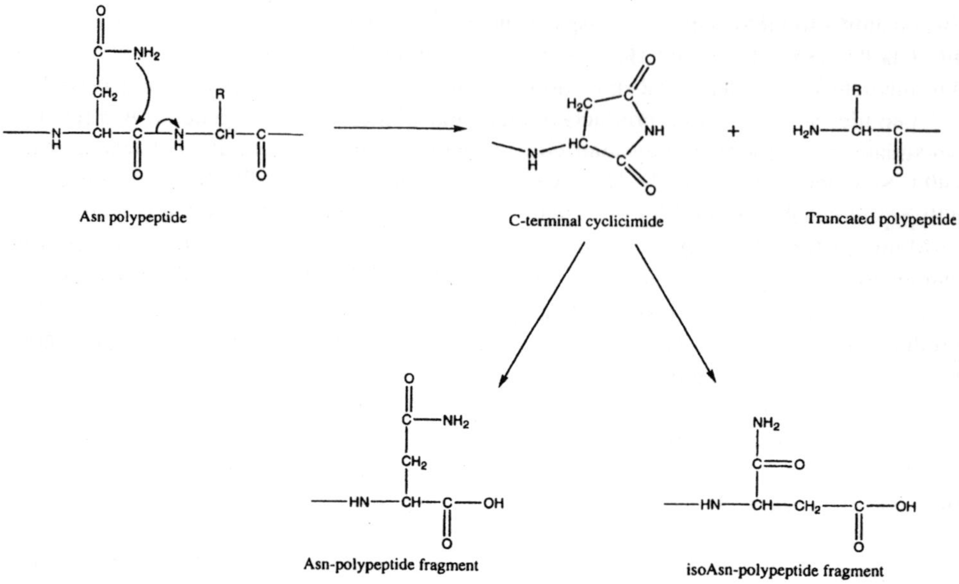

Figure 5.4 Pathways for the spontaneous fragmentation of Asn polypeptides.

Asn residue, while in the neutral-to-basic pH region, Asn residues deamidate via the formation of a cyclicimide intermediate.

Secondary and higher ordered structures can also affect the rates of deamidation of Asn residues in peptides and proteins. Clarke (1987) has shown that native proteins generally exist in conformations in which the nitrogen atom of peptide bond on the C-terminal side of Asn residues cannot approach the side-chain carbonyl carbon atom without large-scale conformational changes. Therefore, certain proteins will only undergo deamidation when they have been denatured and sufficient conformation flexibility exists to allow for the formation of the cyclicimide. Similar observations have been made recently by our laboratory using a series of synthetic growth-hormone-releasing factors (Stevenson *et al.*, 1993) and β-turn peptidomimetics (M. Xie, R.L. Schowen and R.T. Borchardt, unpublished data).

Another pathway of degradation of Asn residues in peptides and proteins involves spontaneous fragmentation (Aswad, 1994). In this process, the side-chain amide nitrogen of the Asn residue attacks the peptide bond carbonyl carbon atom on the C-terminal side, resulting in the cleavage of the peptide bond (Figure 5.4). Of the two fragments produced, one possesses a C-terminal succinimide residue that can undergo further hydrolysis, yielding a mixture of Asn- and isoAsn-peptides. This reaction does not result in the release of ammonia and occurs predominantly, although not exclusively, in peptides in which the Asn residue is followed by a Pro residue.

While extensive research efforts have been devoted to understanding deamidation in aqueous media, little is known about deamidation in the solid state. In the solid state, the kinetics and mechanisms of this reaction are complex, usually involving concomitant existence of other chemical and physical instability processes. Using the hexapeptide Val-Tyr-Pro-Asn-Gly-Ala, Oliyai *et al.* (1994a) have shown that the pH of the lyophilization solution, the moisture level of the lyophilized product, and the storage temperature are important factors in the solid-state stability of this peptide. With a pre-lyophilization solution at pH 3.5, the lyophilized hexapeptide deamidates via direct hydrolysis to produce the Asp-hexapeptide. The Asp-hexapeptide then undergoes hydrolysis at the Asp-Gly amide bond to generate a small quantity of the tetrapeptide Val-Tyr-Pro-Asp. A significant amount of the cyclicimide was also observed. As the pH, temperature and moisture level were increased, the amount of cyclicimide increased. More recently, our laboratory (M.C. Lai, E.M. Topp, R.L. Schowen, M.J. Hageman and R.T. Borchardt, unpublished data) has determined the effect of polymers (e.g. poly(vinylalcohol), poly(vinylpyrrolidone)) and moisture on the rate of deamidation of the hexapeptide Val-Tyr-Pro-Asn-Gly-Ala. These studies have shown that in solid-state preparations, water can be involved in deamidation of this hexapeptide as either a medium, a reactant or a plasticizer. The plasticizing effect of water obviously affects the mobility of the reactants including the polypeptide. In these studies, interesting differences were also observed in the reactivity of the hexapeptide to deamidation in different polymers, suggesting that prediction of solid-state stability from solution stability data would be problematic.

5.2.2 *Degradation of Asp residues*

It has been established that Asp residues are also potential 'hot spots' in peptides and proteins for chemical degradation. The stability of an Asp residue has been shown to be dependent on its location in the primary sequence and the protein's higher-ordered structure (Ota and Clarke, 1989). Asp residues can degrade by two different pathways depending on the pH of the solution. One pathway involves Asp-to-isoAsp interconversion (Figure 5.5). This pathway proceeds through a cyclicimide intermediate similar to that observed in the deamidation of Asn residues. Depending on the pH, this cyclicimide may or may not degrade further to form isoAsp-products. The cyclicimide intermediate is prone to racemization at the α-carbon, resulting in the potential formation of D,L-cyclicimide-, D,L-Asp- and D,L-isoAsp-containing degradation products (Figure 5.5). The other pathway of Asp degradation involves the hydrolysis of Asp-X and/or X-Asp amide bonds (Figure 5.6). In dilute acid, the rate of hydrolysis of the peptide bonds adjacent to the Asp residue is at least 100 times higher than that of other peptide bonds (Schultz, 1967). The enhanced rate of hydrolysis is attributed to intramolecular catalysis by the side-chain carboxyl group where the X-Asp amide bond cleavage proceeds via a six-membered ring intermediate while the Asp-X bond hydrolysis proceeds via a five-membered ring (Figure 5.6).

Many exogenous and endogenous factors can influence the rate of Asp degradation and the distribution of products. In our laboratory, we have studied the effects of exogenous factors such as pH, buffer species and concentration, and temperature on the kinetics and mechanism of degradation of Asp residues using the model hexapeptide Val-Tyr-Pro-Asp-Gly-Ala (Oliyai and Borchardt, 1993). The rate of the Asp-hexapeptide degradation and the product distribution was shown to be highly dependent on the pH of the solution. Under strongly acidic conditions (pH 1.0–2.0), this hexapeptide degraded predominantly via intramolecular hydrolysis of the Asp-Gly amide bond, producing a tetrapeptide, Val-Try-Pro-Asp, and a dipeptide, Gly-Ala. A small amount of the cyclicimide (Asu-hexapeptide) was also generated which was stable under acidic conditions. As the pH of the solution was increased, the formation of the cyclicimide–hexapeptide became more predominant relative to the hydrolysis of the Asp-Gly amide bond. For example, at pH 4.0, most of the peptide (60%) rearranged to produce the cyclicimide–hexapeptide, which then further hydrolysed to form the isoAsp-hexapeptide, whereas the Asp-Gly peptide bond hydrolysis constituted only approximately 10% of the total degradation products. At

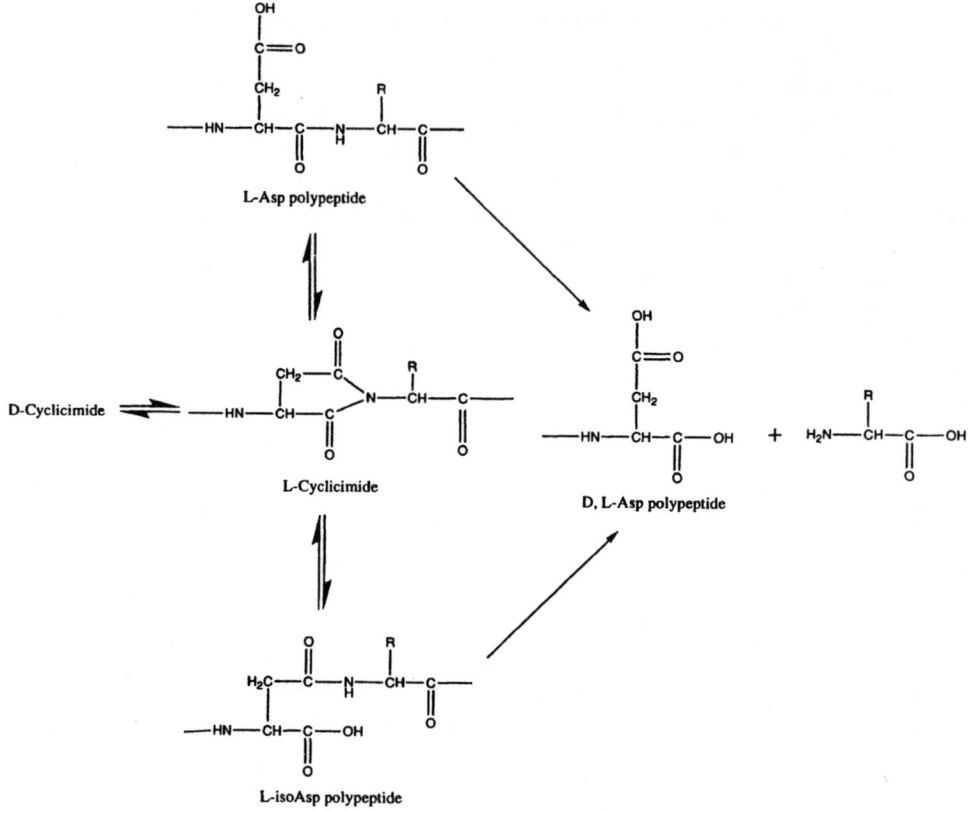

Figure 5.5   Pathways for the degradation of Asp-polypeptides.

pH above 6.0, the contribution from the hydrolysis of the Asp-Gly peptide bond to the overall degradation was negligible. At these pH values, the peptide degraded exclusively to the cyclicimide–hexapeptide, which being extremely unstable in the neutral and basic environment rapidly hydrolysed to the isoAsp- and Asp-hexapeptides. While not determined in this study, the L-cyclicimide hexapeptide should be prone to racemization at the α-carbon, potentially generating the corresponding D-cyclicimide-, D-Asp- and D-isoAsp-hexapeptides (Figure 5.5). Other exogenous factors reported to affect the rates of Asp degradation are temperature (Oliyai and Borchardt, 1993) and dielectric constant of the solvent (Brennan and Clarke, 1993).

The effect of primary sequence on Asp degradation is significant, but not as obvious as that observed for the deamidation of Asn residues. Stephenson and Clarke (1989) have measured the rate of cyclicimide formation at pH 7.4 and 37°C in a series of hexapeptides, Val-Tyr-Pro-Asp-Y-Ala, and observed the following half-lives: Y = Gly, 40.8 days; Y = Ser, 168 days; and Y = Ala, 266 days. Oliyai and Borchardt (1994a) also examined the degradation of a similar series of hexapeptides, Y = Gly, Ser and Val, under acidic (pH 1.1) and alkaline (pH 10.0) conditions. At pH 1.1, the major route of degradation was hydrolysis of the Asp-Y amide bond. They observed that, as the steric bulk of the amino acid Y increased, the rate of hydrolysis decreased, i.e. the Asp-Gly hexapeptide hydrolysed 1.6 times faster than the Asp-Ser peptide and 2.3 times faster than the Asp-Val peptide. A minor amount of cyclicimide–hexapeptide was also observed with the Asp-Gly- and Asp-Ser-hexapeptides but not for the Asp-Val-hexapeptide. At pH 10.0, the

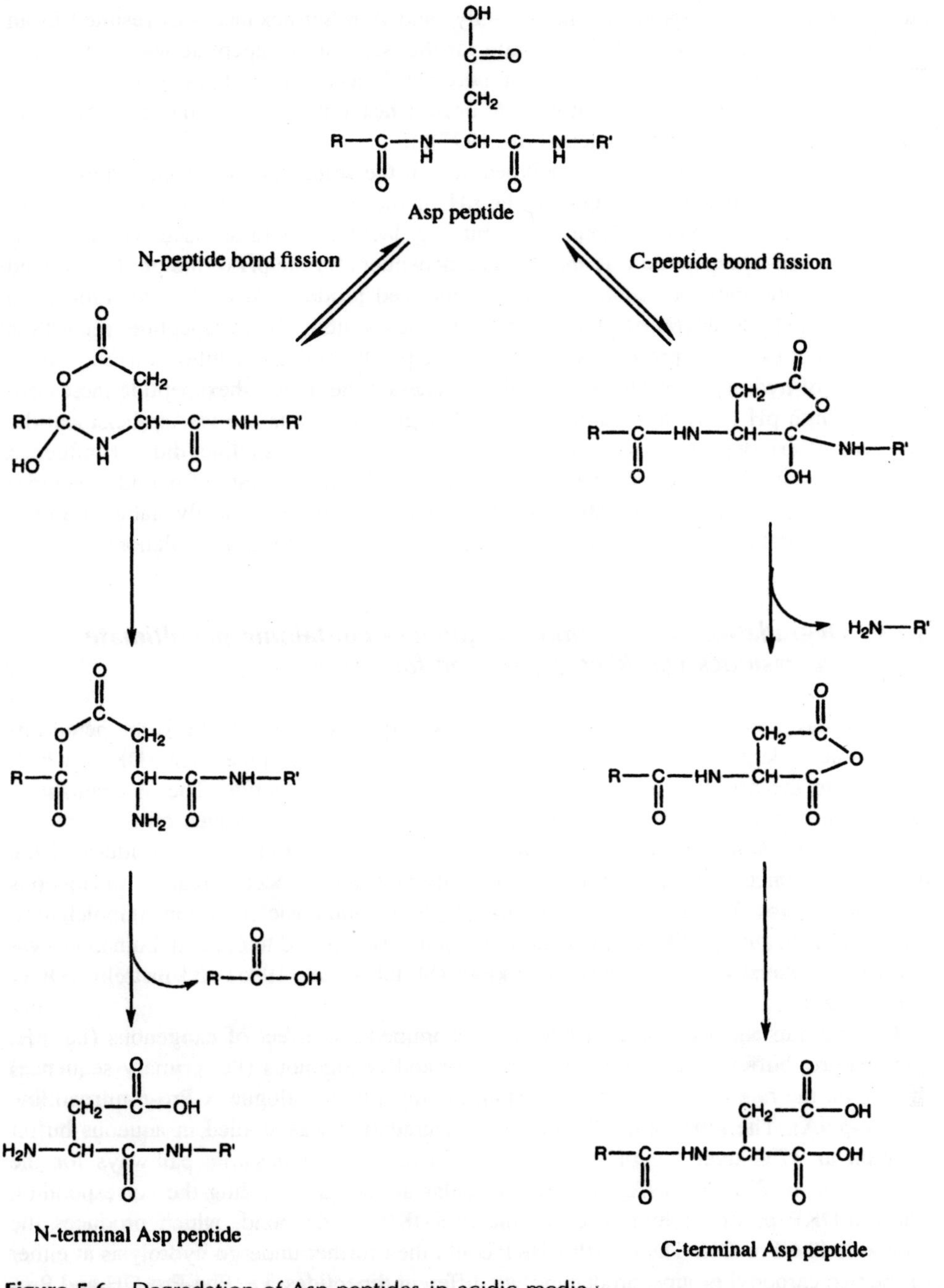

Figure 5.6 Degradation of Asp-peptides in acidic media.

Asp-peptides predominantly degraded to the isoAsp-peptides. As expected, the formation of the cyclicimide was highly sensitive to modifications made in the amino acid located on the C-terminal of Asp. For example, replacing Gly with Ser resulted in a four-fold decrease in the rate of cyclicimide formation. Similarly, substituting Gly with Val further decreased the rate of cyclicimide formation to such an extent that it was not detected.

Furthermore, the degradation of the Asp-Gly- and Asp-Ser-hexapeptides resulted in an isoAsp-to-Asp ratio of 4–5, while the ratio for the Asp-Val-hexapeptide was less than 1. The Asp-Val-hexapeptide also underwent more rapid racemization. In general, the formation of the cyclicimide in an Asn-peptide occurs faster than that of the corresponding Asp-containing peptide (Schöneich et al., 1997).

Degradation of Asp residues in polypeptides in the solid state has been examined by Oliyai et al. (1994b) as a function of the pH of the pre-lyophilized solution, moisture level, temperature, and type of bulking agent (e.g. lactose, mannitol) used. The degradation of the Asp-hexapeptide was shown to be dependent on the pH of the pre-lyophilized solution and the moisture content of the lyophilized product. At acidic conditions, the lyophilized peptide degraded predominantly to the cyclicimide–hexapeptide, regardless of the type of excipient present. As the pH of the pre-lyophilized solution was increased, the extent of hydrolysis of the cyclicimide to generate the isoAsp-hexapeptide increased. Thus, at high pH (8.0), the major product of degradation was the isoAsp-hexapeptide. The type of excipient (lactose vs mannitol), amorphous vs crystalline, did not influence the pathway of degradation of the Asp residue, but affected the rate of peptide degradation; i.e. the amorphous formulations were consistently more chemically stable at all pH values, temperatures and moisture levels than were the crystalline formulations.

5.2.3   *Degradation of N-terminal sequences containing penultimate Pro residues via diketopiperazine formation*

Peptides and proteins that possess an N-terminal sequence in which Pro is the penultimate residue undergo non-enzymatic hydrolysis yielding a diketopiperazine (DKP), which arises from the first two amino acids, and a truncated polypeptide. The mechanism of DKP formation involves nucleophilic attack of the N-terminal nitrogen on the carbonyl carbon of the peptide bond between the second and third amino acid residues in the primary sequence. This intramolecular aminolysis reaction occurs readily in aqueous solution and was shown to be catalysed in both acidic and basic conditions (Goolcharran and Borchardt, 1998). Diketopiperazine formation was reported to occur in human growth hormone (Battersby et al., 1994), bradykinin (Straub et al., 1995) and histrelin (Oyler et al., 1991).

Goolcharran and Borchardt (1998) have examined the effect of exogenous (i.e. pH, temperature, buffer species and concentration) and endogenous (i.e. primary sequence) factors on the rate of DKP formation using the dipeptide analogue X-Pro-*p*-nitroaniline (X-Pro-pNA). The kinetics of Phe-Pro-pNA degradation was studied in aqueous buffer solution at 37°C over the pH range 1–10. There are two possible pathways for the degradation of Phe-Pro-pNA; i.e. intramolecular aminolysis yielding the corresponding Phe-Pro-DKP or direct hydrolysis of the Pro-DKP amide bond, which produces the dipeptide Phe-Pro-OH. The Phe-Pro-DKP could then further undergo hydrolysis at either of the two carbonyl groups, producing two different dipeptides, i.e. Phe-Pro-OH and Pro-Phe-OH. In the pH range studied, Phe-Pro-DKP was the only product generated upon degradation of Phe-Pro-pNA. At pH values between 3 and 8, Phe-Pro-DKP was stable. In contrast, at pH values < 3 and > 8, Phe-Pro-DKP undergoes hydrolysis to the dipeptide, Phe-Pro-OH. The rate of disappearance of Phe-Pro-pNA followed pseudo-first-order kinetics and exhibited significant dependence on pH. Buffer catalysis was observed above pH 5.0, but little or no catalysis was observed below pH 4.5. From these data, a pH–rate profile was generated which suggested that the rate of DKP formation depends on the

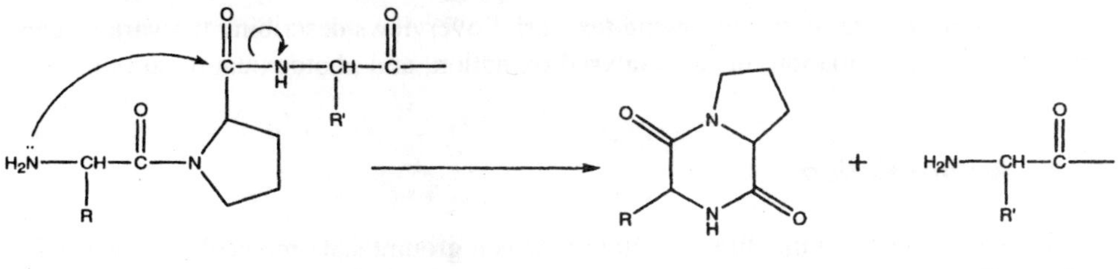

Figure 5.7 Degradation of N-terminal sequences containing penultimate Pro residues via diketopiperazine formation.

degree of ionization of the N-terminal amino group. It is also apparent that the unprotonated N-terminal amino group is much more reactive than the protonated form. At pH < 3, where the reactant exists predominantly in the protonated form, the rate of degradation is independent of pH. From pH 4 to 6, the rate increases with pH, but it is not first-order in hydroxide ion concentration (slope = 0.73). This may be due to the complexity of the mechanism involving different ionic species and buffer catalysis. Between pH 6 and 8 there is a plateau. In this region, the free N-terminal amino group is available for cyclization, but the hydroxide ion concentration is not high enough; thus, degradation occurs by the water-catalysed pathway. In the pH region 9 to 10, the slope of the pH–rate profile is unit positive, indicating specific-hydroxide ion catalysis. A pKa value of 6.1 was kinetically determined for the amino group.

Modification of the amino acid residue located on the N-terminal side of Pro was shown to have a major influence on the rate of DKP formation (Goolcharran and Borchardt, 1998). For example, with the series of dipeptide analogues X-Pro-pNA, the half-lives of DKP formation in 0.05 M phosphate buffer (pH 7.0) at 37°C were reported as follows: X = Gly, 5.1 days; X = Ala, 1.1 days; X = Val, 2.5 days; X = Phe, 0.5 days; X = β-cyclohexylalanine, 0.8 days; and X = Arg, 0.7 days. The effect of alkyl and aryl substituents in intramolecular reactions has been well documented, with the results suggesting that an increase in the bulkiness of the substituent increases the rate of cyclization (Borchardt and Cohen, 1972). This does not seem to be true for the series studied above, as the Ala analogue cycles twice as fast as the bulkier Val analogue. Thus, simple steric bulk cannot be used to explain the effect of the N-terminal residue on the rate of DKP formation. The effect may be due to contributions from several different factors, including the ability of the X-Pro peptide bond to undergo *cis–trans* isomerization.

5.3 Oxidation pathways

Oxidation is one of the most commonly observed chemical degradation pathways of peptides and proteins (Stadtman, 1992; Li *et al.*, 1995; Schöneich *et al.*, 1997). Among the amino acids most susceptible to oxidative modification are those that contain a sulphur (e.g. Met and Cys) or an aromatic ring (e.g. His, Tyr, and Trp) (Stadtman, 1992; Li *et al.*, 1995; Schöneich *et al.*, 1997). There are a number of mechanisms that may result in oxidative modification of amino acids (Stadtman, 1993). Despite the variations in conditions used to induce oxidation, the underlying principle involves the activation of oxygen to generate a few key reactive oxygen species, which ultimately then react with the protein or peptide. In the following text, brief overviews describing the various characteristics of autooxidation, metal-catalysed oxidation, and photooxidation are provided.

5.3.1 Autooxidation

Autooxidation refers to the direct reaction between ground state molecular oxygen and a compound in the absence of any catalytic processes (Schöneich *et al.*, 1997). True autooxidation is of minor importance as a degradation pathway of proteins and peptides for the following reason. Ground state oxygen exists in the triplet state, i.e. there are two unpaired electrons of parallel spin direction in each of the two antibonding orbitals of oxygen. The majority of organic molecules differ from oxygen in that they are in the singlet ground state, i.e. their valence electrons are in paired antiparallel electron spins. Therefore, the direct reaction of oxygen in the triplet state with an organic molecule in the singlet state is a spin-restricted reaction and no significant bond formation is expected (Schöneich *et al.*, 1997). The triplet state configuration of oxygen permits the acceptance of a single electron but not a pair of electrons of antiparallel spin directions. Therefore, any significant oxidation reaction between ground state oxygen and an organic molecule in the singlet state is expected to occur via an outer-sphere electron transfer (Schöneich *et al.*, 1997).

The improbability of such an autooxidation reaction was shown by Schöneich *et al.* (1997) for a hypothetical formulation that contained a standard protein of molecular weight 22 000 in an air-saturated and metal- and peroxide-free aqueous buffer. The rates of autooxidation of the most labile amino acid residues, Cys and Met, were calculated for a pseudo-first-order process. The approximate half-life for the autooxidation of one deprotonated Cys residue was calculated to be 32 090 days. Moreover, the approximate half-life of the Met residue was calculated to be greater than 8.8×10^7 years. Given these calculations for a hypothetical protein formulation, it is apparent that true autooxidation of peptides and proteins is a very slow process that poses no significant threat to the stability of these compounds. However, many so-called cases of autooxidation are observed for peptides and proteins. In fact, it is likely that these cases of oxidation are metal-catalysed or light-induced.

5.3.2 Metal-catalysed oxidation

In the presence of a transition metal ion (i.e. Fe(III) and Cu(II)), oxygen, and an electron donor (reducing agent), metal-catalysed oxidation of proteins and peptides can occur at significant rates (Stadtman, 1993). Pharmaceutical processes such as protein/peptide synthesis, purification, storage of bulk drug, and storage of the dosage form can provide conditions leading to metal-catalysed oxidation reactions (Manning *et al.*, 1989; Cleland *et al.*, 1993). Metals such as Fe(III) and Cu(II) are commonly encountered as contaminants that may originate from reagents and, particularly, from buffers used in processing and formulation of proteins and peptides (Li *et al.*, 1995). Normally, transition metal ions in their oxidized states (i.e. Cu(II) and Fe(III)) do not readily react with oxygen to generate a more reactive oxygen species. An electron donor or reducing agent known as a prooxidant is needed to reduce the transition metal ion; this reduced form may then interact with oxygen to generate reactive oxygen species. The prooxidant in pharmaceutical formulations may arise from contaminants in buffers or may have been added to the formulation, ironically, as an antioxidant (i.e. ascorbic acid) (Schöneich *et al.*, 1997). In addition, the side-chains of certain amino acid residues such as Trp, Cys, and Tyr found in proteins and peptides can act as prooxidants (Timmins *et al.*, 1982; Schöneich *et al.*, 1997). Equation 5.1 illustrates the prooxidant donating an electron to reduce the transition metal ion Fe(III).

$$\text{Fe(III)} + \text{prooxidant} \rightarrow \text{Fe(II)} + \text{oxidized prooxidant}^- \tag{5.1}$$

At this point, it is important to note that the transition metal ion being reduced may be ligated or complexed to appropriate ligands (Stadtman, 1993). Further reaction of the ligated or complexed reduced metal with molecular oxygen O_2 to generate reactive oxygen species occurs at the metal binding site (Stadtman, 1993). The generated reactive oxygen species reacts preferentially with an amino acid residue at the metal binding site before diffusion into the bulk solution. This feature makes metal-catalysed oxidation a site-specific process. That is, given a certain protein, one or at most only a few amino acid residues located at the metal-binding site are specific targets (Stadtman, 1993). As illustrated in equations 5.2 and 5.3, the reduced form of the transition metal ion, complexed by appropriate ligands, may donate an electron to reduce oxygen to form superoxide radical. In these equations, C_x represents a complex that may include a number of species in solution (i.e. amino acid residue(s), buffer, reducing agent, etc.). The pertinent reacting species of the complex (C_x) are illustrated individually.

$$O_2 + C_x\text{Fe(II)} \rightarrow C_x\text{Fe(II)}{-}O{-}O^{\cdot} \rightarrow C_x\text{Fe(III)}{-}O{-}O^- \tag{5.2}$$

$$C_x\text{Fe(III)}{-}O{-}O^- \rightarrow C_x\text{Fe(III)} + O_2^{\cdot-} \tag{5.3}$$

Superoxide anion radical may then react with protons to generate hydrogen peroxide

$$2O_2^{\cdot-} + 2H^+ \rightarrow H_2O_2 \tag{5.4}$$

It is also possible that hydrogen peroxide is directly generated

$$O_2 + C_x\text{Fe(II)} \rightarrow C_x\text{Fe(II)}{-}O{-}O^{\cdot} \tag{5.5}$$

$$C_x\text{Fe(II)}{-}O{-}O^{\cdot} + C_x\text{Fe(II)} + 2H^+ \rightarrow 2C_x\text{Fe(III)} + H_2O_2 \tag{5.6}$$

Hydrogen peroxide can then oxidize susceptible amino acids such as Cys or Met directly or it can undergo various other reactions to form a number of intermediate complexes with the reduced and oxidized forms of the metal. These intermediate complexes have been proposed to oxidize amino acids (Yamazaki and Piette, 1991; Stadtman, 1990). A reactive oxygen species commonly proposed to be the damaging species in metal-catalysed oxidation is the hydroxyl radical or an equivalent bound/complexed form of it. The hydroxyl radical may arise from a Fenton-type reaction where there is a one-electron reduction of hydrogen peroxide

$$C_x\text{Fe(II)} + H_2O_2 \rightarrow C_x\text{Fe(III)} + {}^{\cdot}\text{OH} + {}^-\text{OH} \tag{5.7}$$

The hydroxyl radical or an equivalent bound/complexed form may then react immediately with the amino acid at the metal-binding site before diffusion into bulk solution. It is important to note that other reactive oxygen species may be involved in metal-catalysed oxidation. At this time, only speculation about the exact nature of the reactive oxygen species has been possible because of the site-specific nature of the reaction. The amino acids most susceptible to metal-catalysed oxidation are His, Arg, Lys, Pro, Met, and Cys (Stadtman, 1993). Table 5.1 provides a list of the common degradation products arising from metal-catalysed reaction of these amino acids. As a specific example, the

Table 5.1 Common degradation products formed from metal-catalysed oxidation of amino acid residues

Oxidized residue	Products formed
His	2-oxo-His, aspartic acid, asparagine
Lys	2-amino-adipicsemialdehyde
Cys	–S–S–disulphide crosslinks
Pro	Glutamic acid, glutamic semialdehyde, 2-pyrrolidone, *cis/trans*-4-hydroxyproline
Arg	Glutamic semialdehyde
Met	Met-sulphoxide

reader is referred to a recent paper by Khossravi and Borchardt (1998) dealing with His oxidation by ascorbate/Cu(II)/O_2. Proposed mechanisms of interaction of the reactive oxygen species with susceptible amino acid residues to generate the degradation product can be found in literature references cited in a review by Stadtman (1993).

In metal-catalysed oxidation, the location of the amino acid in the protein, i.e. whether it is exposed or buried, is not a significant factor in its oxidation potential. Amino acids are most susceptible to metal-catalysed oxidation because their side-chains aid in the formation of favourable metal-binding sites in proteins, making them specific targets for oxidation. In contrast to the site-specific nature of metal-catalysed oxidation reactions, oxidation induced by light (photooxidation) or contaminants is of a non-site-specific nature. In such a mechanism, the surface area exposure of the susceptible amino acid becomes an issue in its oxidation potential (Shechter *et al.*, 1975). In other words, the oxidation potential of the amino acid residue depends on whether it is exposed or buried within the protein. An example of a non-site-specific oxidation by contaminants is the oxidation of Met by hydrogen peroxide. In the presence of hydrogen peroxide, Met in peptides and proteins is oxidized to Met sulphoxide (Nguyen *et al.*, 1993). Peroxide contaminants in formulations have been know to arise from surfactants such as polysorbate 80 (Hora *et al.*, 1991) and polyethylene glycols (MacGinity *et al.*, 1975).

5.3.3   *Photooxidation*

Some pharmaceutical proteins are photosensitive and, therefore, undergo oxidative modification when exposed to light. Light-induced oxidation of protein pharmaceuticals may occur during protein processing and storage (Li *et al.*, 1995). The detailed mechanism of photooxidation of protein pharmaceuticals is complex and beyond the scope of this chapter. Therefore, a simple introduction to the conditions that initiate photooxidation reactions is provided. Photooxidation is initiated when a compound absorbs a certain wavelength of light, which provides energy to raise the molecule to an excited state (Halliwell and Gutteridge, 1989). The excited molecule can then transfer the energy to oxygen, converting it to singlet oxygen, while returning to ground state. Alternatively, the excited molecule can react directly with other molecules. This compound, which in essence initiates photooxidation reactions, is often referred to as a photosensitizer (Halliwell and Gutteridge, 1989). Efficient photosensitizers commonly have certain structural characteristics, such as aromatic systems and/or high conjugation with appropriate chromophores, which favour

Table 5.2 Common degradation products formed from reaction of singlet oxygen with amino acid residues

Oxidized residue	Products formed
His	2,5-endoperoxide, aspartic acid, asparagine
Trp	N-formylkynurenine, kynurenine
Cys	–S–S–disulphide crosslinks, sulphonic acid
Met	Met-sulphoxide

the absorbance of light (Halliwell and Gutteridge, 1989). Some compounds that have acted to initiate light-induced reactions are phenothiazines (used as tranquillizers), tetracycline antibiotics, various porphyrins (free and bound to proteins), riboflavin, and bilirubin (Halliwell and Gutteridge, 1989). One of the reactive oxygen species commonly involved in photooxidation reactions is singlet oxygen. The electron configuration of singlet oxygen allows this reactive oxygen species to cause oxidative damage through direct interaction with molecules. The electrons in the highest occupied orbital of singlet oxygen are in antiparallel spin directions, in contrast to triplet oxygen, which has electrons in parallel spin directions. Therefore, singlet oxygen is not spin-restricted in its reaction with an organic molecule. The interaction of singlet oxygen with amino acid residues is investigated in laboratories by initiation of photooxidation reactions using efficient photosensitizer dyes such as methylene blue, rose bengal, and acridine orange (Foote, 1968; Halliwell and Gutteridge, 1989). The amino acids found to be most susceptible to these photooxidation reactions are His, Trp, Met, and Cys (Spikes and Straight, 1967; Halliwell and Gutteridge, 1989). A list of the most common degradation products reported to arise from the reaction of singlet oxygen with these amino acids is provided in Table 5.2 (Spikes and Straight, 1967; Halliwell and Gutteridge, 1989). Further information on the mechanism of photooxidation reactions can be found in references provided in a book by Halliwell and Gutteridge (1989) and a review by Foote (1968).

5.3.4 *Strategies to prevent oxidation*

There are several strategies for stabilization of compounds against oxidation. The three main strategies of stabilization covered here are intrinsic, physical, and chemical.

An intrinsic method of stabilization utilizes site-directed mutagenesis. Stabilization of the protein towards oxidation is achieved by directly replacing the oxidation-labile amino acid. It is important to understand that a major shortcoming of this method of stabilization is that the structural and functional characteristics of the protein may be altered by replacement of the amino acid residue(s).

The physical method relies on the physical state of the formulation (solid vs liquid) as a basis for stabilization. Oxidation may occur in the solid-lyophilized or liquid-solution states of the protein. However, oxidation is usually slower in the solid state (Kenarke and Richards, 1966). Therefore, stabilization of the protein towards oxidation may be achieved by choosing a lyophilized solid-state formulation during storage.

Chemical additives may be utilized as another method of stabilization. Two of the most common are antioxidants and chelating agents. Antioxidant additives tend to end free radical reactions by reacting with radicals to terminate the chain. However, care should be taken that the antioxidant (i.e. thiol, ascorbic acid) does not function as a prooxidant in the presence of trace metals to initiate metal-catalysed oxidation reactions. Chelating agents are used to remove metals such as iron and copper from pharmaceutical formulations, and usually tend to protect proteins from metal-catalysed reactions. However, chelates have been shown to induce rather than prevent oxidation in certain situations (Zhao *et al.*, 1996). For example, Fe(III)–EDTA chelate has been observed to induce peptide oxidation (Zhao *et al.*, 1996). Overall, it is important that special care is taken in choosing the right means of stabilization of proteins and peptides against oxidation.

5.4   Other chemical pathways

Several other chemical reactions have been shown to be important degradation pathways for peptides and proteins. These include β-elimination reactions, which can also lead to racemization, and disulphide exchange reactions, which can result in incorrect pairing of disulphide bonds and, consequently, affect the three-dimensional structure of the protein. Each of these types of reaction is discussed below.

5.4.1   *β-Elimination reactions*

β-Elimination is another pathway of peptide/protein degradation. Amino acid residues that undergo β-elimination reactions include Cys, Ser, Thr, Phe, and Lys (Manning *et al.*, 1989). β-Elimination reactions result from the abstraction of a proton from the α-carbon of an amino acid residue in a polypeptide chain resulting in the formation of a carbanion intermediate. Addition of a proton to the opposite face of the molecule can lead to racemization. Alternatively, the carbanion intermediate can undergo further reaction to form a dehydroalanine residue (R = H) (Figure 5.8). β-Elimination reactions have been observed in a number of proteins, i.e. lysozyme (Nashef *et al.*, 1977) and bovine pancreatic ribonuclease A (Zale and Klibanov, 1986).

5.4.2   *Disulphide exchange reactions*

Disulphide bonds are of great importance to the structural stability of many proteins, as they are the most frequently encountered covalent crosslinks in proteins (Kosen, 1992). The disulphide bond may serve to join two independent polypeptide chains in an intermolecular fashion or may form between two Cys residues of one polypeptide chain in an intramolecular fashion. Thus, the interchange and/or cleavage of disulphide bonds can lead to an altered protein three-dimensional structure, which may result in the loss of biological activity (Kosen, 1992).

The stability of the disulphide bond is threatened by the presence of catalytic amounts of thiols. The thiolate ion is a reactive species that acts as a nucleophile attacking the sulphur atom of the disulphide resulting in disulphide exchange at neutral-to-alkali pH values

$$R^*S^- + R''\text{-}S\text{-}S\text{-}R'' \rightleftharpoons R^*\text{-}S\text{-}S\text{-}R'' + R''S^- \qquad (5.8)$$

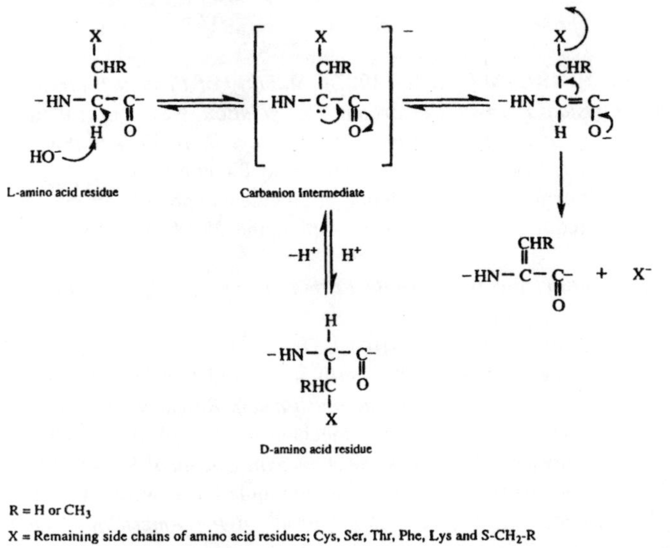

Figure 5.8 Mechanism of β-elimination reactions of amino acid residues at alkali pH.

The rate of reaction is greatest at alkali pH values, since the thiolate ion will predominate at higher pH values. An example of a protein that undergoes thiol disulphide exchange reactions is lysozyme. A more thorough discussion of the conditions leading to this reaction may be found in references in Kosen (1992).

Disulphide exchange reactions may also occur in acidic media. Benesch and Benesch (1958) proposed a mechanism of disulphide exchange in acidic media. The reaction intermediate is the sulphenium cation ($R*S^+$), which may be formed by heterolytic cleavage of a protonated disulphide bond

$$R*\text{-S-S-}R* + H^+ \rightleftharpoons [R*\text{-S-SH}^+\text{-}R*] \rightleftharpoons R*S^+ + R*SH \tag{5.9}$$

The electrophilic $R*S^+$ may then react with the sulphur atom of a disulphide bond leading to disulphide exchange

$$R*S^+ + R'\text{-S-S-}R' \rightleftharpoons R*\text{-S-S-}R' + R'\text{-S}^+ \tag{5.10}$$

Stabilization of the reaction at acidic pH values may be achieved by the addition of an external thiol that can scavenge $R*S^+$ (Benesch and Benesch, 1958). However, the reaction mechanism that occurs at neutral-to-alkali pH values (equation 5.8) is catalysed by the addition of an external thiol. Inhibition of thiol disulphide exchange reactions at neutral pH can be achieved by the addition of efficient thiol scavengers such as *p*-mercuribenzoate and N-ethylmaleimide (Zale and Klibanov, 1986).

5.5 Conclusion

The chemical instability of proteins and peptides poses a problem in their development as pharmaceuticals. A better understanding of the underlying mechanisms of instability of these complex molecules is essential to provide knowledge such that optimum stability conditions can be achieved in the pharmaceutical development processes.

References

AHERN, T.J. and MANNING, M.C. (eds), 1992a, *Stability of Protein Pharmaceuticals. Part A: Chemical and Physical Pathways of Protein Degradation*, New York: Plenum Press.

(eds), 1992b, *Stability of Protein Pharmaceuticals. Part B: In Vivo Pathways of Degradation and Strategies for Protein Stabilization*, New York: Plenum Press.

ASWAD, D.W., 1984, Stoichiometric methylation of porcine adrenocorticotropin by protein carboxyl methyltransferase requires deamidation of asparagine 25, *Journal of Biological Chemistry*, **259**, 10714–10721.

(ed.), 1994, *Deamidation and Isoaspartate Formation in Peptides and Proteins*, Baco Raton: CRC Press.

BATTERSBY, J.E., HANCOCK, W.S., CONNOVA-DAVIS, E., OESWEIN, J. and O'CONNOR, B., 1994, Diketopiperazine formation and N-terminal degradation in recombinant human growth hormone, *International Journal of Peptide and Protein Research*, **44**, 215–222.

BENESCH, R.E. and BENESCH, R., 1958, The mechanism of disulfide interchange in acidic solution; role of sulfenium ions, *Journal of the American Chemical Society*, **80**, 1666–1669.

BORCHARDT, R.T. and COHEN, L.A., 1972, Stereopopulation control. II. Rate enhancement of intramolecular nucleophilic displacement, *Journal of the American Chemical Society*, **94**, 9166–9174.

BRENNAN, T.V. and CLARKE, S., 1993, Spontaneous degradation of polypeptides at aspartyl and asparaginyl residues: effects of the solvent dielectric, *Protein Science*, **2**, 331–338.

CLARKE, S., 1987, Propensity of spontaneous succinimide formation from aspartyl and asparaginyl residues in cellular proteins, *International Journal of Peptide and Protein Research*, **30**, 808–821.

CLELAND, J.L., POWELL, M.F. and SHIRE, S., 1993, The development of stable protein formulations: a close look at protein aggregation, deamidation, and oxidation, *Critical Reviews in Therapeutic Drug Carrier Systems*, **10**, 307–377.

FOOTE, C.S., 1968, Mechanisms of photosensitized oxidation, *Science*, **162**, 963–970.

GEIGER, T. and CLARKE, S., 1987, Deamidation, isomerization, and racemization at asparaginyl and aspartyl residues in peptides, *Journal of Biological Chemistry*, **262**, 785–794.

GOOLCHARRAN, C. and BORCHARDT, R.T., 1998, Formation of diketopiperazine in model peptides, *Journal of Pharmaceutical Sciences*, **87**, 283–288.

HALLIWELL, B. and GUTTERIDGE, J.M., 1989, The chemistry of oxygen radicals and other oxygen-derived species. *Free Radicals in Biology and Medicine*, 2nd edn, pp. 22–85, Oxford: Clarendon Press.

HORA, M.S., RANA, R.K., WILCOX, C.L., HIRTZER, P., WOLFE, S.N. and THOMPSON, J.W., 1991, Development of a lyophilized formulation of interleukin-2, *Developments in Biological Standardization*, **74**, 295–306.

KENARKE, U.W. and RICHARDS, F.M., 1966, The histidyl residues in ribonuclease-S. Photooxidation in solution and in single crystals; the iodination of histidine-12*, *Journal of Biological Chemistry*, **241**, 3197–3206.

KHOSSRAVI, M. and BORCHARDT, R.T., 1998, Chemical pathways of peptide degradation. IX. Metal-catalyzed oxidation of histidine in model peptides, *Pharmaceutical Research*, **15**, 1096–1102.

KOSEN, A.P., 1992, Disulfide bonds in proteins. *Stability of Protein Pharmaceuticals. Part A: Chemical and Physical Pathways of Protein Degradation*, edited by T.J. AHERN and M.C. MANNING, pp. 31–59, New York and London: Plenum Press.

LI, S., SCHÖNEICH, C. and BORCHARDT, R.T., 1995, Chemical instability of protein pharmaceuticals: mechanisms of oxidation and strategies for stabilization, *Biotechnology and Bioengineering*, **48**, 490–500.

MacGINITY, J.W., HILL, J.A. and LA VIA, A.L., 1975, Influence of peroxide impurities in polyethylene glycols on drug stability, *Journal of Pharmaceutical Sciences*, **64**, 356–359.

MANNING, M.C., PATEL, K. and BORCHARDT, R.T., 1989, Stability of protein pharmaceuticals, *Pharmaceutical Research*, **6**, 903–918.

NASHEF, A.S., OSUGA, D.T., LEE, H.S., AHMED, A.I., WHITAKER, J.R. and FEENEY, R.E., 1977, Effects of alkali on proteins. Disulfides and their products, *Journal of Agricultural and Food Chemistry*, **25**, 245–251.

NGUYEN, T.H., BURNIER, J. and MENG, W., 1993, The kinetics of relaxin oxidation by hydrogen peroxide, *Pharmaceutical Research*, **10**, 1563–1571.

OLIYAI, C. and BORCHARDT, R.T., 1993, Chemical pathways of peptide degradation. IV. Pathways, kinetics, and mechanism of degradation of an aspartyl residue in a model hexapeptide, *Pharmaceutical Research*, **10**, 95–110.

1994a, Chemical pathways of peptide degradation. VI. Effect of the primary sequence on the pathways of degradation of aspartyl residues in model hexapeptides, *Pharmaceutical Research*, **11**, 751–758.

1994b, Chemical instability of proteins in solution and lyophilized formulation. *Protein Formulation and Delivery*, edited by J. CLELAND and R. LANGER, pp. 46–48, Washington, DC: American Chemical Society.

OLIYAI, C., PATEL, J., CARR, L. and BORCHARDT, R.T., 1994a, Solid-state chemical instability of an asparaginyl residue in a model hexapeptide, *Journal of Parenteral Science & Technology*, **48**, 167–173.

1994b, Chemical pathways of peptide degradation. VII. Solid state chemical instability of an aspartyl residue in a model hexapeptide, *Pharmaceutical Research*, **11**, 901–908.

OTA, I.M. and CLARKE, S., 1989, Calcium affects the spontaneous degradation of aspartyl/asparaginyl residues in calmodulin, *Biochemistry*, **28**, 4020–4027.

OYLER, A.R., NALDI, R.E., LLOYD, J.R., GRADEN, D.A. and SHAW, C.J., 1991, Characterization of the solution degradation products of histrelin, a gonadotropin releasing hormone (LH/RH) agonist, *Journal of Pharmaceutical Sciences*, **80**, 271–275.

PATEL, K. and BORCHARDT, R.T., 1990a, Chemical pathways of peptide degradation. II. Kinetics of deamidation of an asparaginyl residue in a model hexapeptide, *Pharmaceutical Research*, **7**, 703–711.

1990b, Chemical pathways of peptide degradation. III. Effects of primary sequence on the pathways of deamidation of asparaginyl residues in hexapeptides, *Pharmaceutical Research*, **7**, 787–793.

PEARLMAN, R. and WANG Y.J., 1996, *Formulation, Characterization, and Stability of Protein Drugs. Case Histories*, New York: Plenum Press.

SANDERS, L.M. and HENDREN R.W., 1997, *Protein Delivery: Physical Systems*, New York: Plenum Press.

SCHÖNEICH, C., HAGEMAN, M.J. and BORCHARDT, R.T., 1997, Stability of peptides and proteins. *Controlled Drug Delivery Challenges and Strategies*, edited by K. PARK, pp. 205–228, Washington, DC: American Chemical Society.

SHECHTER, Y., BURSTEIN, Y. and PATCHORNIK, A., 1975, Selective oxidation of methionine residues in proteins, *Biochemistry*, **14**, 4497–4503.

SCHULTZ, J., 1967, Cleavage at aspartic acid, *Methods in Enzymology*, **11**, 255–263.

SPIKES, J.D. and STRAIGHT R., 1967, Sensitized photochemical processes in biological systems, *Annual Review of Physical Chemistry*, **18**, 409–436.

STADTMAN, E.R., 1990, Metal ion-catalyzed oxidation of proteins: biochemical mechanism and biological consequences, *Free Radical Biology & Medicine*, **9**, 315–325.

1992, Protein oxidation and aging. *Science*, **257**, 1220–1224.

1993, Oxidation of free amino acids and amino acid residues in proteins by radiolysis and by metal-catalyzed reactions, *Annual Review of Biochemistry*, **62**, 797–821.

STEPHENSON, R.C. and CLARKE, S., 1989, Succinimide formation from aspartyl and asparaginyl peptides as a model for the spontaneous degradation of proteins, *Journal of Biological Chemistry*, **264**, 6164–6170.

STEVENSON, C.L., FRIEDMAN, A.R., KUBIAK, T.M., DONLAN, M.E. and BORCHARDT, R.T., 1993, Effect of secondary structure on the rate of deamidation of several growth hormone releasing factor analogs, *International Journal of Peptide and Protein Research*, **42**, 497–503.

STRAUB, J.A., AKIYAMA, A., PARMER, P. and MUSSO, G.F., 1995, Chemical pathways of the bradykinin analog, RMP-7, *Pharmaceutical Research*, **12**, 305–308.

TIMMINS, P., JACKSON, I.M. and WANG, Y.J., 1982, Factors affecting captopril stability in aqueous solution, *International Journal of Pharmaceutics*, **11**, 329–336.

WANG, Y.J. and PEARLMAN, R., 1993, *Stability and Characterization of Protein and Peptide Drugs. Case Histories*, New York: Plenum Press.

YAMAZAKI, I. and PIETTE, L.H., 1991, EPR spin trapping study on the oxidizing species formed in the reaction of the ferrous ion with hydrogen peroxide, *Journal of the American Chemical Society*, **113**, 7588–7593.

ZALE, S.E. and KLIBANOV, A.M., 1986, Why does ribonuclease irreversibly inactivate at high temperatures? *Biochemistry*, **25**, 5432–5444.

ZHAO, F., YANG, J. and SCHÖNEICH, C., 1996, Effects of polyaminocarboxylate metal chelators on iron-thiolate induced oxidation of methionine- and histidine-containing peptides. *Pharmaceutical Research*, **13**, 931–938.

6

Physical Stability of Proteins

JENS BRANGE

Brange Consult, Klampenborg, Denmark

6.1   Introduction

Maintenance of the structural integrity of a therapeutic protein is essential for its efficacy in relation to physiological and pharmacological activity. Therefore, a major challenge confronting the pharmaceutical scientist working with protein production and formulation is the instability of the protein drug during processing, handling, storage and use. The structural changes taking place in a protein can be broadly classified as involving physical or chemical alterations of the molecule. In practice, however, these two distinct routes of transformation of the protein structure have a mutual influence, as changes in conformation affect the susceptibility of the protein to chemical reactions, and vice versa. Chemical instability involves covalent modification in the amino acid sequence (primary structure), i.e. bond formation or cleavage, resulting in a new chemical entity. Physical instability refers to any change of the folded state that does not include bond cleavage or formation, i.e. changes in the spatial, three-dimensional conformation (secondary, tertiary and quaternary structure) of the protein. Once partially or wholly unfolded, the protein can undergo further changes by aggregation with other protein molecules, by adsorption to surfaces or by covalent modifications due to chemical reactions. When aggregation leads to macroscopic ensembles, the process is termed precipitation.

For successful application of proteins as pharmaceuticals, it is essential to understand the stability issues relevant to their production, formulation and use. The processes that affect physical stability include denaturation, i.e. loss of the higher-order, globular structure that a protein adopts upon folding to its native structure, aggregation, precipitation, and adsorption to surfaces. The term *aggregation* is here used to describe a non-native 'self-association' of protein molecules resulting in aggregates that remain in solution and are not visible to the naked eye. *Precipitation*, on the other hand, is a macroscopic process producing a visible change of the protein solution in the form of an increase in viscosity or clouding of the solution. Native, folded proteins may precipitate under certain conditions, most notably salting-out and isoelectric precipitation. Such precipitates, however, are readily distinguished from non-native aggregates by their solubility in normal solvents. Both covalent (polymerization) and noncovalent (aggregation) reaction pathways can be responsible for processes leading to formation of insoluble precipitates, and means to differentiate such mechanistically different processes have been suggested (Costantino *et al.*, 1994b). It is emphasized that this chapter will focus only on the aspects of non-covalent transformation of proteins.

This chapter first gives an overview of the structural details of protein architecture and the different forces contributing to its stability, and then gives a general description of factors favouring protein destabilization which can lead to aggregation and adsorption phenomena. Finally, different approaches for physical stabilization of protein drugs are reviewed. Protein stability during downstream processing has recently been reviewed by Hejnaes *et al.* (1998), and will not be dealt with in depth here.

6.2   Protein structure

The relationship between protein structure and stability is one of the important issues of modern protein science. By understanding the native state of a protein and the mechanism by which it is stabilized, the pharmaceutical scientist can more successfully predict the formulation that will result in a stable dosage form. Therefore, in order to discuss protein stability, a brief description of the characteristics and principles of protein structure and its hierarchical order is appropriate.

The apparently complex structure of proteins is the product of an evolution that probably optimized it for function and not necessarily for stability. One of the most important features of protein structure is that many proteins with quite different amino acid sequences have a similar general architecture, meaning that the overall three-dimensional structures are determined not by all the details of amino acid sequences but by some of their key characteristics. The transformation of the linear sequence (primary structure) from a disordered unfolded state into a compact and functional state is governed by a set of relatively simple principles (Chotia, 1984). Local folding of the polypeptide chain into β-strands and helical segments (secondary structure) is stabilized by a hydrogen bonding network, whereas the packing of α-helices and β-strands (or β-sheets) into a compact globular conformation (tertiary structure) mainly is stabilized by van der Waals' interactions and by the hydrophobic effect (resulting from interaction between non-polar amino

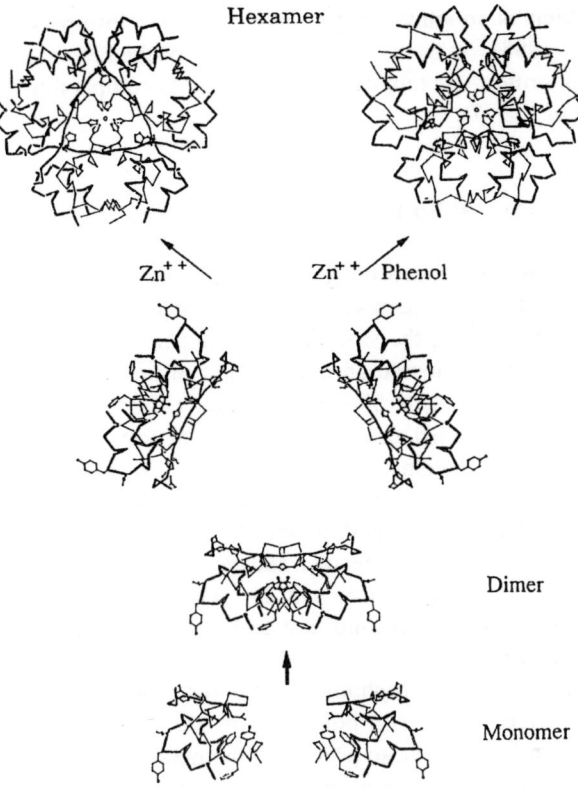

Figure 6.1 Self-association of insulin from the monomer to the dimer, and of three dimers into a hexamer in the presence of zinc ions. If phenol is also present, the gathering of the three dimers is accompanied by a conformational transformation of the B-chain N-terminal from an extended chain into an α-helix.

acid side-chains being more favourable than those they have with water – see section 6.4). Often the fully folded protein molecules will self-associate with one another and form dimers or higher-order subunit associates (quaternary structure). Subunit association is highly specific as it requires the subunits to be in their proper conformation before they coalesce to form the native quaternary structure. Close packing, i.e. optimum complementarity of the subunit interfaces, is the prerequisite of stable self-association, and binding of ligands, often cross-linking the subunits, is further able to stabilize the structure (Figure 6.1).

The distribution of hydrophobic and hydrophilic residues along a sequence is a major determinant of the overall folding of a protein (Miranker and Dobson, 1996). Furthermore, the size, shape and conformation of side-chains determine which particular class of packing occurs and modulate its exact geometry (Chotia, 1984). The protein assumes the most thermodynamically favourable state in a given solvent environment, and close packing strongly restricts the number of possible structures, ensuring a unique organization of the residues for a given protein (Levitt *et al.*, 1997).

Proteins are closely packed both in the solid state and in solution. In fact, they are more tightly packed than almost any other organic matter, as the overall packing efficiency of atoms in the protein core is greater than in crystals of organic molecules. Whereas organic liquids have packing density of less than 0.44, the density of globular proteins is close to 0.75, which is roughly what is expected for close-packing of hard spheres (Privalov and Gill, 1988). At the surface proteins are packed less tightly than in the core, whereas packing at protein–protein interfaces is roughly comparable to that in the protein interior (Levitt *et al.*, 1997).

6.2.1 *Stabilizing interactions*

For the effective and reliable functioning of a protein, the stability of its structure must be great enough for the protein to maintain its native conformation within a certain range of conditions. On the other hand, the structure must not be so rigid as to preclude smaller conformational changes, considered an integral part of protein function.

The stability of the native state of a protein is the difference between the free energy of the folded and the unfolded protein. Due to a delicate balance of stabilizing and destabilizing interactions, proteins are only marginally stable. The large and opposing forces include joint actions of several different energetic and entropic effects, which are dominated by the rigid parts of protein as the flexible, solvent-exposed parts contribute little (Matthews, 1993). The intermolecular and intramolecular forces contributing to the stability of the native structure include van der Waals' interactions, hydrogen bonding and other electrostatic interactions, and, not to be forgotten, the so-called hydrophobic effect (see also section 6.2.2).

It is generally agreed that hydrophobic interaction, representing a combined action of the water-ordering effect by the non-polar groups (the hydrophobic effect) and the van der Waals' interaction between these groups, is the major factor in stabilizing the folded structure of globular proteins. However, recent studies suggest that hydrogen bonding, weak electrostatic interactions in the form of charge–dipole or dipole–dipole interactions, also are major contributors to stability (Creighton, 1991). At least for some proteins, hydrogen bonding and the hydrophobic effect make comparable contributions to the conformational stability (Shirley *et al.*, 1992). In addition, strong electrostatic interactions between oppositely charged groups (salt bridges), weak electrostatic interactions between α-helix dipoles, and between the electrons surrounding the nuclei of atoms of aromatic side-chains contribute to some extent to protein stability (Burley and Petsko, 1989).

Ligands such as cofactors, substrates, specific organic molecules or ions may stabilize or destabilize native structures of proteins, depending on whether the ligands bind with highest affinity for the native or the denatured protein. Metal ions can dramatically increase protein stability when they bind to the flexible fragments of the polypeptide chain, in partiular to the bends (Mozhaev *et al.*, 1988).

The stability of proteins at near-physiological temperatures is not high. The values of free energy of protein stability for most globular proteins are only 5 to 15 kcal/mol (Mozhaev *et al.*, 1988). Amino acids with non-polar side-chains occupy about one half of the total volume of protein molecules (Mozhaev *et al.*, 1988). The burial of non-polar amino acid residues in the interior of a protein represents a very well-fitting three-dimensional jigsaw puzzle, and is a major contributor to stability as 1.3 kcal/mol in stability is gained for each CH_2 group buried by folding (Pace, 1992).

The content of hydrogen bonds in various proteins is almost the same and amounts to 0.75 per amino acid residue. Most intramolecular hydrogen bonds contribute 1.3 kcal/mol to the stability of globular protein structures (Shirley *et al.*, 1992). In a typical protein only one-third of the charged groups are involved in salt bridges, but when present they make a significant contribution to protein stability: up to 5 kcal/mol when they are localized inside the globule, and 1–2 kcal/mol when they are on its surface (Creighton, 1991; Mozhaev *et al.*, 1988). The helix–dipole interactions contribute only about 0.6 kcal/mol to stability (Matthews, 1993).

There is considerable experimental evidence for the stabilizing role of carbohydrate moieties in glycoproteins, whereas only a few cases have been observed in which deglycosylation increases protein stability (Mozhaev and Martinek, 1990).

Protein–protein interactions stabilize the native structure (or the denatured – see section 6.4) by burying hydrophobic clusters localized on the surface of the protein molecule whereby the thermodynamically unfavourable contact of these clusters with water is eliminated. In addition to the shielding of the hydrophobic regions on the surface of the protein, the rigidity of the protein molecule is changed as a result of its multipoint interaction with another molecule. The free energy of stabilization is usually several kcal/mol per protein–protein contact (Mozhaev *et al.*, 1988). Hydrophobic effect is a dominant driving force also in protein–protein association but, in general, hydrophobic amino acids are more frequent in the interior of a globular protein than in the interior of the protein–protein interfaces, and the contribution of the hydrophobic effect to protein self-association is not as strong as to protein folding (Tsai *et al.*, 1997).

The higher-order structure of proteins suggests what forces can be used in stabilization, and the more a formulator knows about the structure of the protein to be formulated, the better the likelihood of stabilizing the protein drug. By understanding the interactions, the formulation chemist can choose excipients to contribute to the thermodynamic stabilization of the protein. It is also important to recognize the potentially denaturing forces that may be applied to the molecule during processing.

6.2.2 *Role of water in structure and stability*

Solvent–protein interactions strongly influence the conformation of the protein molecule. An understanding of proteins requires knowledge of protein hydration as the protein–water interactions play an important role in the folding, stability, and function of proteins. There is a layer of non-covalently bound water molecules on a protein surface that differs essentially from bulk water. The number of bound water molecules is 300–500 for proteins of molecular weight 15–25 kDa, or approximately a monolayer of water, and several states of bound water are distinguished corresponding to the hydration of different types of functional groups (Mozhaev and Martinek, 1990). The layer of bound water associated with a protein in aqueous solution tends to accumulate ionized water molecules oppositely charged to the ionized groups on the surface of the protein molecule.

The role of water in protein structure and stability has been reviewed by Zaks (1992). Protein-bound water molecules can be divided into two main populations: water with hindered rotational freedom that is strongly bound by two or more hydrogen bonds, and weakly bound water with less restriction of rotational freedom. When the number of rotationally free molecules grows, flexibility of the protein structure increases. Hydrophobic effect is, as already mentioned, considered the most important factor contributing to protein stability and such interactions are derived predominantly from an unfavourable entropy decrease of water in the vicinity of non-polar groups. The solvation of a hydrophobic compound decreases its entropy due to increase in ordering of the water molecules surrounding the non-polar groups. To avoid this unfavourable entropy change, non-polar amino acid side-chains are driven into the hydrophobic core, minimizing interaction with the aqueous phase.

In addition to forming a hydration layer around the protein, the presence of water molecules contributes to van der Waals' interactions, salt bridges and, especially, to the hydrogen bonding network within the protein molecule. The molecular events that occur upon binding of water to a protein are well characterized for lysozyme and involve three stages (Zaks, 1992): up to 7% (w/w) of water, the hydration process is dominated by the interaction of water with charged groups; between 7% and 25%, the remaining polar sites are saturated; and finally, the weakly interacting parts of the protein become hydrated and full monolayer coverage of the lysozyme molecule occurs at 38% water corresponding to 300 water molecules per protein molecule.

Organic solvents added to water lower its dielectric constant and reduce polar interactions between the protein and surrounding water molecules, resulting in a decrease in the solubility of the protein. A protein can retain its correct conformation on transfer from a hydrophilic aqueous to a hydrophobic organic solvent system when a layer of bound or 'essential' water remains associated with the folded protein. Dehydration does not seem to disturb the native protein conformation but restricts the mobility. This is facilitated by the ionization of acidic and basic groups of the protein and requires less than 7% water. Lack of protein flexibility associated with dehydration can be reversed by the addition of small amounts of other compounds, such as formamide and glycerol, that have a high potential to form hydrogen bonds (Zaks, 1992). When water is removed from the vicinity of the protein, the protein becomes inert with respect to interactions with hydrophobic solvents, and in dry organic medium (0.02% water) proteins may display a remarkably high stability. Increased resistance to cold denaturation (see section 6.3.2) is another remarkable phenomenon of protein behaviour at low water activity (Zaks, 1992).

6.3   Protein destabilization (denaturation)

To assure protein stability in both liquid and solid formulations, the processes that cause physical destabilization of protein must be understood. Proteins are, as already mentioned, marginally stable and hence readily denatured by various stresses encountered in solution or solid states. Denaturation refers to an alteration of the global fold of a protein molecule, i.e. a disruption of the higher-order structure, such as tertiary and often also the secondary structure. An extreme description of this state is that only random non-covalent interactions are present although conformational preferences do exist. This situation is often referred to as a 'random coil' configuration. However, the random coil model is frequently not an adequate description of denatured proteins in solution. Unfolded states can vary considerably in the extent of residual structure, but some feature that relates to the structure of the native protein will remain and some hydrophobic interactions may persist even under extreme denaturation conditions (Dobson, 1992). Thus, the denatured state can be defined as any non-native conformation and the terms 'denaturation' and 'unfolding' will be used interchangeably in the following, and should be considered equivalent.

Caused by a variety of conditions, the denaturation process is in theory always reversible as the protein molecule, even if aggregated and precipitated, can be returned to its native state after dissolution by 'normalizing' the conditions. Depending on the requirements for reversal of the process, it can be envisioned as reversible or irreversible. However, in pharmaceutical practice, if the unfolded protein cannot easily recover its native state by refolding, for instance by lowering the temperature, then denaturation is considered to be irreversible (Manning *et al.*, 1989).

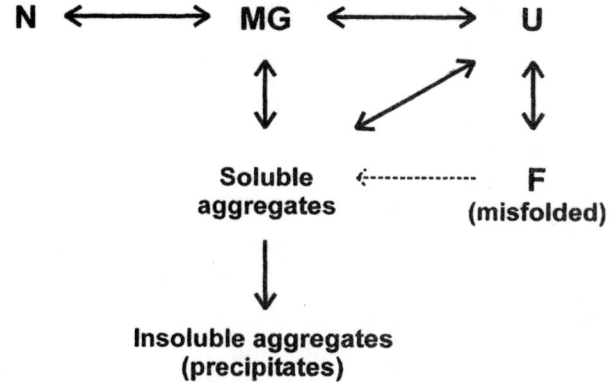

Figure 6.2 Unfolding, aggregation and precipitation. The transition from the native state (N) to the molten globule state (MG) and from MG to the unfolded state (U) are reactions of the first order, i.e. independent of protein concentration, whereas the aggregation reactions are second- or higher-order reactions. In the U state the denatured protein sometimes folds into misfolded (or off-pathway) structures (F) which also may form aggregates.

Proteins often undergo denaturation–renaturation cycles during extraction and purification. Conformational transition of globular proteins into the denatured state, as well as the back-transition into the native state, is a fast process as the protein assumes the most thermodynamically favourable state in a given solvent. Thus both α-helix and β-structure formation show half-lives in the microsecond time range (Jaenicke, 1991). The denaturation of small globular proteins involves a two-state model in which the native state, N, exists in equilibrium with the denatured state D. The denatured state may be an ensemble average of several denatured conformations which frequently exhibit significant amounts of persistent residual structure (Shortle, 1993).

Proteins can be unfolded by various factors, typically by raising or lowering temperature, by extremes of pH, or by addition of organic solvents, surface-active agents or chaotropic denaturants, whose effect is to solvate hydrocarbons more readily. Depending on how denaturation is brought about, the characteristics of the denatured state can differ greatly. For example, the organic solvents and detergents produce a denatured state that still contains some type of highly ordered structure (Shirley *et al.*, 1992).

6.3.1 Unfolding intermediates (molten globule)

In recent years it has become increasingly clear that denaturation of some proteins, especially larger ones, proceeds through a definable thermodynamic state, often native-like, referred to as an unfolding intermediate. There seems to be considerable similarity between folding and unfolding events, and therefore studies of protein folding are providing invaluable insight into the nature of protein stability (Dobson, 1992). A large number of folding studies have during the past decade provided increasing experimental evidence that proteins can exist in states that are intermediate between a fully folded and a highly unfolded state, often referred to as compact denatured states or molten globules (MGs) (Figure 6.2). Many proteins do not show one-step folding but have a distinct, populated intermediate (I state or molten globule state) between the fully unfolded (U state) and the

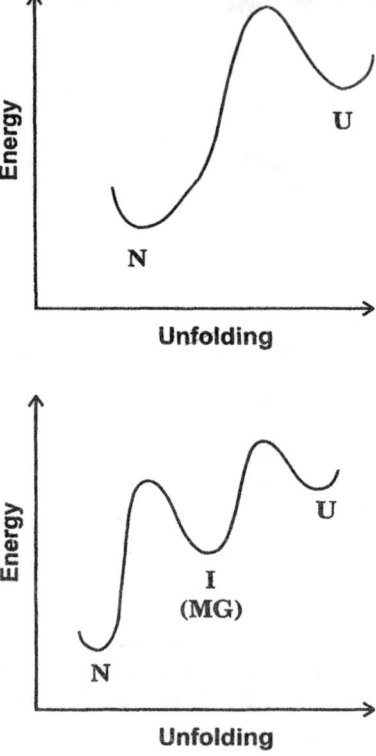

Figure 6.3 Simplified representation of the free-energy profiles for protein unfolding as a result of increasing temperature or increasing concentration of denaturant. Top: One-step unfolding of the native protein (N) into an unfolded structure (U). Bottom: Unfolding via a molten globule intermediate (I).

native state of proteins (N state). The intermediate states are considerably stabilized relative to the U state in the presence of disulphide bonds, and the stability of the intermediate state increases with molecular weight of the protein (Clarke and Waltho, 1997). The MG concept is now widely accepted, and some researchers regard the MG state as a third thermodynamic state of protein molecules (Ptitsyn and Uversky, 1994). This view, however, has recently been challenged (Pfeil, 1998). The free-energy profiles for one-step unfolding and for unfolding via an MG state are shown in Figure 6.3.

The MG intermediates found for different proteins appear to have some common features:

1   they are less compact than the native state, but with Stokes radius close to that of the native protein

2   they contain extensive secondary structure

3   they have loose and disordered tertiary contacts without tight side-chain packing.

The passage of the native state to the MG state involves the breakage of only a small fraction of all intramolecular hydrogen bonds, a fraction similar to that involved in tertiary structure hydrogen bonds (Xie *et al.*, 1991). The more flexible nature of the MG state permits some internal non-polar groups to become exposed to water, making the

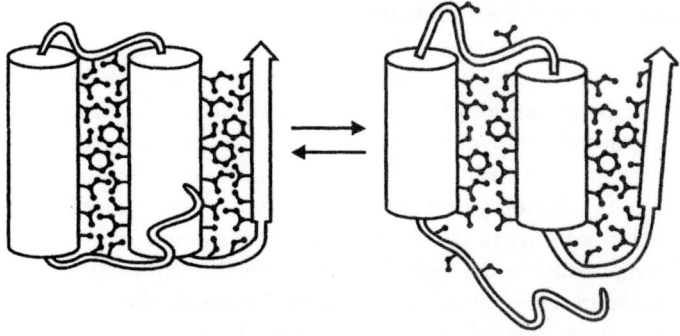

Figure 6.4 Schematic model of the native (left) and the molten globule (right) states of a hypothetical protein molecule. Only hydrophobic side-chains are shown.

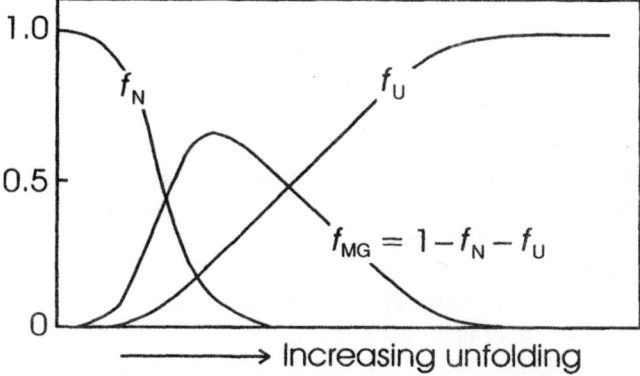

Figure 6.5 Relative population of native (N), molten globule (MG) and unfolded (U) states during unfolding of a hypothetical protein via an intermediate state.

surface of this state more hydrophobic than for the native state, as illustrated in Figure 6.4 (Ptitsyn, 1995), and frequently proteins tend to aggregate in the MG state as measured by equilibrium sedimentation or by light scattering.

Whether two-state or three-state unfolding is observed may depend solely on the stability of the MG state (Baldwin, 1996), and for some small proteins accumulation of intermediates is not observed (Schindler *et al.*, 1995). Recent work has emphasized the complexity of protein folding but has also shed light on folding pathways, the role of intermediates in folding, and the general principles by which proteins adopt their native folds. Many small proteins fold without detectable intermediates, but there are many examples of other proteins whose folding routes pass through partially folded MG-like intermediates (Figure 6.5). However, the presence or absence of intermediates for any given protein is sensitive to folding conditions such as temperature and pH (Pande *et al.*, 1998).

Side-chain interactions (particularly those involving hydrophobic groups) are relatively weak in the intermediate state but form the same pattern as in the native structure, i.e. the I state appears to be neither randomly collapsed nor appreciably misfolded (Clarke and Waltho, 1997). The existence of partially unfolded, stable intermediate conformations has been observed for pharmaceutically relevant proteins, including human growth hormone (Bam *et al.*, 1996) and insulin (Millican and Brems, 1994).

6.3.2   *Temperature-induced changes*

The protein native state is stable in a limited temperature range. When exposed to sufficiently elevated temperatures, all proteins eventually lose native structure. With increasing temperature, hydrogen bonding becomes progressively weaker while the hydrophobic
interactions are strengthened with increasing temperature, at least up to 60°C (Volkin and
Middaugh, 1992). At physiological temperatures hydrophobic interactions provide the
major contribution to stability of the folded state (Franks *et al.*, 1988) but the stability of
proteins at this temperature is not high (approximately 12 kcal/mol).

Upon an increase in temperature, heat ultimately disrupts these non-covalent forces.
At elevated temperatures, proteins become more flexible, leading to partial unfolding, and
their collision frequency increases, resulting in a propensity to form aggregated states.
Thermally induced denaturation is classified as either reversible or irreversible depending
on whether native structure is recovered following return to ambient temperature.

When small proteins are reversibly denatured by heat, a two-state transition is commonly observed where only the folded or the unfolded states are present. The temperature
at which 50% of the molecules are unfolded is called the thermal melting temperature
(T_m). Reversible thermal denaturation is dependent on the precise solution conditions.
Thus, T_m can increase by 20–25 K when the pH of the solution is changed from 3 to 7, or
as a function of the presence and type of neutral salt (Volkin and Middaugh, 1992). One
must exercise caution in extrapolating stability data obtained under accelerated storage
conditions to less extreme temperature conditions. Such extrapolation is valid only if the
kinetics clearly exhibit Arrhenius behaviour (Cleland *et al.*, 1993).

Whereas heat denaturation is driven principally by the loss of polar contacts and the
gain of conformational entropy of the chain, cold denaturation is explained by the weaken-
ing of the hydrophobic effect upon cooling (Dill *et al.*, 1989; Privalov and Gill, 1988).
For most proteins the temperature of cold denaturation is far below the freezing point of
aqueous solutions (Privalov *et al.*, 1986). At the temperature of maximal stability (T_s), generally between −10 and 35°C (Pace and Laurents, 1989), the entropy of protein unfolding
is zero and stabilization of the native structure is achieved only by the enthalpy factor
(Figure 6.6).

As a practical matter, the freezing of a protein solution may also lead to denaturation.
As water molecules crystallize, solutes like salts and buffers are concentrated, causing
potentially dramatic shifts in pH and ionic strength (Volkin and Middaugh, 1992).

6.3.3   *Influence of pH*

The pH of a protein solution is one of the most important factors determining the structural state of the protein. For example, at low pH decreased stability can be observed
as a result of unfavourable electrostatic interactions introduced by the increase in the
positive charge on the protein. A similar argument accounts for the decrease in stability
observed in highly basic solutions due to repulsion of negative charges. Not surprisingly,
most proteins have maximal thermodynamic stability at or near the isoelectric point
where the net charge is zero.

Many proteins form an extensively unfolded denatured state at acid pH (2–3) in the
absence of high concentrations of anions. If anion concentration is increased (or pH
further decreased), and electrostatic repulsion thereby neutralized, the structure becomes
much more compact, a state which earlier was called the A state but now more often is

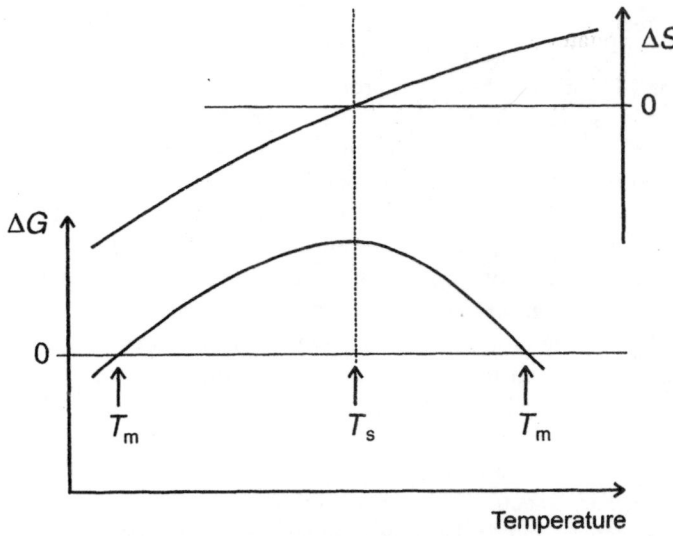

Figure 6.6 Schematic representation of the changes of entropy (ΔS) and Gibbs' free energy (ΔG) of protein unfolding as a result of change in temperature. T_m is the temperature at which the free energy difference between the native and the unfolded state is equal to zero, and represents the temperatures of cold and heat denaturation. T_s is the temperature at which the stability of the native state is maximal and the ΔS is zero.

termed the molten globule state (Shortle, 1993). The effectiveness of an anion in inducing the conformational transition is dependent on the anion species, and the major factor responsible for the transition is electrostatic binding to the positively charged sites of the protein (Goto *et al.*, 1990).

6.3.4 Influence of pressure

Proteins are not very sensitive to hydrostatic pressure, and only at extremely large values of pressure do they exhibit the changes which are very similar to those observed in temperature- and pH-induced denaturation. Whereas moderate pressures of 0.5–2 kbar (500–2000 atm) may induce dissociation of oligomeric proteins, denaturation of the protein monomer requires pressures from 4 to 8 kbar (Mozhaev *et al.*, 1996). Hydrophobic contacts are disfavoured by pressure, and electrostatic interactions become much weaker at elevated pressures. In contrast, hydrogen bonding is stabilized by high pressures (Mozhaev *et al.*, 1996). High pressure can lower the freezing point of water by as much as 20 K; this has been used to facilitate the study of cold denaturation of proteins (Nash and Jonas, 1997).

6.4 Aggregation and precipitation

The formation of soluble or insoluble protein ensembles can be due to different phenomena involving either covalent or non-covalent interactions, as follows.

Chemical reactions (polymerization) between different protein molecules result in the formation of covalently linked protein dimers or polymers. Because the origin of protein aggregates or precipitates is frequently not known, the term 'aggregation' is unfortunately often used in protein literature to describe this process, which should rather be termed 'polymerization'. In practice, however, protein ensembles/precipitates often are mixtures of covalently and non-covalently linked protein molecules (Costantino *et al.*, 1994a, 1994b). Chemical transformation of proteins is dealt with in Chapter 5.

There are two basic types of protein non-covalent interaction, which for clarity here are termed *association* if the process involves protein molecules with native structure, and *aggregation* when denatured protein molecules are involved.

Self-association of the native protein happens as a result of changes in solvent environment (solvent composition, pH, ionic strength, protein concentration, etc.) analogous to the conditions facilitating protein crystallization, isoelectric precipitation, salting out, etc. Thus, if a protein is placed in a high concentration of salt, the surface charges on the protein become masked such that charge–charge repulsion between different native protein molecules does not occur. A similar phenomenon of reduced charge repulsion can occur when the pH of the solution approaches the isoelectric point of the protein. In both cases the surface charge neutralization can result in association of protein molecules with native tertiary structure, and, if the protein concentration is high enough, precipitation of the protein oligomers occurs. In such associates the basic structure of the protein is native-like if not fully native, and the association process is easily reversed by reducing the protein concentration or, if salt and pH changes do not irreversibly alter the conformation of the protein, by adjusting the pH or reducing the salt concentration. Self-association of proteins has often, with insulin as a good example, developed evolutionarily as a requirement for biosynthesis and for ensuring *in vivo* storage stability.

Aggregation of denatured protein molecules was once assumed to arise from ensembles of completely unfolded states of the protein, but recent observations suggest that aggregation is often likely to arise from partially unfolded intermediates. The characteristics and properties of such intermediates may be significantly different from the folded native state as well as the completely denatured protein, and aggregation will be favoured by factors and conditions that favour population of these partly unfolded intermediates (Fink, 1998).

Thus, the term 'association' is here used to characterize ordered, native quaternary structure formation, and the term 'aggregation' is reserved for those cases where a protein forms a regular or non-regular oligomeric structure of non-natively structured proteins. Aggregation may occur during pharmaceutical processing and use of protein drugs by exposure to hydrophobic surfaces, change in temperature (heating or cooling), sonification, or shear stress during filtration or agitation.

6.4.1 *Mechanisms of aggregation*

Protein aggregation can occur from a conformational intermediate produced during folding or denaturation, or from more extensively unfolded (fully denatured) protein molecules. Protein aggregation is most commonly described as a two-step process but if the protein in its native state is oligomeric, an additional initial process of dissociation into monomeric species may apply at high protein concentration where the protein self-associates (Figure 6.7). The first step involves more or less unfolding of the protein molecules whereby buried, hydrophobic residues are exposed to the aqueous solvent. The unfolding is primarily caused by the solvent environment (pH, salt, co-solvents, etc.), temperature, or surface interactions. In the second step, the hydrophobic residues of the

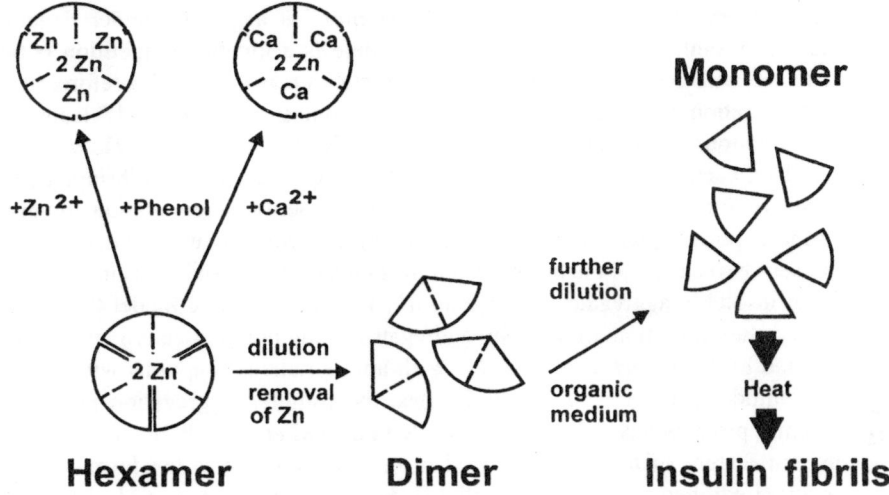

Figure 6.7 Schematic representation of the dissociation of the insulin hexamer as caused by dilution or addition of organic medium. Dissociation into the monomeric species is required for insulin fibril formation, and stabilization of the hexameric state by addition of certain ligands therefore counteracts insulin fibril formation (Brange *et al.*, 1997a).

unfolded protein molecules, now on the surface of the intermediates (or the fully unfolded protein monomers), make intermolecular hydrophobic interactions in order to minimize the unfavourable exposure of the hydrophobic amino acid residues to the solvent. The initial stages of aggregation are quite specific in the sense that they involve the interaction of specific structural subunits of one molecule with 'matching' hydrophobic surface areas of structural subunits of a neighbouring molecule. Two sites can be sufficient, in which case the aggregation most likely propagates in a linear fashion forming long fibres. In any case, the process leads to larger aggregates and eventually their size will exceed the solubility limit.

It has been emphasized that intermediates are more prone to aggregation than the unfolded state, simply because hydrophobicity in the unfolded state is scattered randomly in many small regions, whereas in the intermediates there will be large patches of contiguous surface hydrophobicity with a much stronger propensity for aggregation (Fink, 1998). Some of these surfaces are those that also interact in an intramolecular or intermolecular manner to form the native conformation and associates, respectively (Fink, 1998, Brange *et al.*, 1997b). The hydrophobic interactions in aggregates have often been considered non-specific. However, recent results have provided strong evidence that aggregation occurs by highly specific intermolecular interactions, as a mixture of folding intermediates from two different proteins did not coaggregate with each other, but only with themselves (Speed *et al.*, 1998). The nature of the intermediate state preceding aggregation is the key to understanding the aggregation pathway, but this state has been hypothesized to exist as anything from a small, transient expansion of the native state to a MG structure to a fully unfolded molecule. The sequence of states leading from the unfolded to the native state also allow for the possibility of off-pathway (misfolded) intermediates (see Figure 6.2) (Pande *et al.*, 1998).

Energy transfer – heat, radiation, ultrasound, etc. – is normally required to unfold the protein molecule. Conformational flexibility increases with temperature, and the higher the energy input, induced by, for example, temperature or ultrasound, the more unfolding and subsequent aggregation there will be. Most proteins aggregate under denaturing conditions at concentrations of 0.01–10 mg/ml. There is a kinetic competition between refolding to the native state and higher-order aggregation reactions. Therefore, at high protein concentration aggregation dominates over folding and eventually leads to irreversible formation of insoluble protein aggregates (Kiefhaber *et al.*, 1991; Volkin and Middaugh, 1992). However, as concentration also influences the equilibrium between native protein oligomers and monomers, the relationship between aggregation and protein concentration is not simple. Thus, an increase of insulin concentration under conditions that favour self-association of the hormone into hexamers actually stabilizes insulin against formation of insoluble aggregates (Brange *et al.*, 1997a). A kinetic model that explains quantitatively the formation of protein aggregates from non-associating proteins has been presented by Kiefhaber *et al.* (1991). With appropriate assumptions with regard to relative magnitude of the various reaction rates, the complex aggregation pathways of self-associating proteins have also been modelled (Lencki *et al.*, 1992).

A common feature of thermally-induced protein aggregation is the formation of an intermolecular hydrogen-bonded, antiparallel β-sheet structure (Dong *et al.*, 1995); often the aggregation occurs in a linear fashion eventually resulting in the formation of long fibres. Aggregation is often considered to be an irreversible process, but as long as only a few molecules are forming the individual aggregates, the aggregation process may be reversible. Initially the oligomeric aggregates will be soluble, but once the aggregates reach a certain size and become insoluble, the process is not reversible without dramatic changes in solvent environment.

6.4.2   *Precipitation and fibrillation phenomena*

Precipitation of a protein from solution is a widely used procedure during downstream processing. Solubility of the protein is here lowered by changing the pH to the isoelectric point of the protein (iso-precipitation), by adding high concentrations of salt (salting out) or by inclusion of organic solute such as ethanol, acetone or polyethylene glycol (Hejnaes *et al.*, 1998). Addition of certain metal ions such as zinc or calcium can for some proteins, such as insulin, further reduce solubility (Brange *et al.*, 1987).

However, precipitation during manufacture or storage of a pharmaceutical preparation of a protein is an undesired event. Such precipitation, initially seen as increasing turbidity or viscosity of the solution, can occur for various reasons relating to either chemical (polymerization, often due to disulphide exchange reactions – see Chapter 5) or physical changes of the protein. Sometimes a protein precipitate can be the result of both covalent and non-covalent aggregation pathways (Costantino *et al.*, 1994a, 1994b), and often, when protein aggregation and precipitation is reported in literature, the nature of the precipitate is not known.

Protein precipitation from a solution involves three reactions: formation of stable aggregates of a few molecules (nucleation), development of these centres into larger aggregates or fibres (growth), and floccule formation from soluble aggregates (precipitation). The rate of nucleus formation is normally low and the time period during which stable nuclei build up is called the lag phase. The length of the lag phase depends on the protein concentration and other experimental conditions, but if preformed aggregates are added to the solution, even in small amounts, the lag phase is completely eliminated (Figure 6.8).

Precipitation of non-covalent aggregates happens when the aggregate nuclei have grown into primary particles, typically with a diameter of 100 to 200 nm corresponding

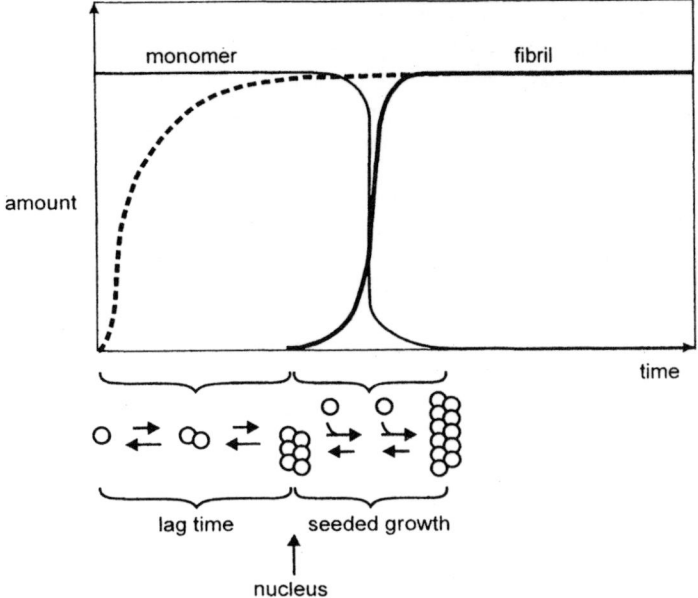

Figure 6.8 Idealized diagram of the formation of protein aggregates (fibrils) as a function of time. The lag time represents the time until a stable nucleus is formed. If fibrils or fibril nuclei are added initially, the lag time disappears and seeded growth starts immediately as represented by the broken curve.

to about 1000 protein molecules (Glatz, 1992). The size and number of primary particles formed depend on the total amount of protein that comes out of solution and the competition between the growth of existing particles and the nucleation of new particles. In the electron microscope, aggregates are often seen as long fibres with diameters ranging from 5 to 150 nm and lengths up to several μm, as in the case of insulin (Brange *et al.*, 1997a). Depending on the exact experimental conditions, the fibres can take different forms (Figure 6.9). The subsequent flocculation of primary particles results in formation of precipitates large enough (10 to 20 μm) to be observed with optical microscopy (Glatz, 1992).

The propensity of insulin, under the influence of heat and exposure to hydrophobic surfaces, to undergo conformational changes resulting in successive, linear aggregation (fibrillation), and formation of a viscous gel or insoluble precipitates, has been one of the most widely studied phenomena in relation to aggregation and precipitation of proteins (for a review see Brange *et al.*, 1997a). The classical method to induce insulin fibrillation has been to heat a concentrated acid solution. The formation of the insulin fibres can be observed as an increase in viscosity or precipitate formation. Whereas nucleation seems to require temperature above normal, the subsequent growth can proceed at ambient or even lower temperature and, depending on the conditions, the growth leads to long fibres resulting in a thixotropic gel or to shorter fibres with a tendency to arrange radially to spherolites with precipitation as the consequence (Waugh *et al.*, 1953; Waugh, 1957). The growth of insulin fibrils has been found to be a function of the surface area of the fibril population and the concentration of insulin in solution (Waugh, 1954), and thus is a highly cooperative process that can remove the insulin quantitatively from solution in the fibrous form. An increase in ionic strength increases the rate of growth and, in particular,

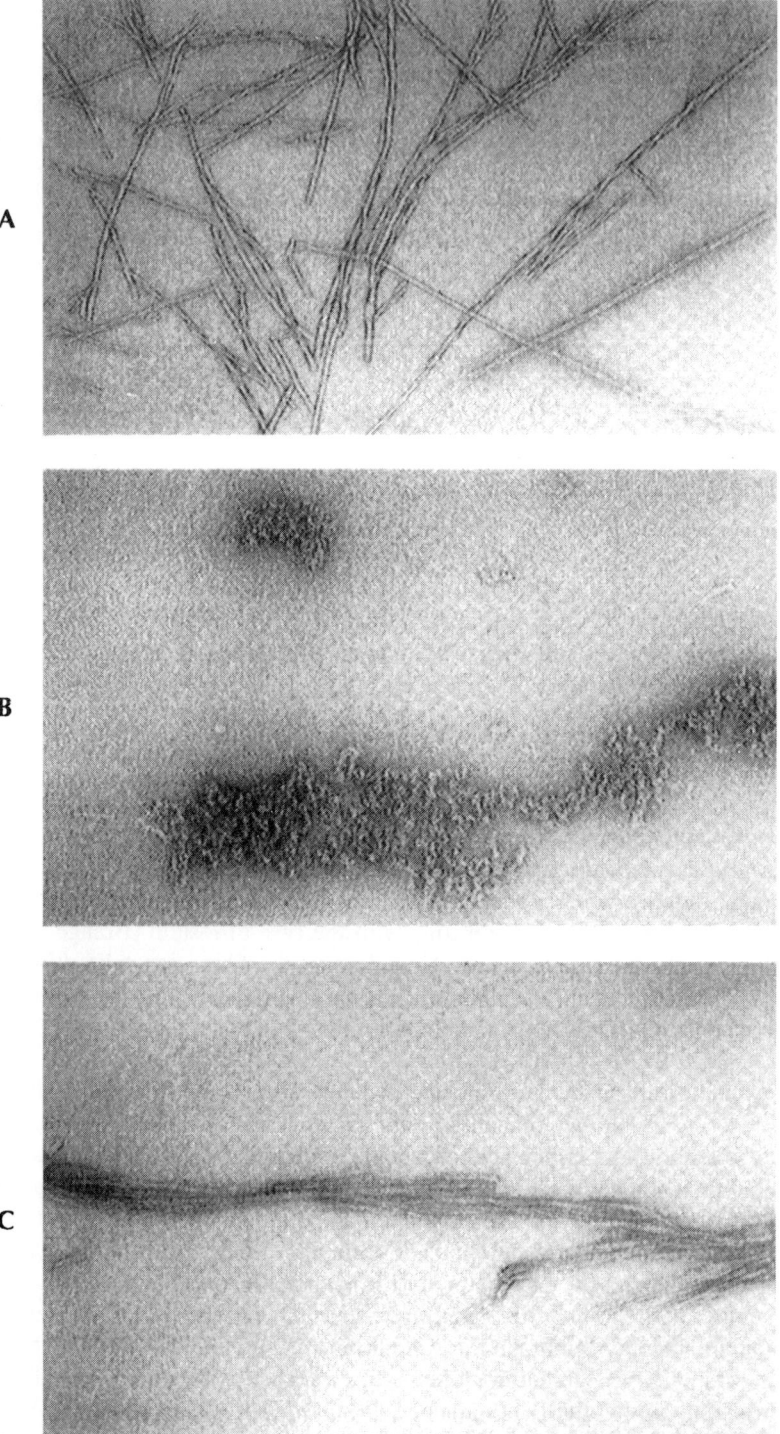

Figure 6.9 Transmission electron micrographs of bovine insulin aggregates (fibrils) formed under various conditions. A: 2% insulin in 0.1 N HCl, pH 1.7, treated at 80°C for 30 min. B: 0.4% insulin, 0.2% phenol, 0.02 M NaCl, pH 7.5, treated by agitation at 37°C for 3 days. C: 2% insulin in 8.3 M acetic acid, 0.1 N NaCl, pH 1.7, treated at 80°C for 2.5 h.

nucleation (Waugh, 1957), whereas high concentrations of organic acids, urea and phenol suppress the nucleation reaction while allowing the growth reaction to proceed (Waugh *et al.*, 1953).

Human insulin does not differ from porcine insulin in fibrillation tendency, whereas bovine insulin, with only two amino acid residues differing in the sequence, surprisingly is significantly more prone to form fibres than the other two species of insulin (Brange *et al.*, 1997a). Apparently the hydrophobicity of the A8 residue (Ala in bovine and Thr in porcine and human insulin), which is on the surface of the molecule, plays a role in the fibrillation process by directing the interactions of insulin with hydrophobic surfaces. With the presence of certain types of insulin-like substances the tendency to fibrillation decreases (Brange *et al.*, 1982). Higher molecular contaminants such as covalent insulin dimer and proinsulin were active in counteracting fibrillation when present in amounts corresponding to a few per cent of the total protein material, whereas more simple derivatives such as desamidoinsulins were ineffective even at 5–10% concentration. Presumably the covalent dimer or proinsulin is able to block further linear aggregation by providing only one of the two complementary interacting surfaces in the insulin fibril.

6.4.3 *Factors influencing aggregation and precipitation*

Environmental conditions that favour a population of partially unfolded intermediates are likely to lead to aggregation. The major factors that determine the extent and rate of *in vitro* protein aggregation are, besides the amino acid sequence of the protein, pH, temperature, ion strength, protein concentration, and the presence of co-solutes (including chaotropes) or ligands that interact selectively with either the native or non-native conformations of the protein. The propensity for a given protein to aggregate may be determined by the lifetime of partially unfolded intermediates (Fink, 1998).

Effect of excipients and cosolvents

Excipients, often referred to as co-solvents because they affect the solvation of the protein, affect the physical stability of proteins by preferential hydration. The driving force for preferential hydration varies for different types of co-solvent compounds. There are two basic mechanisms by which co-solvents are preferentially excluded from the surface of proteins (Timasheff, 1992):

1 the interactions with protein are determined strictly by the properties of the solvent, as the protein remains essentially inert and only presents a surface

2 the interactions (attractive or repulsive) are to a major extent determined by the chemical nature of the protein surface.

The first mechanism comprises steric exclusion of the co-solvent from the surface of the protein (accounting for the preferential exclusion of polyethylene glycol (PEG) from the surface of native proteins) and perturbation of the surface tension of water by the co-solvent, resulting in a depletion of the additive in the surface layer. Lowering of the surface tension appears to be the most prevalent mechanism by which co-solvents are excluded from proteins. Thus, the preferential hydration caused by addition of carbohydrates is related to this effect (Arakawa and Timasheff, 1982).

The second mechanism comprises repulsion from the charges on the protein surface and the solvophobic effect, which is similar in nature to the hydrophobic effect. The solvophobic effect causes co-solvent molecules to migrate away from non-polar regions of the protein and into the bulk solvent because the contacts between these non-polar regions and the water–cosolvent mixture are entropically even more unfavourable than contact between these regions and water. Glycerol and a number of other polyols belong to this latter category although the effect of these compounds is complex. The preferential hydration caused by addition of glycerol is related both to the high tendency of glycerol to repel hydrophobic regions and to its ability to enter the water lattice and strengthen solvent structure (Gekko and Timasheff, 1981a, 1981b). On the other hand, glycerol also has affinity for polar regions of proteins, mainly due to hydrogen bond formation. The net effect is a balance between exclusion and binding, with preferential exclusion generally dominating relative to binding (Timasheff, 1992).

6.5 Surface adsorption

The presence of an air–water or solid–water interface has an important influence on the normal forces stabilizing protein higher-ordered structure in solution. Partial unfolding of the protein can occur at the interface which leads to adsorption to the surface through interaction of hydrophobic amino acid residues and a hydrophobic surface, or through binding of polar amino acid residues to charged surfaces. Therefore, proteins can be adsorbed both to non-polar solid (or air) surfaces and to surfaces with ion-exchange properties (such as glass), although greater adsorption normally occurs at hydrophobic than at hydrophilic interfaces.

Normally a protein in solution reaches a surface by diffusion, and therefore the rate of adsorption is a function of the protein concentration. A characteristic property of adsorption is the saturation effect. The amount of protein on the surface increases steeply with time, followed by a levelling-off to a plateau value which normally falls within the range expected for a close-packed monolayer of protein corresponding to 0.1 to 0.5 $\mu g/cm^2$. Sometimes, however, proteins do not display a well-defined plateau but instead protein concentration continues to rise, although more slowly than initially (Horbett, 1992), probably as a result of protein aggregation at the surface. Eventually, the local protein concentration at the interface may be 1000 times higher than the initial protein concentration in the solution. This means that many partially unfolded protein molecules are in close proximity to each other, often resulting in extensive aggregation of the protein.

Like many solid surfaces, gas interfaces are hydrophobic and can cause protein denaturation and aggregation. At air–water interfaces a large fraction of the protein may be adsorbed to the interface when the solution is agitated, and the turbulence creates additional interface area. In a shaken solution there is a continual creation of new interface for protein to adsorb to, and the large effective surface area can lead to massive protein adsorption and subsequent aggregation. The protein denaturation caused by agitation is the result of both the interfacial and the shear forces on the protein (Cleland *et al.*, 1993).

Binding to hydrophobic surfaces virtually always leads to unfolding of the protein, and exposure of its non-polar interior would be expected to enhance hydrophobic surface interactions. Thus, the more hydrophobic the surface, the greater is the destabilizing effect, probably as a result of stabilization of the denatured form of the protein by the hydrophobic surface. In contrast, when proteins bind to polar surfaces the change in protein structure is small (if any), and the interaction between the surface and the protein may be due to more specific polar contacts between the surface and the protein (Steadman *et al.*, 1992).

Insulin aggregation at solid–water interfaces has been studied by several groups (Brange *et al.*, 1982; Sluzky *et al.*, 1992; Feingold *et al.*, 1984; Chawla *et al.*, 1985). In systematic studies on the influence of different materials, Sluzky *et al.* (1992) showed the fibrillation tendency to increase with the surface area of hydrophobic material, and fibrillation tended to proceed faster with increasing hydrophobicity of the material. They also provided evidence that aggregation caused by agitation is the result of both the interfacial and shear forces on the protein. However, because an insulin solution can be rotated in a glass vial (without any air present) for several months (Brange *et al.*, 1982) or agitated with glass beads for many days (Sluzky *et al.*, 1992) without any fibrillation, the shear forces probably play a minor role compared with the surface interactions. Insulin fibrillation at the agitated air–water interface increases with increasing rate of agitation (Sluzky *et al.*, 1992). The insulin monomer, with its exposed hydrophobic surfaces normally buried by dimer and hexamer formation, has potentially higher binding affinity for a hydrophobic interface than the dimer and hexamer. However, at high insulin concentration the equilibrium is shifted towards the hexamer, which may populate the interface and reduce access for the monomeric species.

6.6 Solid phase stability

Most commercially available therapeutic proteins are available as a lyophilized powder which may undergo physical changes during processing and storage. The stability of a protein in the solid state is highly dependent on the moisture content of the solid and there is an optimum level of residual moisture required to maintain stability (Banga, 1995). Most often moisture induced aggregation is due to covalent reactions but also noncovalent forces seem sometimes to be involved (Costantino *et al.*, 1994a, 1994b). The aggregation reactions and means to prevent them are discussed in detail in Chapter 9.

6.6.1 Lyophilization-induced aggregation

When a protein is not sufficiently stable in aqueous solution, lyophilization is the method most commonly used to ensure stability during long-term storage. While lyophilization may improve storage stability, the process in itself can cause damage of the protein as both freezing and drying stress the protein. During freezing the concentration of all solutes increases dramatically in the non-ice phase. This localized high salt or solute concentration may result in denaturation of the protein, and subsequently, after rehydration, lead to aggregate formation, especially if residual moisture is present. The degree of aggregation noted after rehydration correlates directly with the degree to which the protein is unfolded in the dried state (Dong *et al.*, 1995). Therefore special precautions must be taken when proteins are lyophilized, and it is often necessary to add solutes (cryoprotectants and lyoprotectants) that stabilize against both freezing and drying stresses to ensure protection of the protein. This is discussed in depth in Chapter 9.

6.7 Stabilization of protein drugs

Methods of protein stabilization have been improving steadily in recent years. The various general approaches by which increased physical stability can be achieved are briefly reviewed in the following. A more detailed presentation can be found in Chapters 7 (solution stability) and 9 (solid-state stability). Stabilization of proteins with respect to chemical degradation reactions is discussed in Chapter 5.

The key factors for providing a rational basis for protein stabilization are: the nature of the species responsible for aggregation, the detailed mechanism that leads to aggregation and the underlying kinetic scheme, the structure of the aggregates, the specificity of the intermolecular interaction, and how environmental conditions affect the rate and the amount of aggregation.

6.7.1 *Stabilization strategies*

From the previous sections, it is obvious that rational strategies to improve the practical physical stability of protein drugs include the following main approaches:

1   stabilize the native structure of the protein (i.e. promote self-association, improve native hydrophobic interaction by adding co-solvents or by genetic engineering

2   prevent aggregation of unfolded structures (polyethylene glycols, cyclodextrins?)

3   avoid or block unwanted hydrophobic interfaces (surfactants, avoid head space in container)

4   reduce shear forces (avoid head space in container).

A protein molecule can be stabilized by adding co-solvents or particular ions and salts to the solution; it can also be stabilized by a covalent modification such as glycosylation, or by mutation of its amino acid residues (semisynthesis or genetic engineering). When choosing additives for improving long-term stabilities of proteins in solution or in the solid state, the various factors must be considered on a protein-by-protein basis, as reviewed by Arakawa and Timasheff (1982) and Carpenter *et al.* (1997). Often a combination of stabilizers, working by different mechanisms, may be required for obtaining the desirable stability.

Additives

Solvent environment dictates the ability of the protein to maintain its native state and avoid denaturation from physical forces. It has long been recognized that the addition of a number of polyhydric alcohols, carbohydrates, and some amino acids to aqueous solutions of proteins leads to their stabilization. The mechanism of protein structure stabilization by these compounds was elucidated by Timasheff and coworkers (for reviews see Timasheff, 1992, 1993), who found that the polyols are preferentially excluded from the immediate domain of the protein, which in their presence is surrounded by an environment relatively enriched in water (preferential hydration of the protein). Since this exclusion is thermodynamically unfavourable, the system tends to reduce it by reducing the area of protein–solvent interaction.

Polyethylene glycol (PEG) has unique properties with respect to its interaction with proteins. It is excluded non-specifically from native globular proteins caused by the excluded volume effect. But because PEG also has a strong non-polar character and therefore interacts with newly exposed non-polar residues, it is attracted to the protein in its unfolded state (Timasheff, 1993). Thus, PEG binds to the unfolded form of the protein and thereby stabilizes denatured protein and prevents its aggregation, but once folding has taken place, PEG drives the equilibrium towards the native globular state as it is then excluded from the structure (Lee and Lee, 1987; Banga, 1995). The dual properties of PEG make it difficult to predict its possible effect on a given protein, and the recipe for stabilization must be dictated by each protein individually (Timasheff, 1992).

Cyclodextrins, with a ring structure of six (α-), seven (β-), or eight (χ-cyclodextrin) glucopyranose units, have been proposed as stabilizing excipients for proteins because of their capability to form inclusion complexes due to the hydrophobic nature of the inner core of these carbohydrates (Brewster *et al.*, 1991). The cavity, with a size of 5–8 Å (Banga, 1995), has a particular affinity for aromatic amino acids and may be able to encapsulate them (Cleland *et al.*, 1993). β-cyclodextrin in a concentration of 1.5% has been shown to reduce agitation-induced aggregation of insulin to some degree (Banga and Mitra, 1998).

The major rationale for using surfactants as stabilizers is their ability to bind to hydrophobic interfaces that are potential denaturation sites, i.e. container surfaces and air–liquid interfaces. However, surfactants are also able to bind to exposed hydrophobic domains on proteins whereby aggregation may be reduced (Cleland *et al.*, 1993). The concentration of surfactant must be sufficiently high to cover the surface as a monolayer and the optimum concentration is most likely to be its critical micelle concentration (Banga, 1995). Higher concentrations of surfactants may lead to denaturation of the protein. Non-ionic surfactants such as Tween may act like molecular *in vitro* chaperones in facilitating protein refolding, thus preventing aggregation while not altering the native conformation (Bam *et al.*, 1996).

Genetic engineering

Manipulation of physical stability by protein engineering has been performed on many proteins, and examples of stabilization have been achieved for a number of proteins, in particular the enzymes subtilisin and lysozyme. Stability can be changed by mutating amino acids involved in stabilizing interactions in the folded protein, destabilizing interactions in the unfolded form, or both. Suitable substitutions are those which will increase the hydrophobicity of the protein core and will increase the strength of helical packing (Fersht and Serrano, 1993). Introduction of new S–S bonds in the molecule can either increase or decrease stability (Mozhaev *et al.*, 1988).

Stability of a protein substituted at the interior of the molecule tended to increase linearly with increasing hydrophobicity of the substituted residue, unless the volume of the substituted residue was over a certain limit (Yutani *et al.*, 1987). Stability of some protein molecules has been improved to the level of those from extreme thermophilic organisms. This was achieved by multiple mutations and by utilizing the knowledge gained from protein structures in the extreme thermophiles (Lee and Vasmatzis, 1997).

In order to raise the temperature of thermal denaturation by 10 K it is necessary to add 6 kJ/mol of stabilizing energy, achieved by formation of only one additional ionic or hydrogen bond or a few new hydrophobic contacts (Mozhaev and Martinek, 1990).

References

ARAKAWA, T. and TIMASHEFF, S.N., 1982, Stabilization of protein structure by sugars, *Biochemistry*, **21**, 6536–6544.

BALDWIN, R.L., 1996, On-pathway versus off-pathway folding intermediate, *Folding & Design*, **1**, R1–R8.

BAM, N.B., CLELAND, J.L. and RANDOLPH, T.W., 1996, Molten globule intermediate of recombinant human growth hormone: stabilization with surfactants. *Biotechnol. Prog.*, **12**, 801–809.

BANGA, A.K., 1995, *Therapeutic Peptides and Proteins. Formulation, Processing, and Delivery Systems*, Lancaster, PA: Technomic Publishing Co.

BANGA, A.K. and MITRA, R., 1998, Minimization of shaking-induced formation of insoluble aggregates of insulin by cyclodextrins, *J. Drug Targeting*, **1**, 341–345.

BRANGE, J., HAVELUND, S., HANSEN, P., LANGKJAER, L., SØRENSEN, E. and HILDEBRANDT, P., 1982, Formulation of physically stable neutral insulin solutions for continuous infusion by delivery systems. In: GUERIGUIAN, J.L., BRANSOME, E.D. and OUTSCHOORN, A.S. (eds) *Hormone Drugs*, pp. 96–105, Rockville, MA: United States Pharmacopeial Convention.

BRANGE, J., SKELBAEK-PEDERSEN, B., LANGKJAER, L., DAMGAARD, U., EGE, H., HAVELUND, S., HEDING, L.G., JØRGENSEN, K.H., LYKKEBERG, J., MARKUSSEN, J., PINGEL, M. and RASMUSSEN, E., 1987, *Galenics of Insulin: the Physico-chemical and Pharmaceutical Aspects of Insulin and Insulin Preparations*, Berlin: Springer-Verlag.

BRANGE, J., ANDERSEN, L., LAURSEN, E.D., MEYN, G. and RASMUSSEN, E., 1997a, Toward understanding insulin fibrillation, *J. Pharm. Sci.*, **86**, 517–525.

BRANGE, J., DODSON, G.G., EDWARDS, D.J., HOLDEN, P.H. and WHITTINGHAM, J.L., 1997b, A model of insulin fibrils derived from the x-ray crystal structure of a monomeric insulin (despentapeptide insulin), *Proteins*, **27**, 507–516.

BREWSTER, M.E., HORA, M.S., SIMPKINS, J.W. and BODOR, N., 1991, Use of 2-hydroxypropyl-β–cyclodextrin as a solubilizing and stabilizing excipient for protein drugs, *Pharm. Res.*, **8**, 792–795.

BURLEY, S.K. and PETSKO, G.A., 1989, Electrostatic interactions in aromatic oligopeptides contribute to protein stability, *Tibtech*, **7**, 354–359.

CARPENTER, J.F., PIKAL, M.J., CHANG, B.S. and RANDOLPH, T.W., 1997, Rational design of stable lyophilized protein formulations: some practical advice, *Pharm. Res.*, **14**, 969–975.

CHAWLA, A.S., HINBERG, I., BLAIS, P. and JOHNSON, D., 1985, Aggregation of insulin, containing surfactants, in contact with different materials, *Diabetes*, **34**, 420–424.

CHOTIA, C., 1984, Principles that determine the structure of proteins, *Ann. Rev. Biochem.*, **53**, 537–572.

CLARKE, A.R. and WALTHO, J.P., 1997, Protein folding pathways and intermediates. *Curr. Opin. Biotechnol.*, **8**, 400–410.

CLELAND, J.L., POWELL, M.F. and SHIRE, S.J., 1993, The development of stable protein formulations: a close look at protein aggregation, deamidation, and oxidation, *Crit. Rev. Therapeutic Drug Carrier Systems*, **10**, 307–377.

COSTANTINO, H.R., LANGER, R. and KLIBANOV, A.M., 1994a, Moisture-induced aggregation of lyophilized insulin, *Pharm. Res.*, **11**, 21–29.
1994b, Solid-phase aggregation of proteins under pharmaceutically relevant conditions, *J. Pharm. Sci.*, **83**, 1662–1669.

CREIGHTON, T.E., 1991, Stability of folded conformations, *Curr. Opin. Struct. Biol.*, **1**, 5–16.

DILL, K.A., ALONSO, D.O.V. and HUTCHINSON, K., 1989, Thermal stabilities of globular proteins, *Biochemistry*, **28**, 5439–5449.

DOBSON, C.M., 1992, Unfolded proteins, compact states and molten globules, *Curr. Opin. Struct. Biol.*, **2**, 6–12.

DONG, A., PRESTRELSKI, S.J., ALLISON, S.D. and CARPENTER, J.F., 1995, Infrared spectroscopic studies of lyophilization- and temperature-induced protein aggregation, *J. Pharm. Sci.*, **84**, 415–424.

FEINGOLD, V., JENKINS, A.B. and KRAEGEN E.W., 1984, Effect of contact material on vibration-induced insulin aggregation, *Diabetologia*, **27**, 373–378.

FERSHT, A.R. and SERRANO, L., 1993, Principle of protein stability derived from protein engineering experiments, *Curr. Opin. Struct. Biol.*, **3**, 75–83.

FINK, A.L., 1998, Protein aggregation: folding aggregates, inclusion bodies and amyloid, *Folding & Design*, **3**, R9–R23.

FRANKS, F., HATLEY, R.H.M. and FRIEDMAN, H.L., 1988, The thermodynamics of protein stability. Cold destabilization as a general phenomenon, *Biophys. Chem.*, **31**, 307–315.

GEKKO, K. and TIMASHEFF, S.N., 1981a, Mechanism of protein stabilization by glycerol: preferential hydration in glycerol–water mixtures, *Biochemistry*, **20**, 4667–4676.

1981b, Thermodynamic and kinetic examination of protein stabilization by glycerol, *Biochemistry*, **20**, 4677–4686.

GLATZ, C.E., 1992, Modeling of aggregation–precipitation phenomena. In: AHERN, T.J. and MANNING, M.C. (eds) *Stability of Protein Pharmaceuticals, Part A: Chemical and Physical Pathways of Protein Degradation*, pp. 135–166, New York: Plenum Press.

GOTO, Y., TAKAHASHI, N. and FINK, A.L., 1990, Mechanism of acid-induced folding of proteins, *Biochemistry*, **29**, 3480–3488.

HEJNAES, K., MATTHIESEN, F. and SKRIVER, L., 1998, Protein stability in downstream processing. In: SUBRAMANIAN, G. (ed.) *Bioseparation and Bioprocessing*, Vol. 2, pp. 31–65, Weinheim: Wiley–VCH Verlag.

HORBETT, T.A., 1992, Adsorption of proteins and peptides at interfaces. In: AHERN, T.J. and MANNING, M.C. (eds) *Stability of Protein Pharmaceuticals, Part A: Chemical and Physical Pathways of Protein Degradation*, pp. 195–214, New York: Plenum Press.

JAENICKE, R., 1991, Protein folding: local structures, domains, subunits, and assemblies, *Biochemistry*, **30**, 3147–3461.

KIEFHABER, T., RUDOLPH, R., KOHLER, H.-H. and BUCHNER, J., 1991, Protein aggregation *in vitro* and *in vivo*: a quantitative model of the kinetic competition between folding and aggregation, *Biotechnology*, **9**, 825–829.

LEE, B. and VASMATZIS, G., 1997, Stabilization of protein structures, *Curr. Opin. Biotechnol.*, **8**, 423–428.

LEE, L.L.Y. and LEE, J.C., 1987, Thermal stability of proteins in the presence of poly(ethylene glycols), *Biochemistry*, **26**, 7813–7819.

LENCKI, R.W., ARUL, J. and NEUFELD, R.J., 1992, Effect of subunit dissociation, denaturation, aggregation, coagulation, and decomposition on enzyme inactivation kinetics: I. First-order behavior, *Biotechnol. Bioeng.*, **40**, 1421–1426.

LEVITT, M., GERSTEIN, M., HUANG, E., SUBBIAH, S. and TSAI, J., 1997, Protein folding: the endgame, *Annu. Rev. Biochem.*, **66**, 549–579.

MANNING, M.C., PATEL, K. and BORCHARDT, R.T., 1989, Stability of protein pharmaceuticals, *Pharm. Res.*, **6**, 903–917.

MATTHEWS, B.W., 1993, Structural and genetic analysis of protein stability, *Annu. Rev. Biochem.*, **62**, 139–160.

MILLICAN, R.L. and BREMS, D.N., 1994, Equilibrium intermediates in the denaturation of human insulin and two monomeric insulin analogs, *Biochemistry*, **33**, 1116–1124.

MIRANKER, A.D. and DOBSON, C.M., 1996, Collapse and cooperativity in protein folding, *Curr. Opin. Struct. Biol.*, **6**, 31–42.

MOZHAEV, V.V. and MARTINEK, K., 1990, Structure–stability relationships in proteins: a guide to approaches to stabilizing enzymes, *Adv. Drug Del. Rev.*, **4**, 387–419.

MOZHAEV, V.V., BEREZIN, I.V. and MARTINEK, K., 1988, Structure–stability relationship in proteins: fundamental tasks and strategy for the development of stabilized enzyme catalysts for biotechnology, *CRC Crit. Rev. Biochem.*, **23**, 235–281.

MOZHAEV, V.V., HEREMANS, K., FRANK, J., MASSON, P. and BALNY, C., 1996, High pressure effects on protein structure and function, *Proteins*, **24**, 81–91.

NASH, D.P. and JONAS, J., 1997, Structure of the pressure-assisted cold denatured state of ubiquitin, *Biochem. Biophys. Res. Commun.* **238**, 289–291.

PACE, C.N., 1992, Contribution of the hydrophobic effect to globular protein stability. *J. Mol. Biol.*, **226**, 29–35.

PACE, C.N. and LAURENTS, D.V., 1989, A new method for determining the heat capacity change for protein folding, *Biochemistry*, **28**, 2520–2525.

PANDE, V.S., GROSBERG, A.U., TANAKA, T. and ROKHSAR, D.S., 1998, Pathways for protein folding: is a new view needed? *Curr. Opin. Struct. Biol.*, **8**, 68–79.

PFEIL, W., 1998, Is the molten globule a third thermodynamic state of protein? The example of α-lactalbumin, *Proteins*, **30**, 43–48.

PRIVALOV, P.L. and GILL, S.J., 1988, Stability of protein structure and hydrophobic interaction, *Adv. Protein Chem.*, **39**, 191–234.

PRIVALOV, P.L., GRIKO, Y.V. and VENYAMINOV, S.Y., 1986, Cold denaturation of myoglobin, *J. Mol. Biol.*, **190**, 487–498.

PTITSYN, O.B., 1995, Molten globule and protein folding, *Adv. Protein Chem.*, **47**, 83–229.

PTITSYN, O.B. and UVERSKY, V.N., 1994, The molten globule is a third thermodynamical state of protein molecules, *FEBS Lett.*, **341**, 15–18.

SCHINDLER, T., HERRLER, M., MARAHIEL, M.A. and SCHMID, F.X., 1995, Extremely rapid protein folding in the absence of intermediates, *Nature Struct. Biol.*, **2**, 663–673.

SHIRLEY, B.A., STANSSENS, P., HAHN, U. and PACE, C.N., 1992, Contribution of hydrogen bonding to the conformational stability of ribonuclease T1, *Biochemistry*, **31**, 725–732.

SHORTLE, D., 1993, Denatured states of proteins and their roles in folding and stability, *Curr. Opin. Struct. Biol.*, **3**, 66–74.

SLUZKY, V., KLIBANOV, A.M. and LANGER, R., 1992, Mechanism of insulin aggregation and stabilization in agitated aqueous solutions, *Biotechnol. Bioeng.*, **40**, 895–903.

SPEED, M.A., WANG, D.I.C. and KING, J., 1998, Specific aggregation of partially folded polypeptide chains: the molecular basis of inclusion body composition, *Nature Biotechnol.*, **14**, 1283–1287.

STEADMAN, B.L., THOMPSON, K.C., MIDDAUGH, C.R., MATSUNO, K., VRONA, S., LAWSON, E.Q. and LEWIS, R.V., 1992, The effect of surface adsorption on the thermal stability of proteins, *Biotechnol. Bioeng.*, **40**, 8–15.

TIMASHEFF, S.N., 1992, Stabilization of protein structure by solvent additives. In: AHERN, T.J. and MANNING, M.C. (eds) *Stability of Protein Pharmaceuticals. Part B: In Vivo Pathways of Degradation and Strategies for Protein Stabilization*, pp. 265–285, New York: Plenum Press.

1993, The control of protein stability and association by weak interactions with water: how do solvents affect these processes? *Annu. Rev. Biophys. Biomol. Struct.*, **22**, 67–97.

TSAI, C.-J., LIN, S.L., WOLFSON, H.J. and NUSSINOV, R., 1997, Studies of protein–protein interfaces: a statistical analysis of the hydrophobic effect, *Protein Sci.*, **6**, 53–64.

VOLKIN, D.B. and MIDDAUGH, C.R., 1992, The effect of temperature on protein structure. In: AHERN, T.J. and MANNING, M.C. (eds) *Stability of Protein Pharmaceuticals, Part A: Chemical and Physical Pathways of Protein Degradation*, pp. 215–247, New York: Plenum Press.

WAUGH, D.F., 1954, Protein–protein interactions, *Adv. Protein Chem.*, **9**, 325–437.

1957, A mechanism for the formation of fibrils from protein molecules. *J. Cell. Comp. Physiol.*, **49** (suppl. 1), 145–164.

WAUGH, D.F., WILHELMSON, D.F., COMMERFORD, S.L. and SACKLER, M.L., 1953, Studies on the nucleation and growth reactions of selected types of insulin fibrils, *J. Am. Chem. Soc.*, **75**, 2592–2600.

XIE, D., BHAKUNI, V. and FREIRE, E., 1991, Calorimetric determination of the energetics of the molten globule intermediate in protein folding: apo-α-lactalbumin, *Biochemistry*, **30**, 10673–10678.

YUTANI, K., OGASAHARA, K., TSUJITA, T. and SUGINO, Y., 1987, Dependence of conformational stability on hydrophobicity of the amino acid residue in a series of variant proteins substituted at a unique position of tryptophan synthase α subunit, *Proc. Natl Acad. Sci. USA*, **84**, 4441–4444.

ZAKS, A., 1992, Protein–water interactions. Role in protein structure and stability. In: AHERN, T.J. and MANNING, M.C. (eds) *Stability of Protein Pharmaceuticals, Part A: Chemical and Physical Pathways of Protein Degradation*, pp. 249–271, New York: Plenum Press.

7

Peptides and Proteins as Parenteral Suspensions: an Overview of Design, Development, and Manufacturing Considerations

MICHAEL R. DEFELIPPIS AND MICHAEL J. AKERS

Lilly Research Laboratories, Indianapolis, Indiana, USA

7.1   Introduction and scope

As research in biotechnology continues to identify and elucidate the function of new peptides and proteins, it will become necessary to devise viable formulations and delivery modes so that these potential therapeutics can be marketed. It is widely recognized that peptides and proteins are unique in terms of chemical and physical properties, therefore the types of formulation approaches that can be applied to these molecules are limited. This chapter is concerned with the formulation of peptides and proteins as parenteral suspension preparations. Suspensions are perhaps the most difficult formulations to develop because of the need to balance a number of variables to achieve an acceptable product. Thus, the complexity of the molecules being considered and the type of formulation desired doubly complicate the problem of developing a peptide or protein suspension.

There are at least two general approaches that the formulation scientist can take to preparing a parenteral suspension of a peptide or protein: (1) crystals or amorphous particles are produced *in situ* by combination of sterile solutions, or (2) sterile drug particles are produced separately, then suspended in an appropriate sterile vehicle. This review will focus primarily on concepts related to the preparation of peptide and protein suspensions for parenteral administration with emphasis on particle formation, excipient selection, optimization of properties, characterization techniques, and manufacturing. Our goal is to provide ideas and summarize principles for pharmaceutical scientists who might be involved with designing and developing parenteral peptide or protein suspensions. We will also highlight the difficulties associated with such formulations and identify special design aspects that must be considered.

7.2   Rationale for suspension development

The basic reasons for preparing small molecule pharmaceutical suspensions (Akers *et al.*, 1987) can also be applied to peptides and proteins. These include the following.

1   The solubility of the peptide or protein prohibits solution formulation.

2   The stability of the peptide or protein is improved in a suspension formulation.

3   There is a desire to control or retard the release profile of the peptide or protein.

In considering these reasons, one might question the practicality of preparing a suspension over some other dosage form given the inherent complexity associated with such formulations. For example, peptides or proteins can readily be solubilized by adjustment of solution conditions such as pH or by the addition of appropriate excipients. Assuming that the resulting preparation has appropriate chemical, physical, and microbiological stability over shelf-life and for the intended in-use period, such a formulation approach would clearly be desirable. Many small peptides are stable enough to be prepared as solutions for various routes of administration (Ganderton, 1991). However, as highlighted in other chapters in this book, the susceptibility of peptides and proteins to various forms of chemical and physical degradation means that the option of solution preparations is not always advantageous.

One might then consider a suspension formulation as a means to circumvent stability problems. Indeed, microcrystalline suspensions of insulin demonstrate considerably reduced deamidation at position B3 compared to amorphous or soluble preparations (Brange *et al.*, 1992c); however, this is only one of the many possible degradation reactions known for insulin (Brange (1994) and the references therein). In reality, the formulation scientist would probably opt for freeze-drying the peptide or protein over a suspension product to obtain needed stability despite the inconvenience of requiring a suitable diluent for reconstitution. This point is supported by the number of freeze-dried peptide or protein preparations on the market (e.g. human growth hormone, tissue-type plasminogen activator, erythropoietin, and glucagon).

Thus, the need to retard or control release of the drug provides the best reason for choosing to develop a suspension preparation of a peptide or protein. However, there are

Table 7.1 Examples of peptide or protein suspensions described in this chapter

Peptide or protein	Suspension characteristics	Status
Insulin Ultralente	Crystalline	Commercial product
Insulin NPH	Crystalline	Commercial product
Lys^{B28}, Pro^{B29} human insulin protamine suspension	Crystalline	Patent literature, clinical data reported
Asp^{B28} insulin protamine suspension	Crystalline	Patent literature, clinical data reported
Insulin Semilente	Amorphous	Commercial product
Insulin Lente	Amorphous/crystalline mixture	Commercial product
Regular/NPH insulin mixtures	Soluble/crystalline mixture	Commercial product, preparations with various ratios of soluble/crystalline material are available
25% soluble porcine insulin, 75% crystalline bovine insulin	Soluble/crystalline mixture	Rapitard® commercial product
Insulinotropin	Crystalline	Scientific literature
Interleukin-4	Crystalline	Patent literature
Zinc–interferon alpha-2B	Crystalline	Scientific literature
Alpha interferon	Protamine complex	Patent literature
Insulinotropin	Protamine complex	Scientific literature
Somatostatin	Protamine complex	Scientific literature
Glucagon	Protamine complex	Scientific literature
Gonadotropins	Protamine complex	Patent literature
Bovine somatotropin	Oleaginous suspension	Patent literature
Porcine somatotropin	Oleaginous suspension	Patent literature
Growth hormone releasing hormones	Oleaginous suspension	Patent literature
Adrenocorticotropic hormone	Oleaginous suspension	Scientific literature
Luteinizing hormone-releasing hormone	Suspension of degradable microspheres	Commercial product, Lupron® Depot
Human somatotropin	Suspension of degradable microspheres	Scientific literature

very few examples of marketed peptide or protein suspension formulations (Table 7.1), with the long-acting insulin preparations being the best characterized of this type. We will rely heavily on the body of knowledge related to insulin suspensions as we discuss general concepts in later sections of this review.

7.3   Types of suspensions and particle formation

7.3.1   In situ *particle formation*

Based on insulin preparations that have been produced for the market, there are at least four approaches for producing particles *in situ* that can be considered for development of peptide or protein suspension formulations.

1   The suspension is composed exclusively of crystalline material in vehicle.

2   The suspension is composed exclusively of amorphous material in vehicle.

3   The suspension contains a mixture of crystalline and amorphous material in vehicle.

4   The suspension contains active ingredient in both the suspension and solution phases.

We shall discuss each of these categories in more detail in this section, and highlight fundamental aspects and special considerations which may be generally applied to other peptides and proteins being considered as suspension products.

Crystalline suspensions

The topic of macromolecular crystallization as it pertains primarily to X-ray crystallography has been reviewed in great detail in the literature (McPherson, 1982; Gilliland and Bickham, 1990; Durbin and Feher, 1996 and references therein); however, it is worth briefly summarizing this information as some basic knowledge of protein crystallization is necessary in order to prepare suspensions composed of crystalline material. One advantage of preparing crystals for pharmaceutical suspensions as opposed to X-ray crystallographic studies is that large, perfect single crystals with dimensions of $0.3 \times 0.3 \times 0.3$ mm, 0.027 mm^3 (Drenth, 1994) are not essential. In fact, a suspension composed of microcrystalline material in the size range of about 1–40 µm will probably have desirable pharmaceutical properties such as resuspendability, syringeability, and injectability based on our experiences with marketed insulin preparations. Even with this slight edge, we shall see that the task of preparing peptide or protein crystals for pharmaceutical preparations is not trivial. We assume in our discussion that highly purified bulk material is available for formulation and do not cover isolation and purification methods. However, the importance of having a bulk drug substance with reproducible purity cannot be overemphasized, since impurities will probably influence the crystallization outcome.

There are no theoretically-based rules that can be followed for preparing peptide or protein crystals, and most of what is known is based on empirical observations and experience. The complexity of structure and diversity of chemical groups characterizing these molecules necessitates the elucidation of appropriate conditions that are fairly unique to a given peptide or protein. The objective is to devise conditions that result in supersaturation of the solution so that crystal growth is promoted. Conditions such as pH, ionic strength, temperature, peptide or protein concentration, and the presence of auxiliary ions or other molecules can all affect solubility. Crystallization is governed by a precise balance among many of these parameters such that in most instances there will be a limited set of conditions that will yield the desired crystals as opposed to the formation of amorphous material. Furthermore, slight variations in the conditions that produce crystals can strongly influence the final polymorph (McPherson, 1982).

Prior to initiating studies to determine crystallization conditions, it is essential to have a good understanding of the precipitation behaviour of the molecule of interest. This can

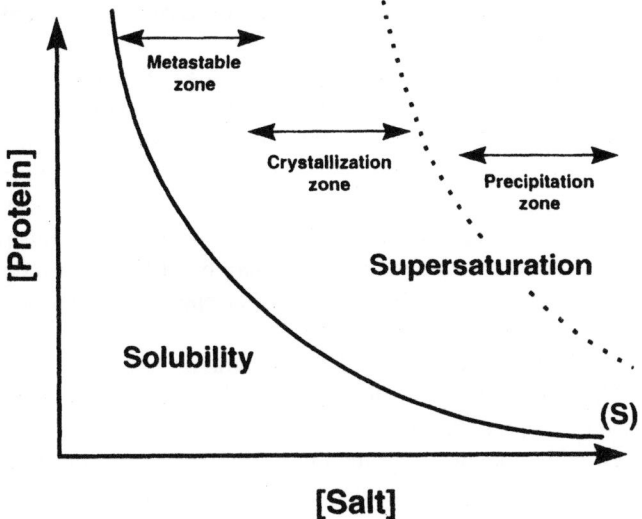

Figure 7.1 Solubility diagram for a hypothetical protein. Figure adapted from Ducruix and Ries-Kautt (1990). © Academic Press Inc., reproduced with permission.

be accomplished by generating solubility curves (Ducruix and Ries-Kautt, 1990), like the one depicted in Figure 7.1, for various conditions. Using this information, one can then screen various conditions at small scale that might result in crystal growth. Once crystallization conditions are identified, optimization studies can be undertaken to improve yield or achieve a desired polymorph. For example, a factorial design evaluating salt concentration, pH, and temperature in appropriate ranges could serve as a reasonable starting point. Carter (1990) provides additional information concerning the use of statistically designed experiments for screening crystal growth conditions.

In general, the best-quality crystals will form slowly from clear solutions (McPherson, 1982). Amorphous precipitates usually occur when aggregation occurs too rapidly, but crystals may grow from this material over time in some cases. As described later in more detail for insulin suspensions, slow crystal growth and prevention of amorphous precipitation are not critical requirements for preparing crystalline pharmaceutical suspensions. The two commercial microcrystalline insulin suspensions are prepared in batch by rapid formation of an amorphous precipitant that slowly converts to microcrystals. Other crystallization aids suitable for preparing pharmaceutical suspensions include seeding and cocrystallization.

Crystallization recipes used to prepare samples for X-ray crystallographic studies are readily available from established databases (e.g. NIST/NASA/CARB Biological Macromolecule Crystallization Database[1]) and these can serve as a starting point in the development of crystalline peptide or protein suspensions. However, it is worth mentioning some practical limitations of this information when applied to the preparation of pharmaceutical crystalline suspensions. Typical crystallizing solutions often contain small organic molecules, various salts, detergents, and/or divalent ions as precipitants or simply because they are necessary ingredients. Since these additives may be difficult to remove completely, they can be expected to become part of the final preparation. All ingredients must

---

[1] The NIST/NASA/CARB Biological Macromolecule Crystallization Database is available on the World Wide Web (http://ibm4.carb.nist.gov:4400/bmcd/bmcd.html).

therefore be pharmaceutically acceptable and proven to have no significant toxicological effects.

In addition to ingredient limitations, techniques such as dialysis, vapour diffusion, or evaporation commonly used to prepare crystals for X-ray crystallography will be difficult to apply in a large-scale manufacturing setting where crystallization may need to be accomplished at volumes ranging from tens to thousands of litres. Complicated sample manipulation procedures, easily performed at the bench, may be impractical or impossible to accomplish in a sterile manufacturing facility. Crystallization by temperature changes such as slow cooling from high temperature again may be difficult to accomplish at large scale because efficient sterile processing requires that operations be accomplished in as short a time as possible.[2]

Using the commercially available insulin NPH (neutral protamine Hagedorn) (Hagedorn *et al.*, 1936) and Ultralente (Hallas-Møller *et al.*, 1952) suspensions as model systems, two practical methods for preparing crystals can be described. In one approach, crystals are grown from a solution containing all the ingredients at the proper concentrations that make up the final formulation. NPH insulin is an example of this type. The other approach involves preparing a concentrated crystal suspension which is then diluted with a suitable, aqueous suspension vehicle to produce the final formulation. This type of process is used for preparing Ultralente insulin. While NPH and Ultralente insulin preparations have been thoroughly reviewed elsewhere (Schlichtkrull, 1961; Brange, 1987), it is worth summarizing the crystallization processes here as the concepts have been applied to other peptides or proteins. Regardless of the technique used, it is important to devise a process that both is reproducible and produces crystalline material having appropriate physical and chemical properties.

A schematic representation of the NPH insulin crystallization process is shown in Figure 7.2. NPH suspension is actually prepared by cocrystallizing insulin with the basic peptide protamine. Two solutions are used for the crystallization. One solution contains insulin and protamine dissolved in water at acidic pH, and the other contains dibasic sodium phosphate adjusted to slightly basic conditions. Both solutions also contain the additional ingredients necessary to complete the crystallization and final formulation that we will discuss in greater detail in the section on excipient selection. Precipitation is initiated by combination of the solutions in a 1:1 ratio causing a rapid change to neutral pH conditions. Amorphous material forms immediately which then transforms over time (approximately 24 hours) to form rod-shaped crystals about 3–6 μm long and 1–1.5 μm wide.

Krayenbühl and Rosenberg (1946) have described systematic studies evaluating the requirements for producing NPH crystals. These conditions can generally be applied to prepare NPH suspensions of human, bovine, or porcine insulin. Recently, protamine cocrystals were prepared with monomeric insulin analogues Lys^{B28}, Pro^{B29}-human insulin (DeFelippis, 1995) and Asp^{B28} insulin (Balschmidt, 1996) for the purpose of preparing NPH-type suspensions. In both cases, modifications to the standard crystallization procedure were necessary to produce appropriate crystals. Lys^{B28}, Pro^{B29}-human insulin•protamine crystallization is complete within 24 hours, but requires temperature control (15°C) as well as additional modifications to the precipitation procedure. For the Asp^{B28} analogue, crystals are reported to form only after extended storage (6 days) at refrigerated conditions

---

[2] Time limitations are necessary because of concerns over inadvertent microbiological contamination and, for solutions prior to filtration sterilization, the potential accumulation of endotoxins. No minimum time is suggested here; however, such processing should be accomplished as quickly as possible.

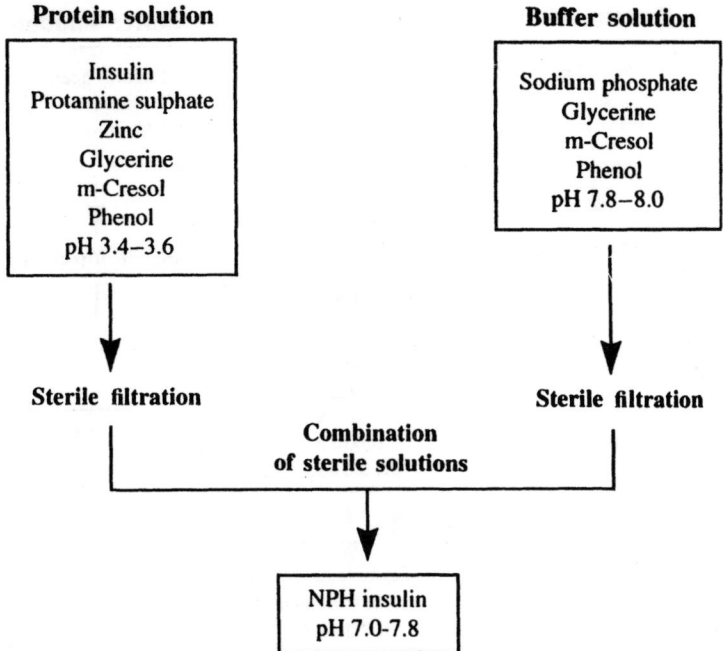

Figure 7.2   Schematic representation of the NPH insulin crystallization process.

These recent examples highlight that even when dealing with known processes and related proteins, optimization of appropriate crystallization conditions is often necessary. The modifications to the NPH procedure necessary to prepare Lys^{B28}, Pro^{B29}-human insulin•protamine crystals have been argued to result from the altered self-association properties of the analogue (DeFelippis *et al.*, 1998). Pharmacological studies have been reported for both analogue suspensions (DeFelippis *et al.*, 1996; Radziuk *et al.*, 1996; Weyer *et al.*, 1997), and the time-actions were shown to be similar to those of protamine suspensions prepared with other insulin species.

Protamine cocrystallization may be an option for preparing other peptide and protein suspensions, since complex formation is driven by electrostatic interactions and most proteins will not be as positively charged as protamine. Binding can be expected to result in precipitation, but the extent to which crystal growth is observed, if at all, depends on the system under investigation. Another reason for considering protamine cocrystallization as an option is the fact that protamine sulphate has been proved to be an acceptable excipient in marketed preparations, although immunogenicity is a consideration (see discussion in section 7.4). Examples of other peptides and proteins for which protamine suspensions have been described include: alpha interferon (Yim, 1988), insulinotropin (glucagon-like-peptide-1-(7-37)) (Kim and Rose, 1995), somatostatin (Brazeau *et al.*, 1974; Martin *et al.*, 1974; Tannenbaum and Colle, 1980), glucagon (Naets and Guns, 1980; Buch and Buch, 1983), and gonadotropins (Donini, 1974). The motivation for the preparation of these protamine•protein complexes was to achieve delayed activity, and testing was limited to investigational studies.

In contrast to the NPH crystallization procedure, Ultralente suspensions are prepared without using protamine sulphate (Figure 7.3). Precipitation is initiated by adjusting conditions to the isoelectric point of insulin in the presence of zinc ions, sodium chloride, and sodium acetate, accomplished by mixing an acidic solution of insulin with buffer

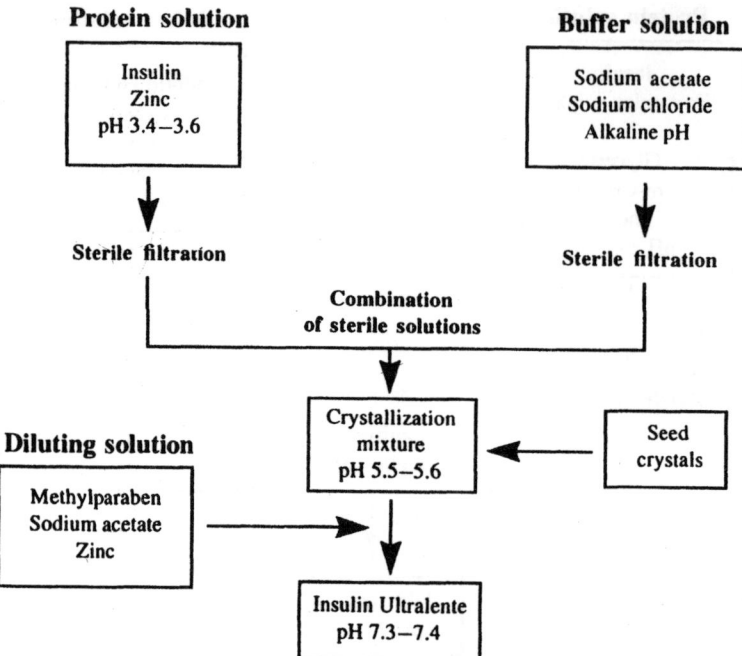

Figure 7.3 Schematic representation of preparation of Ultralente suspension.

such that the appropriate pH is achieved. Most of the ingredients required for the final preparation are present during crystallization except for preservative. Concentrations for insulin and other ingredients are 10-fold higher during crystal growth, and a diluent containing preservative is used to dilute the concentrated suspension to produce the final preparation. Because a monodisperse particle size distribution is desired for the final preparation, predetermined amounts of seed crystals are added during the crystallization phase. Seeding also effectively eliminates self-nucleation. For more detailed information regarding the preparation of seed crystals see Schlichtkrull (1957). Commercial Ultralente preparations contain rhombohedral crystals in the approximate size range of 20–30 μm.

The method described for making Ultralente crystals is specific to insulin and results from its unique self-association and ligand-binding properties, therefore we cannot expect the exact procedures to be generally applicable to other peptides or proteins. However, this example demonstrates viable possibilities for other molecules. In cases where crystallization can only be accomplished under defined conditions not including all required excipients for pharmaceutically acceptable preparations, a separate crystal suspension can be manufactured first followed by dilution with vehicle containing any additional required ingredients. The addition of seed crystals is also possible to initiate crystallization or control size distribution. There are important considerations in either case. Studies evaluating the compatibility of crystals and ingredients in the suspending vehicle are necessary to ensure that degradation of the particles or changes in morphology do not occur. Incorporation of seed crystals into a pharmaceutical process presents an additional burden for establishing suitable control specifications for the seed preparation and validating its use and intended purpose.

Crystallization experiments reported (Kim and Haren, 1995) for insulinotropin (glucagon-like-peptide-1-(7-37)) are illustrative of the types of studies that are necessary to devise a process for preparing a pharmaceutically acceptable crystalline suspension, and further substantiate one of the considerations mentioned above. Analogous to Ultralente insulin, a microcrystalline suspension of insulinotropin was desired to achieve a sustained release preparation. Kim and Haren (1995) demonstrated that insulinotropin can be crystallized from a salt solution adjusted to a pH near its isoelectric point, but these crystals transform to amorphous material when mixed with m-cresol. However, the crystals remain intact if they are first treated with a zinc soaking procedure prior to the addition of phenolic preservative. Note the similarity to the Ultralente insulin process of controlled addition of the necessary ingredients without compromising crystalline morphology.

Other examples of suspensions containing crystalline or precipitated peptides or proteins have been described. A crystalline form of human interleukin-4 has been described and proposed to have potential application as a slow-release pharmaceutical preparation (Hammond, 1991). The introduction of seed crystals, grown by techniques typically used to prepare crystals for X-ray crystallography, was suggested as a method for accelerating large-scale crystallizations. It was further implied that the pharmaceutical preparation would require complexation of the protein with metal ions, and no method for introducing preservatives was identified. A potential controlled-release, crystalline suspension of zinc–interferon alpha-2B has also been described (Reichert *et al.*, 1996). In this example, crystals grown in microgravity were formulated as a suspension and evaluated in primates. The pharmacological data on the crystalline suspension showed a 4–6-fold increase in the serum half-life of the protein compared to a soluble formulation. Donini (1974) described long-acting suspensions containing gonadotropins precipitated by the addition of divalent metal ions and appropriate pH adjustments.

Amorphous suspensions

Semilente insulin is an example of a suspension composed of flocculated amorphous particles. The suspension is prepared by performing the Ultralente crystallization under less than optimum pH conditions. Physical stability has been correlated to the degree of flocculation (Brange, 1987) suggesting that optimization of a final preparation will be necessary even if amorphous precipitation is easily accomplished. While it is difficult to predict whether amorphous suspensions of a given peptide or protein will have the desired pharmacological properties and necessary stability, Semilente insulin preparations highlight the fact that development of a completely crystalline suspension may not be an absolute requirement.

Crystalline and amorphous mixtures

The Lente insulin preparation is an example of a suspension containing a mixture of particles having different morphology derived by combination of Ultralente and Semilente to produce a 3:7 mixture of amorphous to crystalline material. It should be recognized that the suspension is intentionally prepared in this manner and does not result from incomplete crystallization. There was a specific therapeutic rationale related to injection frequency for which this insulin preparation was developed. Whether a similar justification can be identified for other potential peptides or protein suspensions clearly depends on the system under investigation. If a suspension composed of a specific ratio of amorphous to crystalline material is desired, conditions for both precipitation and crystallization need to be elucidated along with establishing necessary compatibility. The Lente insulin preparation also demonstrates that crystalline suspensions containing some defined level of amorphous material are possible alternatives. For example, this might be a useful option in cases where complete crystallization is not possible. Key requirements for a suspension of this type include: a reproducible degree of crystallinity batch-to-batch and demonstration that the crystalline–amorphous ratio remains constant over shelf-life and during the intended use period.

Crystalline and solution mixtures

Biphasic preparations containing mixtures of soluble and suspension phases of insulin provide examples of this category. Once again, the impetus for developing such formulations was based on optimization of injection therapy. The soluble component of the preparation provides the initial insulin effect while the suspension phase provides protracted activity, all in a single injection. Insulin biphasic mixtures containing either Ultralente or NPH as the suspension component have been described.

The mixture containing Ultralente crystals (described in Brange, 1987) is a special case whereby the solubility characteristics of different insulin species was exploited. Rapitard® is composed of 75% beef insulin Ultralente and 25% soluble porcine insulin. Because bovine insulin is less soluble than porcine insulin, the ratio of suspension to soluble insulin is relatively stable, with only a small portion of bovine insulin dissolved in solution. Biphasic mixtures composed of NPH and soluble regular insulin are more common, and numerous ratios 10:90 through 50:50 (soluble–suspension) are commercially available. These preparations must contain the same insulin species in the suspension and the solution phases (homogeneous), as exchange can occur in heterogeneous mixtures over long-term storage (Edwards *et al.*, 1996).

Depending on the treatment regimen intended for a specific peptide or protein, such a formulation approach offered by biphasic insulin mixtures might be a useful option. As an alternative to deliberately designing a biphasic mixture, one could also envision a suspension that by nature of the crystallization procedure is not completely crystalline and contains a known and constant concentration of soluble protein or peptide. This situation may occur in cases where crystal growth is not 100% complete. For example, Yim (1988) has indicated that the conditions for producing an insoluble complex of alpha interferon and protamine can be adjusted to allow for a specific portion of soluble interferon for immediate effect. For it to be commercially viable, however, the solid–solution ratio in such a suspension must be reproducible and stable, and the preparation should have some practical pharmacological purpose, especially if a significant amount of soluble material exists.

7.3.2 Combination of particles and vehicle

Preparation of particles

In this method of suspension manufacture, the drug is prepared as a dry solid independently from the vehicle. As with small-molecule processing procedures, there are at least four methods of producing the sterile powder. These include crystallization, lyophilization, spray-drying, and supercritical fluid particle formation (SCFPF).

- *Crystallization:* Many of the concepts relating to crystallization described above can be applied to produce crystals for this formulation approach. A key difference involves harvesting the crystals from mother liquor and drying them.

- *Lyophilization:* The proper application of this technique to peptides and proteins has been reviewed elsewhere (Pikal, 1990a, 1990b). Bulking or stabilizing agents can be incorporated into the process as needed.

- *Spray-drying:* The active ingredient is dissolved in solvent and sprayed into a drying chamber. Rapid solvent evaporation is accomplished using a hot stream of sterile gas, resulting in the formation of uniform spherical particles (Gölker, 1993; Wendel and Çelik, 1997).

- *Supercritical fluid particle formation (SCFPF):* Supercritical carbon dioxide is used as either solvent or antisolvent to achieve supersaturation and subsequent generation of particles. Based on equipment configuration and methodology employed, three SCFPF techniques have been described: rapid expansion of supercritical solution (RESS), gas antisolvent (GAS), and precipitation with compressed antisolvents (PCA). Detailed descriptions and comprehensive reviews comparing the various SCFPF techniques can be found in Gallagher *et al.* (1989), Tom and Debenedetti (1991), Yeo *et al.* (1993), Phillips and Stella (1993) and Subramaniam *et al.* (1997).

Given the propensity of peptides and proteins to denature under high-stress conditions, crystallization and lyophilization are more generally applicable to this class of compounds. These two methods also have proven ability to maintain the sterility of the dried material. The spray-drying process might result in denaturation either at the liquid–air interface or from the high temperature required to evaporate solvent (Yeo *et al.*, 1993); however, the technique may be appropriate for small peptides that lack higher-order structure or selected proteins. For example, Mumenthaler *et al.* (1994) have observed that spray-drying caused denaturation of human growth hormone, but similar processing did not adversely affect tissue-type plasminogen activator. Increasing attention is focusing on the potential of SCFPF for pharmaceutical processing, in particular its suitability for peptide and protein particle formation. A microparticulate insulin powder (<5 μm) prepared by GAS expansion of dimethylsulphoxide was reported (Yeo *et al.*, 1993), presaging the utility of SCFPF techniques for generating protein particles.

Regardless of the method employed to produce particles, some additional biophysical characterization studies should be performed to confirm that particle processing procedures do not adversely affect the properties of the molecule. Particle size reduction of the solid material by milling and sieving may be required after drying, depending on the technique employed and the size of the particles initially produced. The impact of these additional operations on the integrity of the peptide or protein needs to be assessed.

Preparation of vehicle and combination

Aqueous or nonaqueous vehicle containing any necessary excipients is prepared separately from particle formation. Examples of nonaqueous vehicles include any highly purified natural or synthetic oil such as sesame, peanut, or other vegetable oils. Depending on the solubility of the constituents and overall viscosity of the vehicle, sterilization can be accomplished by either filtration or autoclaving. The sterile combination approach offers more flexibility in the choice of vehicle (aqueous or nonaqueous), since particle growth is accomplished independently.

Once processing of each section is completed, the dry particles and vehicle are aseptically combined. Some form of agitation is required to achieve a homogeneous dispersion of particles. In the case of peptides or proteins, appropriate controls should be in place to ensure that the dispersion process does not result in denaturation or other physical changes.

Peptide and protein suspensions prepared by combination of particles and vehicle

An example of a protein suspension prepared by the combination of particles and vehicle is bovine somatotropin (bST) (Bramley *et al.*, 1989; Ferguson *et al.*, 1990; Mitchell, 1991; Ferguson, 1997). This highly viscous suspension product is marketed under the trademark Prosilac® and is used for increasing milk production in dairy cows. The thick suspension results in protracted release of bST, thereby minimizing injection frequency. The vehicle in this case is sesame oil. The oil is thickened by incorporating a wax such as beeswax, although a variety of other agents can serve this purpose. Dry powders of the growth hormone are prepared by lyophilization and then dispersed into the vehicle. Mitchell (1991) describes a similar suspension, but the growth hormone is complexed with metal ions.

Preparations closely matching the bST suspension design have been described for other growth hormones (Martin, 1986) and growth hormone releasing factors (Brooks and Needham, 1994). A long-acting parenteral suspension of adrenocorticotropic hormone (ACTH) in sesame oil gelled with aluminium monostearate has also been described (Chien, 1981). It should be recognized that the highly viscous nature of suspensions such as the one described for bST will require the use of large gauge (14–16 g) needles, making such preparations generally unattractive for parenteral administration to humans.

7.4 Excipient selection

It would be impossible to identify all possible examples of excipients or auxilliary substances used in parenteral suspensions, as selection is highly dependent on the properties of the active ingredient and the desired final preparation. Some excipients are integral to the production of particles and their presence is essential for maintaining specific properties. This is especially true for the insulin preparations described earlier. Other excipients, such as preservatives, are included to meet various regulatory requirements for parenteral pharmaceutical products. Thus, the choice and concentration of excipients is a major consideration in the design of suspension preparations. Not only should these ingredients perform intended functions, but.optimization studies and thorough evaluations for compatibility with the other constituents of the preparation and the container/closure system are essential. These requirements impose limitations on excipient choices in addition to the prerequisite that they be acceptable for use in pharmaceutical products for injection.

The various excipients that are used in parenteral suspensions are categorized as buffering agents, isotonicity modifiers, preservatives, stabilizers, complexing agents, or other auxiliary agents. In some cases these ingredients may have dual functions, as highlighted below.

- *Buffering agents:* Physiologically tolerated buffers are added to maintain pH in a desired range; examples include sodium phosphate, sodium bicarbonate, sodium citrate, and sodium acetate. The addition of a buffer is not absolutely necessary if it can be demonstrated that the formulation maintains the desired target pH range. In certain cases, these agents are present as a result of the process for achieving particle formation, yet have no significant buffering capacity at the pH of the final preparation. Ultralente insulin is an example of such a situation. The sodium acetate is present during crystal growth at pH 5.5, but the final suspension is adjusted to neutral pH conditions where the buffering capacity is minimal. Potential interactions between buffers and metal ions must be considered, as reaction products can lead to compromised stability.

- *Isotonicity modifiers:* These agents are added to minimize pain that can result from cell damage due to osmotic pressure differences at the injection depot. Glycerin and sodium chloride are examples used in insulin suspensions. Effective concentrations can be determined by osmometry using an assumed osmolality of 285 mOsmol/kg for serum as a guide (Siegel, 1990). Typical concentrations of 7 and 16 mg/ml are used for sodium chloride and glycerine, respectively. Which agent is chosen may be dictated by the need to have a particular ingredient present during particle formation, as is the case for sodium chloride in the Lente insulin preparations. The two examples of isotonicity modifiers differ in ionic strength (sodium chloride, high ionic strength; glycerin, low ionic strength), and these properties might influence the choice of one over the other depending on compatibility and stability considerations.

- *Preservatives (antimicrobial agents):* Multidose parenteral preparations require the addition of preservatives at sufficient concentration to minimize risk of patients becoming infected upon injection. Regulatory requirements for antimicrobial effectiveness have been established that take account of whether the formulation has inherent bacterial growth inhibition properties. The Ph. Eur. criteria are more stringent than USP, and this fact should be considered when deciding on the intended markets for the pharmaceutical preparation (Akers and Taylor, 1990). Another important point regarding antimicrobial effectiveness testing concerns the use of oleaginous suspension vehicles that may complicate test methodology because of the immiscibility with aqueous microbiological media (Workman and Clayton, 1996). Typical preservatives for parenteral suspensions include: m-cresol, phenol, methylparaben, ethylparaben, propylparaben, butylparaben, chlorobutanol, benzyl alcohol, phenylmercuric nitrate, thimerosol, sorbic acid, potassium sorbate, benzoic acid, chlorocresol, and benzalkonium chloride (Nash, 1988; Boyett and Davis, 1989; Denyer and Wallhaeusser, 1990). Toxicology issues will probably impose limitations on the use of other chemicals, especially for chronic use applications. The type of preservative and concentration chosen may also be influenced by factors related to crystal growth, maintaining acceptable suspension stability, or compatibility with already grown crystals in addition to achieving necessary antimicrobial effectiveness. For example, insulin Ultralente cannot be formulated with phenol as the crystal morphology is destroyed over time (Brange, 1987), but methylparaben does not exhibit this effect. The insulinotropin crystallization is also conducted in the absence of preservative, and formed crystals will transform to amorphous material if m-cresol is added before a zinc soaking treatment (Kim and Haren, 1995). In contrast, insulin NPH crystals require phenolic preservative for crystal growth (Krayenbühl and Rosenberg 1946), and a mixture of m-cresol and phenol in a defined ratio is present in commercial preparations.

- *Stabilizers:* Stabilizers include a variety of agents that impart stability to particles themselves or the entire suspension. General categories include: metal ions (zinc, calcium, etc.), salts used to produce crystals, and organic molecules. Divalent metal ions play a pivotal role in insulin self-assembly and bringing about crystallization. It has been demonstrated that the addition of excess zinc ions to human insulin Ultralente suspensions extends the duration of activity (Hoffman, 1994). Zinc ions were shown to impact the physical stability of NPH insulin (Massey and Sheliga, 1988; Massey *et al.*, 1988), and were found to be critical for stabilizing insulinotropin crystals (Kim and Haren, 1995). The various salts necessary to achieve crystal growth may also serve as stabilizers in the final suspension. Insulin Ultralente suspensions cleverly exploit the requirement for sodium chloride for crystal growth by also using the ingredient as a tonicity modifier. Organic ligands such as phenolic preservatives, in addition to serving a role as antimicrobial agents, may function as stabilizers.

- *Complexing agents:* Protamine sulphate is an example of a complexing agent used to prepare suspensions. As excess protamine is undesirable from an immunogenicity standpoint (Horrow, 1985; Nell and Thomas, 1988; Ellerhorst *et al.*, 1990) and may impact the stability of biphasic (solution–suspension) mixtures by complexing some of the soluble component, the exact ratio required to complex completely all the available peptide or protein needs to be determined. Under appropriate conditions, no detectable free protamine or peptide/protein remains in the supernatant. This condition is defined as the isophane ratio (Hagedorn *et al.*, 1936). Originally described using a turbidimetric analysis, HPLC was more recently used to evaluate the isophane ratio for an insulintropin•protamine complex (Kim and Rose, 1995).

- *Other auxiliary agents:* Depending on the type of suspension and its properties and intended use, various auxiliary agents might be included in the preparation. The types of ingredients that might also be added can be classified as: wetting agents, flocculating/suspending agents (surfactants, hydrophobic colloids, or electrolytes), viscosity modifiers, antibiotics and antioxidants. Specific examples of each class of agent can be found (Nash, 1988; Boyett and Davis, 1989). Acids and bases such as hydrochloric acid and sodium hydroxide are auxiliary agents necessary for pH adjustment during particle formation and of the final suspension. The choice of auxiliary agent and its concentration may be highly dependent on whether the suspension is for acute or chronic applications. Furthermore, the compatibility of auxiliary agents with drug particles must be demonstrated.

It should be apparent to the pharmaceutical scientist involved in development of suspensions that the formulation and process are integrally related, especially in the case of *in situ* particle growth. Therefore, excipient selection cannot be considered independently of development aspects relating to the manufacturing process.

7.5 General requirements for suspension products

In addition to demonstrating appropriate chemical, physical, and microbiological stability over shelf-life and during its intended in-use period, a well-formulated suspension should have the following characteristics (Akers *et al.*, 1987).

1 Resuspension of particles is accomplished with reasonable agitation.

2 Rapid settling of dispersed particles does not occur.

3 Particles can be homogeneously dispersed such that consistent doses are obtained repeatedly.

4 The particles do not cake or pack at the bottom of the container over the shelf-life period making it difficult to redisperse the product.

5 The suspension manufacturing process reproducibly produces particles with properties that are maintained batch-to-batch and during the preparation's intended shelf-life and in-use periods.

6 The suspension product must be easily drawn up into a syringe through a 20–25 gauge needle and readily expelled.

Optimization of these characteristics is an essential part of the development process for suspension products.

Characteristics 1–5 are concerned with special requirements for suspensions relating to elegance, physical attributes, and stability. We shall address these items in greater detail in section 7.6.2. Requirement 5 is especially important for suspensions intended to have specific delayed or controlled release profiles, as the properties of the particles will govern pharmacology.

Characteristic 6 relates to suspension properties essential for administration of proper doses. Needles for parenteral injections have become increasingly small in diameter in an attempt to minimize pain and increase patient compliance. While 20–25 gauge needles are acceptable, certain insulin injector devices are currently utilizing 28 or 30 gauge needles. Suspensions composed of large particles could clog these narrow gauge needles, affecting the ability both to withdraw a proper dose from a container into a syringe (syringeability) and to inject the dose into the patient (injectability). In addition to these particle size considerations, suspensions that are composed of particles that tend to agglomerate or employ highly viscous vehicles can also affect syringing and injection operations or patient acceptability. Exceptionally viscous suspensions are not necessarily out of the question, as the bST product demonstrates; however, the preparation requires 14–16 gauge needles for subcutaneous administration. Clearly, such a suspension is of limited practicality for human health products.

During development of a suspension, both syringeability and injectability need to be examined with appropriate *in vitro* studies. Such studies may involve withdrawals and ejections using the intended needles and syringes and/or delivery devices. Dose potency of the active component is then confirmed by a suitable analytical assay. Chien *et al.* (1981) have devised an apparatus that allows quantitative measurements of syringeability. The apparatus was shown to be appropriate for parenteral solutions and suspensions, especially those that are nonaqueous.

7.6 Testing and optimization of chemical, physical, and microbiological properties

As with any pharmaceutical product, there is a need both to optimize and to test the chemical, physical, and microbiological properties of the preparation and demonstrate appropriate stability over shelf-life and during the intended in-use period. Suspensions are somewhat more complicated in this regard, because optimization of physical properties is extremely challenging and additional testing is usually required. However, many of the concepts relating to other dosage forms also apply to suspensions, so that in some cases only a general overview is provided. The reader is referred to other chapters in this book for additional details.

7.6.1 *Chemical properties*

Peptides and proteins are subject to a variety of chemical modifications resulting in the formation of specific degradation products, regardless of the type of pharmaceutical preparation. The kinds of degradation product that can form depend on the amino acid sequence of the peptide or protein, formulation conditions, manufacturing operations, and product storage and handling. Since it will be impossible to prevent degradation entirely, the goal of chemical optimization is to identify the degradation products, determine the conditions under which they form most readily, and take steps to minimize their rates of formation.

Reversed-phase, ion-exchange, and size-exclusion HPLC assays are particularly useful for peptide and protein products. It should be recognized that such methods must also include an appropriate procedure for solubilizing solid particles. Whenever possible, the most prominent degradation products observed by HPLC should be isolated and subjected to other chemical analyses such as mass spectrometry or sequencing to determine identity, and may further require activity or toxicology testing. Stability programmes should include assays developed to monitor the rates of change of degradation products that can affect product quality and possibly impact the end-user. For example, assays for drug potency, other related substances, and higher molecular weight species are generally good chemical stability indicators. A series of research papers on the chemical stability of insulin (Brange, 1992; Brange and Langkjaer, 1992; Brange et al., 1992a, 1992b, 1992c) provide an excellent overview of the experimental approaches to be followed when evaluating protein-based pharmaceutical preparations, including suspensions.

Excipients present in the formulation can chemically degrade, interact with various surfaces during manufacturing, interact with the container or closure, or interact with the peptide or protein, thereby negatively affecting critical properties of the preparation. Degradation products derived from excipients can also react with the peptide or protein. Therefore, chemical optimization work should also include an evaluation of the excipient properties. The covalent insulin protamine polymer formed in insulin NPH suspensions (Brange, 1992; Brange and Langkjaer, 1992; Brange et al., 1992a, 1992b) is an example of a protein–excipient chemical interaction producing an additional degradation product.

Depending on the type of suspension, additional assays may be needed to evaluate the preparation. For example, biphasic mixtures composed of solution and suspension phases may require a method to quantitate the proportion of soluble to solid material because the time-action of the two components is different. Simply centrifuging the formulation to separate the phases followed by assay of the supernatant may not be appropriate, as the results may be influenced by adsorption of soluble protein onto solid particles. This phenomenon has been observed for biphasic NPH–soluble insulin mixtures (Brange, 1987; Dodd et al., 1995). To overcome this limitation, Rolim and Bristow (1995) have devised a method whereby treatment of the insulin suspension with a Tris buffer effectively desorbs soluble protein, affording estimates for the amount of soluble and solid material.

Other general chemical testing may be undertaken in addition to the routine stability evaluations. Examples include: photostability, inverted container stability, packaging compatibility, and aberrant temperature conditions. Photostability testing is required by ICH guidelines, as phenolic preservatives and possibly peptides or proteins (Holt et al., 1977) can degrade when exposed to light for extended periods of time. Inverted container stability is applicable for vial presentations having head space between the closure and suspension. By including studies on inverted samples, information can be obtained on the potential incompatibility of formulation with the closure. Packaging compatibility with the formulation should be evaluated by performing stability studies in the intended container/closure system for the product. The effect of aberrant temperatures (freezing and heating) on the product should be determined, as unpredictable and rather extreme conditions may be encountered during shipping, during warehouse storage, or in the hands of the consumer.

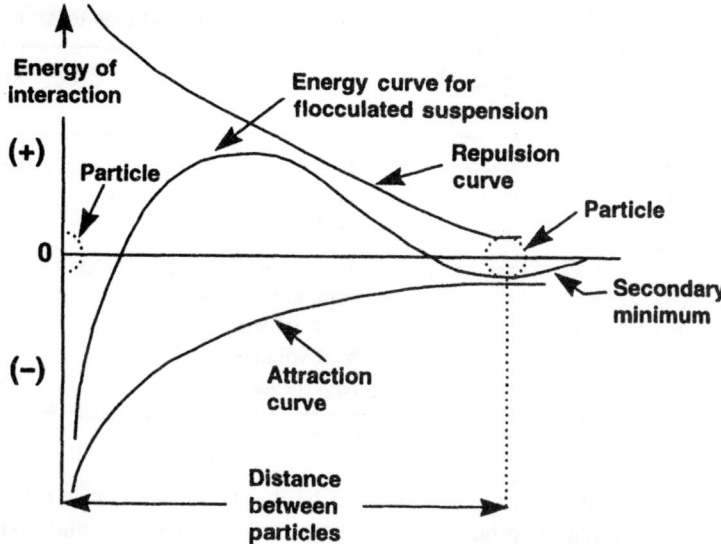

Figure 7.4 Potential energy curves for particle interactions in coarse suspensions. Figure adapted from Martin (1961). Reproduced with permission of the American Pharmaceutical Association.

7.6.2 Physical properties

General concepts and basic theory: optimization of physical properties of dispersed systems

Suspensions are thermodynamically unstable systems, and the goal is to design a preparation that is kinetically stable for a sufficient period of time (i.e. shelf-life) so that product performance is not compromised by gross changes in physical properties. Two common physical instability problems are caking and crystal growth. In order to understand the behaviour of suspensions and methods for optimization, a basic understanding of theoretical concepts explaining these physical transformations is required. However, a detailed discussion on theory is beyond the scope of this review, and the reader is referred to appropriate texts on this subject for additional details.

The Derjaguin, Verwey, Landau, and Overbeek (DVLO) theory was originally devised to explain the stability of colloidal systems, but the principles have also been invoked to explain particle interactions in coarse dispersions such as suspensions (Martin, 1961; Hiestand, 1964; Matthews and Rhodes, 1968; 1970; Kayes, 1977; Schneider *et al.*, 1978). According to DVLO theory, the forces on particles in a dispersion are due to electrostatic repulsion and van der Waals' attraction. Note that other forces are usually included to explain interactions in dispersed systems adequately. Potential energy curves for particle interactions are shown in Figure 7.4. The forces at particle surfaces will affect the degree of flocculation and agglomeration observed for suspensions. Thus, DVLO theory provides a framework for understanding the interactions of particles which control physical properties of suspensions.

With reference to the composite curve in Figure 7.4, the collision of particles will be opposed if the repulsion energy is high (e.g. low electrolyte concentration in aqueous suspensions). Such a system is referred to as 'deflocculated'. When the particles settle, the energy barrier is overcome and strong attractive forces in the potential well cause a

Table 7.2 Relative properties of deflocculated and flocculated particles in suspension

State	Particle characteristics	Sedimentation rate	Appearance	Cake	Resuspension
Deflocculated	Exist as separate entities	Slow	Initially suspended, but settles to a small volume	Yes	Difficult to redisperse
Flocculated	Exist as weak aggregates (floccules)	Fast	Settling results in the presence of an obvious, clear vehicle region; final volume may be large or small	No	Easy to redisperse

densely packed sediment to form. Eventually, a hard cake results that is difficult to disperse using normal agitation procedures for resuspension. Such a condition is highly undesirable, as a non-uniform dispersion of particles can impact dosing reliability.

For coarse dispersions which are flocculated, the potential energy barrier is still too large to be surmounted by approaching particles (Martin, 1961; Schneider *et al.*, 1978). However, weaker attractive forces occur at significant interparticle distances at the region referred to as the secondary minimum in Figure 7.4. Particle interactions in this case result in the formation of loose aggregates (floccules). Flocculation can be induced in a suspension by the addition of a flocculating agent such as an electrolyte. Suspensions that are flocculated are considered pharmaceutically stable because sedimented material is readily redispersed upon normal agitation procedures.

The properties of flocculated and deflocculated suspensions are compared in Table 7.2 (Nash, 1988; Zografi *et al.*, 1990). For flocculated suspensions, the sedimentation properties may result in a preparation that appears to contain a greater part of clear vehicle upon settling. This condition is not a serious problem provided caking does not occur making the particles difficult to disperse with minor agitation. Ultralente insulin is an example of a suspension that displays a very small sedimentation volume, but the particles are easily resuspended to homogeneity with gentle shaking of the vial. Because Ultralente insulin crystal growth conditions are defined and the suspension has appropriate stability, there is little value in making adjustments to improve its physical appearance upon settling. Thus, suspension formulation design may require compromises between aesthetic aspects and other desirable physical attributes of the preparation.

Sedimentation volume and zeta potential measurements are useful for optimizing the physical properties of suspensions by providing information on the degree of flocculation. Sedimentation volume is determined by measuring the equilibrium volume of settled particles relative to the total suspension volume after resuspension. The quantity is expressed as a ratio:

$$F = V_u/V_o \tag{7.1}$$

where V_u is the equilibrium volume of sediment and V_o is the total suspension volume.

Zeta potentials are determined to estimate surface charges as discussed in section 7.7. The relationship between sedimentation volume and zeta potential is illustrated in Figure 7.5 (Martin, 1961). The addition of a flocculating agent causes a progressive reduction in zeta potential and changes in sedimentation volume. There exists a region where the

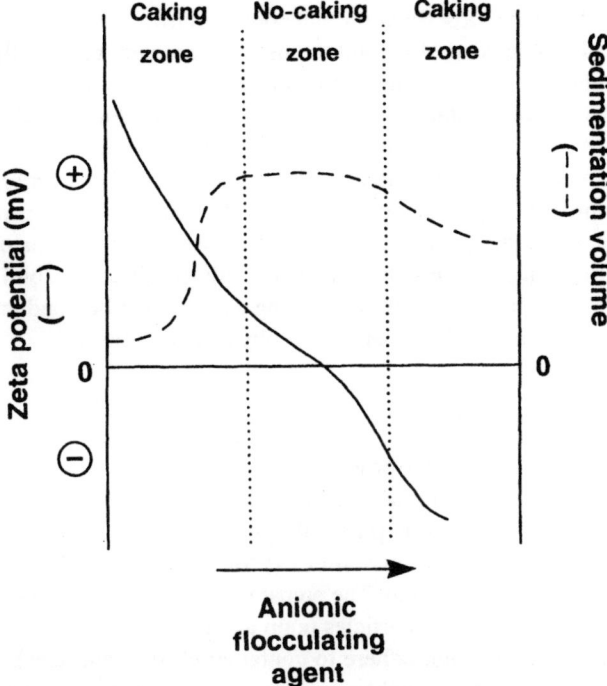

Figure 7.5 Caking diagram showing the effect of a flocculating agent on zeta potential and sedimentation volume. Figure adapted from Martin (1961). Reproduced with permission of the American Pharmaceutical Association.

sedimentation volume is maximized (flocculated) and no caking is observed. Note that if too much flocculating agent is added, overflocculation and caking can occur. Exposing suspensions to extremes in temperature or mechanical stress can also produce this effect (Hem *et al.*, 1986). This example indicates that the zeta potential must be controlled in order to produce a suspension with desirable physical properties.

Most examples of the application of these principles reported in the literature describe small-molecule systems (Matthews and Rhodes, 1970; Kayes, 1977). However, proteins carry charge due to ionization of surface groups and specific adsorption of ions (anions or cations), and these concepts are applicable to protein suspensions as well. For example, Kim *et al.* (1995) describe the effect of chloride ions and pH on sedimentation volume and zeta potential of zinc insulin suspensions. Suspensions were prepared at three sodium chloride concentrations and adjusted to different pH values. Sedimentation volume increased while the zeta potential became more negative with increasing pH. At the highest sodium chloride concentration, maximum values for sedimentation volume and zeta potential were observed at pH 6.9, indicating that a flocculated, pharmaceutically stable suspension can be prepared when zinc insulin is precipitated with 120 mM sodium chloride at pH 6.9.

The method for preparing stable suspensions described above is referred to as controlled flocculation'. Another method for achieving optimal physical stability of suspensions is termed the 'structured vehicle approach'. In this case, viscosity-imparting suspending agents are added to the vehicle to reduce sedimentation and maintain the particles in suspended state. Depending on the application, structured vehicles might not be appropriate for parenteral preparations because the high viscosity would affect syringeability (Nash, 1988).

Akers *et al.* (1987) have described the phenomenon of crystal growth in pharmaceutical suspensions. Particle size distribution, dissolution and recrystallization, pH and temperature changes, and polymorphism and solvate formation are factors that can affect crystal growth. The close contact of particles in settled deflocculated suspensions will favour crystal growth by a process referred to as 'Ostwald ripening' (Patel *et al.*, 1986).

One method to retard crystal growth involves the addition of viscosity-imparting agents, but again this approach may not be appropriate for parenteral suspensions. The best approaches for minimizing crystal growth in peptide or protein suspensions are to control the particle size distribution, select the correct polymorph (if applicable), or use the controlled flocculation approach. Appropriate testing should also be conducted to determine the impact of pH and temperature changes on physical stability.

Testing of physical properties

As highlighted above, the physical properties of suspensions differ from those of other pharmaceutical preparations, and much of the testing will be focused in this area. One of the simplest attributes to evaluate is physical appearance. This qualitative assessment is performed in stability programmes and after exposing the suspension to various stresses of temperature or mechanical agitation. The preparation must remain 'elegant' (no clumping of particles, uniform dispersion of particles upon agitation, particle characteristics remain unchanged, and material does not adhere to container/closure surfaces) after exposure to reasonable conditions. In instances where product elegance is compromised, the information can be used to draft instructions for proper storage and use of the product.

Other, more quantitative evaluations of physical stability involve measurement of sedimentation volume and rate. The procedure for determining sedimentation volume is described above. Sedimentation rate is used to estimate the rate of particle settling, and can be done in conjunction with sedimentation volume determinations by measuring the top boundary of particles progressing downwards as a function of time. In addition to determining sedimentation rate for samples exposed to extreme conditions, this important parameter must be evaluated for flocculated suspensions where particles sediment rapidly. After proper resuspension manipulations, the particles must remain dispersed in vehicle for sufficient time to allow injections of proper doses.

To simulate the various stresses a suspension might encounter during shipping or patient handling, thermomechanical testing is routinely performed to evaluate product performance under these conditions. Methods of this type include some agitation of the suspension in its container induced by either shaking or rotation (e.g. Massey and Sheliga, 1988; Massey *et al.*, 1988). Temperatures are generally elevated during mechanical stress testing as well. Such treatment can result in a rather unsightly appearance of the suspension due to clumping or in damage to the particles, changes in sedimentation properties, adherence of solids to the container/closure system, and loss or change of activity, depending on the duration of exposure. These conditions can result from extended exposure to high temperatures alone, as shown for insulin suspensions (Brange, 1987). All these aspects should be evaluated for potential significance to the final product.

In the same manner that aberrant temperature studies are performed to evaluate chemical stability, the impact of extreme conditions on physical attributes must be determined. The effect of excessively high temperatures has been mentioned, but physical changes can also occur at extreme cold temperatures. Graham and Pomeroy (1978) have studied the effects of freezing and thawing on insulin suspensions. Freezing and thawing caused an increase in sedimentation rate attributed to aggregation of the crystals. Microscopy also revealed some crystal damage, but bioactivity was unaffected.

As highlighted earlier, syringing studies are generally required for suspension preparations stored in vials to ensure that the preparation can be used reliably. Syringing can also be performed to evaluate physical stability. Procedures involve performing daily resuspension manipulations and withdrawal of doses from product containers under conditions that mimic the intended use profile. Evaluations are routinely conducted at room temperature and continue for a length of time corresponding to the proposed in-use dating period. The physical appearance of the suspension is examined throughout, and a thorough chemical analysis of the material remaining is conducted at the end of testing. A common occurrence for suspensions evaluated in this manner is the adherence of solids to the wall of the container, primarily in the region closest to the stopper. This material forms as the suspension flows downwards upon settling. Vehicle drains, leaving solid particles on the interior glass surface.

7.6.3 *Microbiological properties*

In designing a suspension formulation and a manufacturing process for producing it, steps must be taken to ensure that the microbiological properties are optimized and maintained in the final product. The process must include appropriate procedures for sterilization, taking into consideration the way in which particles are produced. In certain cases, sterile filtration of solutions prior to initiating particle growth is appropriate, while other processes involve aseptic combination of preformed sterile particles with a sterilized vehicle. Because of the sensitivity of peptides and proteins to extreme conditions, careful thought must be given to the manner in which sterilization is accomplished. For example, terminal sterilization is unlikely to be tolerated by peptides and proteins.

Many of the issues associated with identifying the appropriate antimicrobial preservative for a suspension have been described in section 7.4; however, there are additional points to consider regarding optimization. Antimicrobial effectiveness testing should be conducted once all the ingredients and concentrations are established, as the results will reflect the overall composition of the preparation including final pH conditions. Therefore, it should not be assumed that a given concentration of preservative providing sufficient antimicrobial efficacy for one preparation will be suitable for another. Furthermore, the efficacy of preservatives varies, some being more potent than others. The properties and relative antimicrobial effectiveness of some commonly used preservatives were compared by Allwood (1982), O'Neill and Mead (1982), and Denyer and Wallhaeusser (1990).

Phenolic preservative concentrations can be decreased due to absorption into tubing during mixing and recirculation operations. Permeation through the closure and chemical degradation of these agents is also possible. To account for possible degradation over shelf-life, formulations varying in preservative concentration should be subjected to antimicrobial effectiveness testing to define the effective range wherein efficacy is maintained and to establish specifications. In addition, preservative excesses may be necessary to account for loss during manufacturing and to achieve final target concentrations.

7.7 Techniques for characterizing and optimizing suspensions

A number of techniques can be applied throughout development of peptide and protein suspensions to achieve the optimal properties described in sections 7.5 and 7.6. Characterization and optimization efforts can be directed towards the particle formation process, the particles themselves, or the suspension properties of the preparation. This section will introduce the various techniques available, but our intent is to provide general insight into what is available and its application rather than to give a detailed description of each method. The reader is referred to appropriate and readily available textbooks for more detailed information on a given technique. Our survey is by no means comprehensive, as it focuses primarily on solid state applications. It should be emphasized that an understanding of the solution state properties of the peptide and protein being formulated can be extremely valuable, if not essential, to suspension development efforts. Furthermore, chemical analyses on the solid peptide or protein particles should be performed to establish the composition of the material.

Microscopy: This is perhaps one of the most basic techniques, and its importance cannot be overstated. White-light microscopy can be used to visualize particles in the size range of 0.4–100 μm (Martin, 1993), and the technique is particularly useful for characterizing nonspherical or amorphous particles. Microscopy can also be used to monitor particle growth over time.

Particle sizing: Information on particle size is essential throughout all stages of suspension development. In addition, particle size can be an important property for evaluating product stability and ensuring process control. A number of techniques are available to determine particle size and size distributions. Regardless of the methodology employed in the various commercial instruments, measured parameters are usually reported in terms of the diameter of spherical particles. In many cases, particles may not be spherical or may be irregularly shaped so that sizes are approximate and will vary among techniques. Particle size distributions, usually included as part of the analysis, may be more useful as they provide information on the population of species present in the sample. The choice of technique will depend on the nature of the sample to be measured and appropriateness of the methodology.

Orr and Spence (1977) report on the use of the Coulter counter for a variety of pharmaceutical applications. Of particular interest is their description of a quantitative assay for determining the amount of amorphous material in crystalline insulin suspensions. In similar studies, laser-light diffraction spectroscopy was used to estimate the particle sizes and distributions of commercially available insulin suspensions (Komatsu et al., 1996). Komatsu et al. (1996) also demonstrated stress-induced (sonication) reduction of particle size in certain cases, presumably due to crystal damage. This observation highlights how particle size analysis can be useful for examining the effects of physical stress on suspension products such as mixing and recirculation typically used in manufacture, and agitation caused by shipping and consumer use.

Electrophoretic light scattering (ELS): ELS provides a direct measure of the velocity of particles moving in an electric field. Velocity is obtained by measuring the Doppler shift of laser light scattered from a moving particle electrophoresing in liquid. The electrophoretic mobility (U), which is proportional to the surface charge density of the particle, is determined from

$$U = V/E \tag{7.2}$$

where V is the electrophoretic velocity and E is the applied electric field. The zeta potential (ζ) is derived from U using

$$U = \zeta\varepsilon/\eta \tag{7.3}$$

where ζ is zeta potential, η is viscosity, $\varepsilon = \varepsilon_o D$, D is the dielectric constant, and ε_o is permittivity of free space. Zeta potentials are usually expressed in millivolts.

As surface charge governs particle interactions, zeta potentials are useful to determine during development of pharmaceutical suspensions, since the quantities can be correlated with physical attributes and stability of coarse dispersions (see section on 'General concepts and basic theory' above). Detailed information on the application of ELS for controlling the physical properties of suspensions is given by Martin (1961), Kayes (1977), Schneider *et al.* (1978) and Kim *et al.* (1995). In another relevant application of Doppler ELS, the technique has been used to study the pH-dependent surface charge of NPH insulin (Dodd *et al.*, 1995). The results were used to explain the adsorption of soluble insulin onto the NPH crystals in biphasic mixtures.

Dynamic light scattering: Although primarily applicable to solution studies, the technique can be employed during suspension development as a tool to study the potential for systems to form certain particles. It is widely recognized that protein crystallization begins with aggregation of individual molecules in solution. Aggregates formed during the early prenucleation phase will determine the potential of the system to form crystals. The goal of dynamic light scattering experiments is to determine whether precipitation or crystallization will occur based on aggregation behaviour in solution. Such information can assist with screening activities for appropriate crystallization conditions. Kadima *et al.* (1991) have shown that differences in aggregate size can be used to define conditions where solubility is maintained, precipitation occurs, or crystals form.

Calorimetry: Differential scanning calorimetry (DSC) has been applied to study the crystal growth mechanism of a protein suspension as demonstrated by Ooshima *et al.* (1997). DSC curves were interpreted in terms of a population of species differing in the degree of self-association. Such information might be useful for predicting crystallization behaviour. In another example of the application of calorimetry, isothermal titrating calorimetry (ITC) has been used to study adsorption of soluble insulin onto NPH crystals and obtain estimates for the thermodynamic parameters associated with this process (Dodd *et al.*, 1995).

Scanning probe microscopy: The scanning probe microscopy (SPM) techniques of scanning tunnelling microscopy (STM) and atomic force microscopy (AFM) with specific applications to pharmaceutical systems have been reviewed by Shakesheff *et al.* (1996). AFM has many advantages for the study of biological molecules, including: no requirement for sample treatment allowing imaging in the natural state, high resolution (nanometre) imaging of surface topography, and the ability to image in liquid. Related to the subject matter of this review, the AFM technique is appropriate for characterizing protein crystal packing and growth mechanisms. For example, Yip and Ward (1996) have used tapping-mode AFM to identify polymorphs of bovine insulin. The *in situ* imaging capabilities of the technique allow the direct visualization of the effects of additives and other parameters as crystal growth occurs. AFM can also be used to examine surface morphology of polymeric materials for drug delivery applications.

Infrared and Raman spectroscopy: Vibrational spectroscopy techniques are particularly useful for obtaining structural information on solid peptide or protein particle samples produced by the various methods discussed in 'Preparation of particles' above. Because some of these processing procedures can result in denaturation, characterization studies should be performed on the material to confirm structural integrity. Examples of the application of these techniques include the use of infrared spectroscopy to detect denaturation induced by lyophilization (Prestrelski *et al.*, 1993a, 1993b), and application of Raman spectroscopy to characterize various crystal forms of insulin (Tensmeyer and Shields, 1990).

X-ray crystallography: As mentioned earlier, X-ray crystallographic data on crystalline particles comprising suspensions are not essential and in most cases are not feasible due to the small size or lack of crystal quality. The practical limitations of applying the technique to commercial suspensions is illustrated nicely in the work of Balschmidt *et al.* (1991). In order to obtain the crystal structure of porcine insulin cocrystallized with clupeine-Z (NPH insulin), the addition of 4 M urea to the crystallization mixture was required. While it can be argued that the difference in crystallization methodology may not induce major structural differences, the relevance of the finding to commercial crystals can still be questioned. However, no other technique can provide equivalent detailed information at the molecular level.

In vitro dissolution: In the development of sustained or controlled-release suspensions it is useful to have an *in vitro* assay available for quickly approximating dissolution properties. Analogous to dissolution testing for solid dosage forms, the procedure requires some detection method to continuously monitor release of drug. For example, Graham and Pomeroy (1984) have developed a continuous-flow spectrophotometric method that can categorize insulin suspension preparations based on clinical time-action classifications.

7.8 Suspension manufacture

7.8.1 Scale-up

The scale-up of parenteral suspensions to commercially viable manufacturing processes presents a significant challenge. As pointed out earlier, crystallization of peptides and proteins at small scale is not simple, but the difficulty of the problem is magnified by virtue of the large volumes needed and the strict controls required for the preparation of pharmaceutical products. Generally, incremental increases in scale are attempted starting from the bench process and progressing upwards in volume to the required batch size. Changes in container composition (e.g. glass vs stainless steel) and geometry will occur during the transition, and could impact the crystallization. In addition, one must consider how certain operations performed with ease in the laboratory such as additions, mixing, transfers, and temperature control will be conducted under aseptic conditions of a manufacturing facility.

The other methods of particle generation are no easier to scale-up (Akers *et al.*, 1987), especially when peptides or proteins are the target compound. Seemingly sound lab- or pilot-scale procedures can produce undesirable outcomes at larger scale. Therefore, steps must be taken to ensure that particle generation and size reduction operations are accomplished without affecting the properties of the molecule or reproducibility of the process. Milling operations for dry peptide or protein particles must be conducted aseptically because no practical means of resterilization is feasible if sterility is compromised. Non-aqueous (oil) vehicles require special consideration if they are to be sterilized by filtration, as filter composition, pore size, and flow rates can impact capture efficiency. Finally, an appropriate strategy needs to be devised for aseptically combining the particles and vehicle.

7.8.2 *Manufacturing controls: special considerations for peptide and protein suspensions*

A regulatory workshop report on scale-up of liquid and semisolid disperse systems has defined the primary finished product attribute to be controlled during scale-up as the degree of 'sameness' of the finished product from batch to batch (van Buskirk *et al.*,

1994). Many of these concepts apply to coarse dispersions as well. Once scale-up is accomplished, the same requirements must continue to be met for each batch manufactured for the market. Four criteria are used to evaluate sameness:

1   adherence to raw material controls

2   adherence to in-process controls

3   adherence to finished product specifications

4   bioequivalence to previous lots.

These criteria are applicable to any pharmaceutical dosage form; however, we will consider criteria 2 and 3 in more detail since the requirements are rather unique for suspensions.

   In-process controls are those operations within a manufacturing process where a critical parameter is controlled to a proven acceptable range or an analytical assay is conducted to determine proper interim conditions prior to progressing to the next step. Examples of possible in-process controls for suspension processes include: order of addition, rate of addition, location of addition, temperature, heat gain/loss (rate and overall time), agitation (type, rate, intensity, and duration), particle size control, suspension homogeneity, and morphology. In addition to these specific in-process controls, drug potency, concentrations of key excipients, pH, bioburden, and filter integrity will be evaluated at various stages during manufacture.

   All in-process controls need to be supported by studies designed to establish appropriate specifications that will ensure batch-to-batch reproducibility. The order, rate, and location of additions and temperature control can be extremely important, especially for producing the desired crystalline form of a peptide or protein. Heat gain or loss can result in denaturation or influence particle generation if strict controls are not in place. Agitation is another critical parameter requiring precise control when dealing with peptide or protein suspensions. Excessive foaming or entrapped air should be avoided, as denaturation at the air–liquid interface is possible. If foam is generated and remains at the surface, crystal nucleation can occur resulting in increases in the number of crystals and changes in size distribution as observed by Schlichtkrull (1961). It may not be possible to include wetting agents or surfactants with the suspension, so steps must be taken in manufacturing to prevent foaming. Particle damage can also occur if mixing and recirculation operations are too severe. However, a balance must be established so that adequate mixing is conducted to achieve suspension homogeneity without impacting the particles. The influence of mixing on particle generation, control, and stability has been discussed in a report by Genck (1995).

   As with any finished dosage form, the product specifications must define key attributes of the preparation that ensure safety, identity, strength, purity, and quality. Besides potency, purity and stability, suspension preparations will require, at a minimum, specifications for content uniformity, particle morphology, and physical appearance. Specifications might also be established for other parameters such as particle size and distribution, sedimentation rate and sedimentation volume, or rheological properties depending on the nature of the final suspension. We have described approaches and techniques that can be applied to measure these parameters and define appropriate specifications in other sections of this chapter. Content uniformity in the context of suspensions relates to demonstration of uniform potency throughout a filled lot of product. For peptide- or protein-based preparations, nitrogen determinations, colorimetric assays, or an HPLC assay for potency can be employed. Regardless of the method used, assays are conducted on a statistical sampling of the filled lot.

7.9 Other related systems

The suspensions described so far are composed of solid particles of the peptide or protein dispersed in vehicle plus excipients. Recall that these preparations were devised with the intention of retarding or controlling activity. While these suspensions display extended activity, their duration may still not be long enough to preclude frequent daily injections. For example, the time-action of Ultralente insulin, although variable, is generally greater than 24 hours, but multiple injections of insulin are still required every day. This particular suspension, which is the longest-acting insulin preparation, highlights a major short-coming of the systems we have described. Specifically, is the pharmaceutical effect that is ultimately achieved worth the effort involved in developing such a suspension? The bST suspension is one exception, displaying sustained release of bioactive somatotropin for 7–28 days (Ferguson, 1997); the practical limitations of such preparations have been discussed earlier.

How can the activity of therapeutically useful peptides and proteins be extended to longer durations, and can release of drug be better controlled? One approach involves encapsulation of these molecules into degradable microspheres. The focus of this text is not concerned with drug delivery, and we will not provide a detailed description of this technology as it has received a great deal of attention in the literature (Langer, 1990; Mackay, 1990; Pitt, 1990; Sanders, 1990; Kuo and Saltzman, 1996). The subject relates to the present discussion, however, because these systems represent examples of inject-able suspensions. The particles have diameters of 1–100 μm and are supplied as dried powders. Prior to injection, the particles are mixed with an appropriate vehicle, dispersed, and administered. Release kinetics are controlled by polymer degradation and diffusion of the drug, and the duration can be adjusted from days to months (Langer, 1996).

Microencapsulation involves rather harsh conditions which may involve high shear, organic solvents, or high temperatures. In addition, the encapsulated molecules will be exposed to high body temperature over extended periods of time. As a result of these processing requirements and potential stability issues, the technology was not thought to be appropriate for peptides and proteins. However, examples of encapsulation procedures for peptides and proteins have been described in the literature (Langer, 1990; Mackay, 1990; Pitt, 1990; Sanders, 1990). In general, peptides lacking complex, higher-order structure are amenable to the microencapsulation approach. An example of a peptide that has been encapsulated is leuprolide acetate, a synthetic nonapeptide analogue of LHRH (luteinizing hormone-releasing hormone). The microencapsulated peptide is marketed as Lupron® Depot and is used for the treatment of advanced prostatic cancer. Reconstitu-tion of the dried particles with vehicle results in a suspension that is administered intramuscularly at monthly intervals. Clearly, such a suspension design is not subject to many of the physical instability issues addressed in the section on 'General concepts and basic theory' above.

The microencapsulation approach can also be reliably applied to selected proteins if the unique properties of these molecules are considered as part of process development. This fact is illustrated nicely in the procedure devised for microencapsulating human growth hormone (Johnson *et al.*, 1996). By exploiting the stabilizing effect of zinc ion complexation and using a low-temperature method for incorporation during encapsula-tion, degradable microspheres were prepared containing structurally intact human growth hormone. The dried microspheres were suspended in vehicle containing carboxymethyl cellulose, polysorbate, and sodium chloride and injected into monkeys to evaluate the pharmacological properties of the preparation. Elevated serum concentrations of human growth hormone were sustained for greater than one month. Johnson *et al.* (1996) propose that their encapsulation process may be applicable to other proteins; therefore, this technology might result in other sustained-release protein-based preparations in the future. Of course, the success of such systems depends on demonstrating the desired pharmacological properties in humans, and the ability to adapt the procedure into a viable commercial manufacturing process.

7.10 Conclusions

Suspensions are generally considered the most complicated type of formulation to develop because of the multitude of factors which must be considered from design through manufacture. Similarly, the unique chemical and physical properties of peptides and proteins make them one of the most difficult classes of molecules to develop as pharmaceuticals. Therefore, the development of commercially viable peptide or protein suspensions is a significant challenge. This is demonstrated by the limited number of peptide or protein suspensions on the market. Because peptides and proteins are natural therapeutic agents and generally have high potencies, however, the benefits of such preparations often outweigh the difficulties associated with formulation development. In particular, the extended release properties afforded by suspensions are extremely attractive from the standpoint of patient compliance and convenience.

Building on the concepts devised for small molecule suspensions and the years of accumulated knowledge obtained on marketed insulin suspensions, we have attempted to summarize approaches for designing, developing, and manufacturing suspensions of other peptides and proteins. We have described methods for producing particles and suitable vehicle and identified the various excipients that can be used to prepare peptide and protein suspensions having optimal chemical, physical, and microbiological properties. Techniques and procedures for evaluating these properties to meet the various requirements for suspension preparations have also been addressed. In addition, we have briefly touched on the issues and challenges associated with the manufacture of parenteral suspensions. The numerous examples and practical information provided herein will assist formulation scientists in the development of high-quality suspension preparations of peptides or proteins.

Acknowledgements

The authors are grateful to Dr Mary Stickelmeyer and Dr Henry Havel for careful review of this work and for providing useful suggestions. Dr Joseph Rinella is thanked for providing some of the reference material cited in this chapter.

References

AKERS, M.J. and TAYLOR, C.J., 1990, Official methods of preservative evaluation and testing, in DENYER, S.P. and BAIRD, R.M. (eds) *Guide to Microbiological Control in Pharmaceuticals*, pp. 292–303, Chichester: Ellis Horwood.

AKERS, M.J., FITES, A.L. and ROBISON, R.L., 1987, Formulation design and development of parenteral suspensions, *J. Parenteral Sci. Technol.*, **41**, 88–96.

ALLWOOD, M.C., 1982, The effectiveness of preservatives in insulin injections, *Pharm. J.*, **229**, 340.

BALSCHMIDT, P., HANSEN, F.B., DODSON, E.J., DODSON, G.G. and KORBER, F., 1991, Structure of porcine insulin cocrystallized with clupeine-Z, *Acta Cryst.*, **B47**, 975–986.

BALSCHMIDT, P., 1996, *Asp^{B28} insulin crystals*, United States Patent, Patent number: 5,547,930.

BOYETT, J.B. and DAVIS, C.W., 1989, Injectable emulsions and suspensions, in LIEBERMAN, H.A., RIEGER, M.M. and BANKER, G.S. (eds) *Pharmaceutical Dosage Forms Disperse Systems*, Vol. 2, pp. 379–416, New York: Marcel Dekker.

BRAMLEY, M.R., CARTER, A.B. and DUNWELL, D.W., 1989, *Somatotropin formulations*, European Patent Application: Publication number 0 314 421.

BRANGE, J., 1987, *Galenics of Insulin: the Physico-chemical and Pharmaceutical Aspects of Insulin and Insulin Preparations*, Berlin, Heidelberg: Springer-Verlag.

    1992, Chemical stability of insulin 4. Mechanisms and kinetics of chemical transformations in pharmaceutical formulation, *Acta Pharm. Nord.*, **4**, 209–222.

    1994, *Stability of Insulin: Studies on the Physical and Chemical Stability of Insulin in Pharmaceutical Formulation*, Dordrecht: Kluwer Academic Publishers.

BRANGE, J. and LANGKJAER, L., 1992, Chemical stability of insulin 3. Influence of excipients, formulation, and pH, *Acta Pharm. Nord.*, **4**, 149–158.

BRANGE, J., HALLUND, O. and SØRENSEN, E., 1992a, Chemical stability of insulin 5. Isolation, characterization and identification of insulin transformation products, *Acta Pharm. Nord.*, **4**, 223–232.

BRANGE, J., HAVELUND, S. and HOUGAARD, P., 1992b, Chemical stability of insulin. 2. Formation of higher molecular weight transformation products during storage of pharmaceutical preparations, *Pharm. Res.*, **9**, 727–734.

BRANGE, J., LANGKJAER, L., HAVELUND, S. and VØLUND, A., 1992c, Chemical stability of insulin. 1. Hydrolytic degradation during storage of pharmaceutical preparations, *Pharm. Res.*, **9**, 715–726.

BRAZEAU, P., RIVIER, J., VALE, W. and GUILLEMIN, R., 1974, Inhibition of growth hormone secretion in the rat by synthetic somatostatin, *Endocrinology*, **94**, 184–187.

BROOKS, N.D. and NEEDHAM, G.F., 1994, Injectable extended release formulations and methods, United States Patent, Patent number: 5,352,662.

BUCH, J. and BUCH, A., 1983, Sustained effect of zinc–protamin–glucagon in hyperlipidaemic patients, *Acta Pharmacol. Toxicol.*, **53**, 188–192.

CARTER, C.W., JR, 1990, Efficient factorial designs and the analysis of macromolecular crystal growth conditions, in CARTER, C.W., JR (ed.) *Methods: a Companion to Methods in Enzymology Protein and Nucleic Acid Crystallization*, Vol. 1, pp. 12–24, Duluth, MN: Academic Press.

CHIEN, Y.W., 1981, Long-acting parenteral drug formulations, *J. Parenteral Sci. Technol.*, **35**, 106–139.

CHIEN, Y.W., PRZYBYSZEWSKI, P. and SHAMI, E.G., 1981, Syringeability of nonaqueous parenteral formulations – development and evaluation of a testing apparatus, *J. Parenteral Sci. Technol.*, **35**, 281–284.

DEFELIPPIS, M.R., 1995, *Monomeric insulin analog formulations*, United States Patent, Patent number: 5,461,031.

DEFELIPPIS, M.R., BAKAYSA, D.L., YOUNGMAN, K.M., RADZIUK, J. and FRANK, B.H., 1996, Preparation and characterization of neutral protamine lispro (NPL) suspension, *Diabetes*, **45**, Suppl. 2, 74A.

DEFELIPPIS, M.R., BAKAYSA, D.L., BELL, M.A., HEADY, M.A., LI, S., PYE, S., YOUNGMAN, K.M., RADZIUK, J. and FRANK, B.H., 1998, Preparation and characterization of a cocrystalline suspension of [Lys^{B28}, Pro^{B29}]-human insulin analog, *J. Pharm. Sci.*, **87**, 170–176.

DENYER, S.P. and WALLHAEUSSER, K.-H., 1990, Antimicrobial preservatives and their properties, in DENYER, S.P. and BAIRD, R.M. (eds) *Guide to Microbiological Control in Pharmaceuticals*, pp. 251–273, Chichester: Ellis Horwood.

DODD, S.W., HAVEL, H.A., KOVACH, P.M., LAKSHMINARAYAN, C., REDMON, M.P., SARGEANT, C.M., SULLIVAN, G.R. and BEALS, J.M., 1995, Reversible adsorption of soluble hexameric insulin onto the surface of insulin crystals cocrystallized with protamine: an electrostatic interaction, *Pharm. Res.*, **12**, 60–68.

DONINI, P., 1974, *Long-active gonadotropins*, United States Patent, Patent number: 3,852,422.

DRENTH, J., 1994, *Principles of Protein X-ray Crystallography*, New York: Springer-Verlag.

DUCRUIX, A.F. and RIES-KAUTT, M.M., 1990, Solubility diagram analysis and the relative effectiveness of different ions on protein crystal growth, in CARTER, C.W., JR (ed.) *Methods: a Companion to Methods in Enzymology Protein and Nucleic Acid Crystallization*, Vol. 1, pp. 25–30, Duluth, MN: Academic Press.

DURBIN, S.D. and FEHER, G., 1996, Protein crystallization, *Annu. Rev. Phys. Chem.*, **47**, 171–204.

EDWARDS, S.L., DEFELIPPIS M.R., FRANK, B.H., KILCOMONS, M.A., SHELIGA, T.A., STICKELMEYER, M.P., YOUNGMAN, K.M. and HAVEL, H.A., 1996, Assessment of the stability of insulin lispro mixtures with human insulin NPH, presentation at *The 211th National Meeting of the American Chemical Society*, New Orleans, 24–28 March.

ELLERHORST, J.A., COMSTOCK, J.P. and NELL, L.J., 1990, Protamine antibody production in diabetic subjects treated with NPH insulin, *Am. J. Med. Sci.*, **299**, 298–301.

FERGUSON, T.H., 1997, Peptide and protein delivery for animal health applications, in PARK, K. (ed.) *Controlled Drug Delivery Challenges and Strategies*, pp. 289–308, Washington, DC: American Chemical Society.

FERGUSON, T.H., HARRISON, R.G. and MOORE, D.L., 1990, *Injectable sustained release formulation*, United States Patent, Patent number: 4,977,140.

GALLAGHER, P.M., COFFEY, M.P., KRUKONIS, V.J. and KLASUTIS, N., 1989, Gas antisolvent recrystallization: new process to recrystallize compounds insoluble in supercritical fluids, in JOHNSTON, K.P. and PENNINGER, J.M.L. (eds) *Supercritical Fluid Science and Technology*, pp. 334–354, Washington, DC: American Chemical Society.

GANDERTON, D., 1991, The development of peptide and protein pharmaceuticals, in HIDER, R.C. and BARLOW, D. (eds) *Polypeptide and Protein Drugs Production, Characterization and Formulation*, pp. 211–227, Chichester: Ellis Horwood.

GENCK, W.J., 1995, Mixing's influence on precipitations, *Chemical Processing*, 46–50.

GILLILAND, G.L. and BICKHAM, D.M., 1990, The biological macromolecule crystallization database: a tool for developing crystallization strategies, in CARTER, C.W., JR (ed.) *Methods: a Companion to Methods in Enzymology Protein and Nucleic Acid Crystallization*, Vol. 1, pp. 6–11, Duluth, MN: Academic Press.

GÖLKER, C.F., 1993, Final recovery steps: lyophilization, spray-drying, in STEPHANOPOULOS, G. (ed.) *Biotechnology*, Vol. 3, *Bioprocessing*, 2nd edn, pp. 696–714, Weinheim: VCH.

GRAHAM, D.T. and POMEROY, A.R., 1978, The effects of freezing on commercial insulin suspensions., *Int. J. Pharm.*, **1**, 315–322.

1984, An *in-vitro* test for the duration of action of insulin suspensions, *J. Pharm. Pharmacol.*, **36**, 427–430.

HAGEDORN, H.C., JENSEN, B.N., KRARUP, N.B. and WODSTRUP, I., 1936, Protamine insulinate, *J. Am. Med. Assoc.*, **106**, 177–180.

HALLAS-MØLLER, K., PETERSEN, K. and SCHLICHTKRULL, J., 1952, Crystalline and amorphous insulin–zinc compounds with prolonged action, *Science*, **116**, 394–399.

HAMMOND, G., 1991, *Crystalline interleukin-4*, International Patent Application, Publication number: WO 91/19742.

HEM, S.L., FELDKAMP, J.R. and WHITE, J.L., 1986, Basic chemical principles related to emulsions and suspension dosage forms, in LACHMAN, L., LIEBERMAN, H.A. and KANIG, J.L. (eds) *The Theory and Practice of Industrial Pharmacy*, pp. 100–122, Philadelphia: Lea & Febiger.

HIESTAND, E.N., 1964, Theory of coarse suspension formulation, *J. Pharm. Sci.*, **53**, 1–18.

HOFFMAN, J.A., 1994, *Insulin formulation*, European Patent Application, Publication number: 0 646 379 A1.

HOLT, L.A., MILLIGAN, B., RIVETT, D.E. and STEWART, F.H.C., 1977, The photodecomposition of tryptophan peptides, *Biochim. Biophys. Acta*, **499**, 131–138.

HORROW, J.C., 1985, Protamine: a review of its toxicity, *Anesth. Analg.*, **64**, 348–361.

JOHNSON, O.L., CLELAND, J.L., LEE, H.J., CHARNIS, M., DUENAS, E., JAWOROWICZ, W., SHEPARD, D., SHAHZAMANI, A., JONES, A.J.S. and PUTNEY, S.D., 1996, A month-long effect from a single injection of microencapsulated human growth hormone, *Nature Med.*, **2**, 795–799.

KADIMA, W., MCPHERSON, A., DUNN, M.F. and JURNAK, F., 1991, Precrystallization aggregation of insulin by dynamic light scattering and comparison with canavalin, *J. Cryst. Growth*, **110**, 188–194.

KAYES, J.B., 1977, Pharmaceutical suspensions: relation between zeta potential, sedimentation volume and suspension stability, *J. Pharm. Pharmacol.*, **29**, 199–204.

KIM, Y. and HAREN, A.M., 1995, The application of crystal soaking technique to study the effect of zinc and cresol on insulinotropin crystals grown from a saline solution, *Pharm. Res.*, **12**, 1664–1670.

KIM, Y. and ROSE, C.A., 1995, Precipitation of insulinotropin in the presence of protamine: effect of phenol and zinc on the isophane ratio and the insulinotropin concentration in the supernatant, *Pharm. Res.*, **12**, 1284–1288.

KIM, Y., CUFF, G.W. and MORRIS, R.M., 1995, Effect of chloride ion on the sedimentation volume and zeta potential of zinc insulin suspensions in neutral pH range, *J. Pharm. Sci.*, **84**, 755–759.

KOMATSU, H., KITAJIMA, A. and OKADA, S., 1996, Estimation of average particle sizes and size distributions of commercially available human-insulin aqueous suspensions using laser-light diffraction spectroscopy, *Chem. Pharm. Bull.*, **44**, 1966–1969.

KRAYENBÜHL, C. and ROSENBERG, T., 1946, Crystalline protamine insulin, *Rep. Steno Mem. Hosp. Nord. Insulinlab*, **1**, 60–73.

KUO, P.Y.P. and SALTZMAN, W.M., 1996, Novel systems for controlled delivery of macromolecules, *Crit. Rev. Eukaryotic Gene Expression*, **6**, 59–73.

LANGER, R., 1990, New methods of drug delivery, *Science*, **249**, 1527–1533.

1996, Controlled release of a therapeutic protein, *Nature Med.*, **2**, 742–743.

MACKAY, M., 1990, Delivery of recombinant peptide and protein drugs, *Biotech. Genet. Eng. Rev.*, **8**, 251–278.

MCPHERSON, A., 1982, *Preparation and Analysis of Protein Crystals*, New York: Wiley.

MARTIN, A., 1993, *Physical Pharmacy*, 4th edn, Philadelphia: Lea & Febiger.

MARTIN, A.N., 1961, Physical chemical approach to the formulation of pharmaceutical suspensions, *J. Pharm. Sci.*, **50**, 513–517.

MARTIN, J.B., RENAUD, L.P. and BRAZEAU, P., JR, 1974, Pulsatile growth hormone secretion: suppression by hypothalamic ventromedial lesions and by long-acting somatostatin, *Science*, **186**, 538–540.

MARTIN, J.L., 1986, *Prolonged release of growth promoting hormones*, Australian Patent Application, Application number: AU-A-61092/86.

MASSEY, E.H. and SHELIGA, T.A., 1988, Human insulin (HI) isophane suspension (NPH) with improved physical stability, *Pharm. Res.*, **5**, S-34.

MASSEY, E.H., TENSMEYER, L.G. and SHELIGA, T.A., 1988, Aggregation of human insulin (HI) in isophane suspension formulations (NPH), presentation at *The 196th ACS National Meeting*, Los Angeles, September 25–30.

MATTHEWS, B.A. and RHODES, C.T., 1968, Some studies of flocculation phenomena in pharmaceutical suspensions, *J. Pharm. Sci.*, **57**, 569–573.

1970, Use of the Derjaguin, Landau, Verwey, and Overbeck theory to interpret pharmaceutical suspension stability, *J. Pharm. Sci.*, **59**, 521–525.

MITCHELL, J.W., 1991, *Prolonged release of biologically active somatotropin*, United States Patent, Patent number: 5,013,713.

MUMENTHALER, M., HSU, C.C. and PEARLMAN, R., 1994, Feasibility study on spray-drying protein pharmaceuticals: recombinant human growth hormone and tissue-type plasminogen activator, *Pharm. Res.*, **11**, 12–20.

NAETS, J.P. and GUNS, M., 1980, Inhibitory effects of glucagon on erythropoiesis, *Blood*, **55**, 997–1002.

NASH, R.A., 1988, Pharmaceutical suspensions, in LIEBERMAN, H.A., RIEGER, M.M. and BANKER, G.S. (eds) *Pharmaceutical Dosage Forms Disperse Systems*, Vol. 1, pp. 151–198, New York: Marcel Dekker.

NELL, L.J. and THOMAS, J.W., 1988, Frequency and specificity of protamine antibodies in diabetic and control subjects, *Diabetes*, **37**, 172–176.

O'NEILL, J.J. and MEAD, C.A., 1982, The parabens: bacterial adaptation and preservative capacity, *J. Soc. Cosmet. Chem.*, **33**, 75–84.

OOSHIMA, H., URABE, S., IGARASHI, K., AZUMA, M. and KATO, J., 1997, Mechanism of crystal growth of protein: differential scanning calorimetry of thermolysin crystal suspension, in BOTSARIS, G.D. and TOYOKURA, K. (eds) *Separation and Purification by Crystallization*, pp. 18–27, Washington, DC: American Chemical Society.

ORR, N.A. and SPENCE, J., 1977, Applications of particle size analysis in the pharmaceutical industry, *Analyst*, **102**, 466–472.

PATEL, N.K., KENNON, L. and LEVINSON, R.S., 1986, Pharmaceutical suspensions, in LACHMAN, L., LIEBERMAN, H.A. and KANIG, J.L. (eds) *The Theory and Practice of Industrial Pharmacy*, pp. 479–501, Philadelphia: Lea & Febiger.

PHILLIPS, E.M. and STELLA, V.J., 1993, Rapid expansion from supercritical solutions: application to pharmaceutical processes, *Int. J. Pharm.*, **94**, 1–10.

PIKAL, M.J. 1990a, Freeze-drying of proteins part I. Process design, *BioPharm*, **3**, 18–27.
1990b, Freeze-drying of proteins part II. Formulation selection, *BioPharm*, **3**, 26–30.

PITT, C.G., 1990, The controlled parenteral delivery of polypeptides and proteins, *Int. J. Pharm.*, **59**, 173–196.

PRESTRELSKI, S.J., ARAKAWA, T. and CARPENTER, J.F., 1993a, Separation of freezing- and drying-induced denaturation of lyophilized proteins using stress-specific stabilization, *Arch. Biochem. Biophys.*, **303**, 465–473.

PRESTRELSKI, S.J., TEDESCHI, N., ARAKAWA, T. and CARPENTER, J.F., 1993b, Dehydration-induced conformational transitions in proteins and their inhibition by stabilizers, *Biophys. J.*, **65**, 661–671.

RADZIUK, J., BRADLEY, B., WELSH, L., DeFELIPPIS, M.R. and ROACH, P., 1996, Profiles of biological activity after subcutaneous administration of mixtures of Lys^{B28}-Pro^{B29} human insulin (lispro) in soluble and neutral protamine formulations, *Diabetes*, **45**, Suppl. 2, 218A.

REICHERT, P., NAGABHUSHAN, T.L., LONG, M.M., BUGG, C.E. and DELUCAS, L.J., 1996, Macroscale production and analysis of crystalline interferon alpha-2B in microgravity on STS-52, in EL-GENK, M.S. (ed.) *Space Technology & Applications International Forum (STAIF-96)*, Vol. 361, Pt 1, pp. 139–148, Woodbury, NY: American Institute of Physics.

ROLIM, C. and BRISTOW, A.F., 1995, Determination of insulin-in-solution in biphasic isophane insulin formulations, *Pharmeuropa*, **7**, 22–25.

SANDERS, L.M., 1990, Drug delivery systems and routes of administration of peptide and protein drugs, *Eur. J. Drug Metab. Pharmacokin.*, **15**, 95–102.

SCHLICHTKRULL, J., 1957, Insulin crystals IV. The preparation of nuclei, seeds and monodisperse insulin crystal suspensions, *Acta Chem. Scand.*, **11**, 299–302.
1961, *Insulin Crystals*, Copenhagen: Novo.

SCHNEIDER, W., STAVCHANSKY, S. and MARTIN, A., 1978, Pharmaceutical suspensions and the DVLO theory, *Am. J. Pharm. Educ.*, **42**, 280–289.

SHAKESHEFF, K.M., DAVIES, M.C., ROBERTS, C.J., TENDLER, S.J.B. and WILLIAMS, P.M., 1996, The role of scanning probe microscopy in drug delivery research, *Crit. Rev. Therapeutic Drug Carrier Systems*, **13**, 225–256.

SIEGEL, F.P., 1990, Tonicity, osmoticity, osmolality and osmolarity, in GENNARO, A.R. (ed.) *Remington's Pharmaceutical Sciences*, 18th edn, pp. 1481–1498, Easton, PA: Mack Publishing.

SUBRAMANIAM, B., RAJEWSKI, R.A. and SNAVELY, K., 1997, Pharmaceutical processing with supercritical carbon dioxide, *J. Pharm. Sci.*, **86**, 885–890.

TANNENBAUM, G.S. and COLLE, E., 1980, Ineffectiveness of protamine zinc somatostatin as a long-acting inhibitor of insulin and growth hormone secretion, *Can. J. Physiol. Pharmacol.*, **58**, 951–955.

TENSMEYER, L.G. and SHIELDS, J.E., 1990, The Raman spectra of crystalline 4Zn, 2Zn, and Na insulin, in ADAR, F. and GRIFFITHS, J.E. (eds) *Raman and Luminescence Spectroscopies in Technology II*, Vol. 1336, pp. 222–234, Bellingham, WA: Proc. SPIE.

TOM, J.W. and DEBENEDETTI, P.G., 1991, Particle formation with supercritical fluids – a review, *J. Aerosol Sci.*, **22**, 555–584.

VAN BUSKIRK, G.A., SHAH, V.P., ADAIR, D., ARBIT, H.M., DIGHE, S.V., FAWZI, M., FELDMAN, T., FLYNN, G.L., GONZÁLEZ, M.A., GRAY, V.A., GUY, R.H., HERD, A.K., HEM, S.L., HOIBERG, C., JERUSSI, R., KAPLAN, A.S., LESKO, L.J., MALINOWSKI, H.J., MELTZER, N.M., NEDICH, R.L., PEARCE, D.M., PECK, G., RUDMAN, A., SAVELLO, D., SCHWARTZ, J.B., SCHWARTZ, P., SKELLY, J.P., VANDERLAAN, R.K., WANG, J.C.T., WEINER, N., WINKEL, D.R. and ZATZ, J.L., 1994, Workshop III report: Scale-up of liquid and semisolid disperse systems, *Eur. J. Pharm. Biopharm.*, **40**, 251–254.

WENDEL, S. and ÇELIK, M., 1997, An overview of spray-drying applications, *Pharm. Technol.*, **21**, 124–156.

WEYER, C., HEISE, T. and HEINEMANN, L., 1997, Premixed formulation of B28Asp and NPH-insulin: pharmacodynamic properties of a 30/70-stable mixture, *Diabetologia*, **40**, Suppl. 1, A350.

WORKMAN, W.E. and CLAYTON, R.A., 1996, Microbial sterility testing of oil-formulated bovine somatotropin using Tween® 80 dispersion, *J. Pharm. Biomed. Anal.*, **15**, 193–200.

YEO, S.-D., LIM, G.-B., DEBENEDETTI, P.G. and BERNSTEIN, H., 1993, Formation of microparticulate protein powders using a supercritical fluid antisolvent, *Biotech. Bioeng.*, **41**, 341–346.

YIM, Z., 1988, *Stable interferon complexes*, European Patent Application, Publication number: 0 281 299.

YIP, C.M. and WARD, M.D., 1996, Atomic force microscopy of insulin single crystals: direct visualization of molecules and crystal growth, *Biophys. J.*, **71**, 1071–1078.

ZOGRAFI, G., SCHOTT, H. and SWARBRICK, J., 1990, Disperse systems, in GENNARO, A.R. (ed.) *Remington's Pharmaceutical Sciences*, 18th edn, pp. 257–309, Easton, PA: Mack Publishing.

8

Peptides and Proteins as Parenteral Solutions

MICHAEL J. AKERS AND MICHAEL R. DEFELIPPIS

Lilly Research Laboratories, Indianapolis, Indiana, USA

8.1   Overview and introduction

The purpose of this chapter is to provide practical guidance to formulation scientists charged with the development of stable, manufacturable, and elegant solution dosage forms of peptides and proteins. The chapter will cover the basics of chemical stabilization, physical stabilization, and microbiological quality of proteins and peptides in solution. We will place more emphasis on the *approaches* used to solve protein/peptide solution formulation problems than on discussing the nature of degradation mechanisms which are covered elsewhere in this text and in many other excellent publications. We also will have some coverage of packaging and manufacturing of protein solution dosage forms, in the spirit of emphasizing that scientists developing these dosage forms must be equally concerned with the formulation, the package, and the manufacturing process.

There are at least 22 protein products on the market, four of which are stored as ready-to-use solutions, and the rest of which are stored as freeze-dried powders, then reconstituted into solutions by adding a diluent before administration. Approximately 200 peptides and proteins are being studied in the clinic, most of which are freeze-dried products. It is reasonable to assume that nearly every one of these peptide and protein products, commercial or in clinical study, has had to overcome and control stability issues in solution. The type of stability issue and the degree of complexity of the degradation mechanism differ from protein to protein, but approaches to resolve instability issues in solution are relatively universal.

There are some basic guidelines to consider in the development of parenteral solutions of proteins and peptides. These are summarized as follows.

1   A thorough understanding of the physical and chemical properties of the protein or peptide bulk drug substance is necessary. Well-documented analytical techniques are now available for studying these properties in solution. Effects of temperature, pH, shear, oxygen, buffer type and concentration, ionic strength, and protein/peptide concentration must be understood. From preformulation studies, protein/peptide chemical and physical degradation pathways will be better understood so that the final formulation, manufacturing process, and packaging system will be rationally developed.

2   The route of administration must be known in order to select the final dosage form, vehicle, volume, and tonicity requirements for the product. For example, if the primary route of administration is intravenous, the vehicle has to be water although some water-miscible co-solvents can be used. The volume can be limitless (unless an antimicrobial preservative is part of the formulation, in which case the volume is limited to 15 ml), and the tonicity does not necessarily have to be isotonic because the injected solution will be rapidly diluted. However, if the route of administration will be subcutaneous or intramuscular, then the vehicle can be aqueous or nonaqueous, the volumes are limited (usually no more than 2 ml for subcutaneous, 3 ml for intramuscular), and the tonicity of the product needs to be more tightly controlled since the product is not quickly or readily diluted. The rate of injection is also a factor to be considered in selection of final formulation ingredients in that some ingredients, including the protein/peptide itself, can be irritating and even cause local inflammatory reactions if injected too quickly and/or at too high a concentration.

3   Careful screening and choice of solutes for solubilization, stabilization, preservation, and tonicity adjustment must take place. These aspects will be the thrust of this chapter.

4   Potential effects of the manufacturing process on the stability of the protein/peptide in the final formulation must be understood. Proteins/peptides cannot withstand terminal sterilization techniques (heat, gas, radiation) and, thus, must be sterilized by aseptic filtration. The filter used must be qualified so that it does not bind the protein/peptide. The effect of flow rate during filtration and filling on solution stability must be studied. Also, the effect of shear (mechanical stress) that is encountered during manufacturing must be known. Time limitations must be established from the time the protein solution is compounded until it is sterile-filtered in order to avoid any increase in endotoxin levels from whatever the bioburden, however small, may be in the nonsterile solution. Harwood *et al.* (1993) and Nail and Akers (2000) are excellent

references that deal thoroughly with all aspects of the manufacturing of sterile protein and peptides dosage forms.

5   Selection of the most compatible container/closure system is tremendously important. Formulation scientists must appreciate that the container and closure system is just as important as the final solution formulation in assuring long-term stability and maintenance of sterility and other quality parameters of the product. Proteins and peptides are well known to adsorb to glass, so experiments must be designed to study this possibility and, if adsorption occurs significantly, additives such as albumin must be considered to reduce the adsorption. Glass leachates and particulates are possible, and the formulator must be aware of this. Experiments must be conducted to assure elimination of this potential problem. The choice of rubber closure is particularly important because of known potential for the closure to leach some of its own ingredients into a solution, to adsorb components of the protein/peptide formulation, to core (rubber particulates) when penetrated by a needle, to generate particulates, and to leak due to problems with the fitment on the glass vial or resealability of the elastomer after needle penetration. Studies on adsorption of the protein to plastic surfaces will be necessary if the final product will be a plastic container. Even if plastic is not part of the primary container, protein–plastic compatibility studies should be done since plastic tubing such as silicone or polyvinyl chloride will be used in pharmaceutical process equipment (e.g. filling machines), and the final dosage form might be added to large-volume parenteral solutions contained in plastic bags.

6   Studies must be conducted to understand the effects of distribution and storage on the stability of the final product. Temperature excursions during shipping, mechanical stress, exposure to light, and other simulated shipping and storage conditions must be studied. From these studies, appropriate procedures for distribution and long-term storage of these relatively unstable dosage forms can be developed.

Table 8.1 provides of summary of the key steps in the development of solution dosage forms of peptides and proteins.

8.2   Optimizing hydrolytic stability

The effect of solution pH on stability is a very important factor to study in early protein solution development. Figure 8.1 schematically depicts expected stability problems of proteins as a function of pH. Preformulation stability studies are conducted very early in the product development cycle to elucidate relative protein solubility and stability over an appropriate pH range (normally pH 3 to pH 10). The relationship of stability and solubility at various pH values usually follows a pattern of higher solubility, lower chemical stability; or lower solubility, lower physical stability. Protein solubility is minimum generally at its isoelectric point. Insulin, for example, has an isoelectric point of 5.4, and at this pH it is quite insoluble in water (<0.1 mg/ml). Adjusting the solution pH to less than 4 or greater than 7 greatly increases insulin solubility (>30 mg/ml, depending on zinc concentration and species source of insulin), but also increases the rate of deamidation at the pH ranges (Brange *et al.*, 1992b). An example of the effect of pH on deamidation and polymerization of insulin is shown in Figure 8.2 (Brange and Langkjaer, 1993). In dosage form development, the scientist must first determine what pH range provides acceptable solubility of the protein for proper dosage, then determine whether this pH range also provides acceptable stability. There is usually a give-and-take relationship between solubility and stability, and it is up to the scientist to identify what pH is optimal for both.

Table 8.1 Development strategy for protein and peptide parenteral solution dosage forms

Formulation and package development studies	Process studies

Final strategy/objectives
Development of final formulation
 	
• Justification of choice of excipients, pH, specifications
 	
Selection of container/closure
 	
• Extractables
• Container/closure integrity
• Glass leachates, particulates
 	
Stability and compatibility studies
 	
• Effects of light, oxygen, high temperature, freezing
• Interaction of excipients with active components
• Long-term stability studies of final container formulation in final container/closure system
• Temperature/shipping excursions
 	
Microbiological characteristics
 	
• Antimicrobial properties
• Preservative efficacy
• Endotoxin control
	Optimization studies of excipients, pH, other possible variations
	Process development
	
	• Process control (e.g. time, temperature during each processing step
	• Filter selection/validation
	
	  – Microbial retention
	  – Adsorption
	  – Extractables
	
	• Effect of terminal sterilization
	• Justification of excess
	• Process validation
	
	  – Sterilization of components
	  – Aseptic process
	  – Cleaning
	  – Filling
	
	Establishment of critical process parameters
	Establishment of control strategy

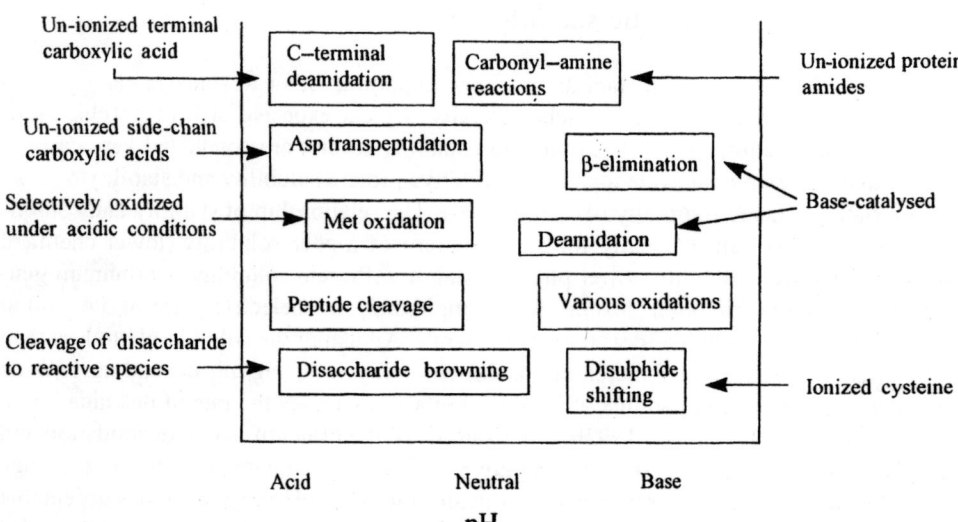

Figure 8.1 Protein reactions as a function of pH (courtesy of Dr Lee Kirsch).

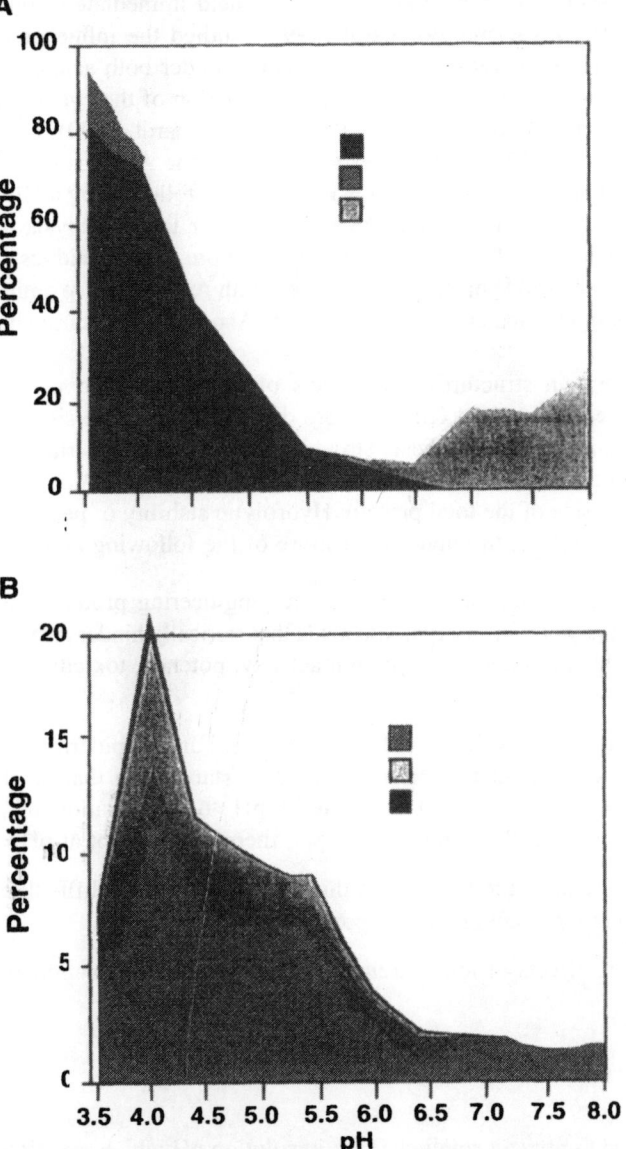

Figure 8.2   Effect of pH on deamidation and polymerization of insulin: chemical transformation during storage (25°C, 12 months) of rhombohedral bovine insulin crystals (0.7% NaCl, 0.2% phenol) as a function of pH. (A) Formation of the hydrolysis products monodesamido and didesamido insulins and the insulin split product (A8–A9). (B) Formation of covalent dimers and oligomers. Reprinted with permission from Brange and Langkjær (1993); ©Plenum Press.

Hydrolysis or deamidation occurs with peptides and proteins containing susceptible Asn and Gln amino acids, the only two amino acids that are primary amines. The side-chain amide linkage in a Gln and Asn residue may undergo deamidation to form free carboxylic acid. Deamidation can be promoted by a variety of factors including high pH, temperature and ionic strength (Manning *et al.*, 1989). The rate of deamidation is affected by amino acid sequence, particularly the amino acid immediately following the Asn or Gln amino acids. Oliyai and Borchardt (1994) studied the influence of primary amino acid sequence on the degradation of Asp residues under both acidic and alkaline conditions. As expected, the rate of intramolecular formation of the cyclic imide, the first step in the hydrolytic degradation pathway (Patel and Borchardt, 1990), was most affected by the size of the amino acid on the C-terminal side of the Asp residue. Deamidation rates for peptides will be highest when Asn is immediately followed by a Gly amino acid since Gly has no side-chain, thus no opportunity to hinder the hydrolysis reaction sterically. C-terminal substitution of Gly with increasingly more bulky residues (Ser, Val) inhibits the amount of cyclic imide produced. However, with respect to Asp amide bond hydrolysis with adjacent amino acids either before or after Asp, such structural changes had little or no effect.

For larger protein structures, the effects of adjacent amino acid sequences on the deamidation rates of Asn and Gln are more difficult to estimate simply due to the three-dimensional complexities of these structures. However, it certainly is intuitive that adjacent amino acids and their size will have some effect on Asn and Gln deamidation regardless of the size of the total protein. Hydrolytic stability of peptides and proteins can be minimized, therefore, through one or more of the following approaches.

1   Optimization of amino acid sequence, i.e. engineering protein structures to remove unstable amino acids or insert amino acids that sterically hinder Asn or Gln deamidation, as long as this does not affect protein activity, potency, toxicity, or any other quality attribute.

2   Formulate at optimal solution pH. For example, human epidermal growth factor 1–48 demonstrates some interesting pH-dependent stability in that at pH <6 succinimide formation at Asp[11] is favoured, while at pH >6 deamidation of Asn[1] is favoured (Senderoff *et al.*, 1994). The optimal pH, therefore, is right at pH = 6.

3   Store at low temperatures, although this will always create difficulties during distribution and long-term storage of the product.

4   Optimize the effects of ionic strength (to be discussed in section 8.2.2).

8.2.1 *Buffers*

Buffers are used to prevent small changes in solution pH which can affect protein solubility and stability. Buffers are composed of salts of ionic compounds, the most common of which are acetate, citrate, and phosphate. Buffer systems acceptable for use in parenteral solutions are listed in Table 8.2.

The proper selection of buffer type and concentration is determined by performing solubility and stability studies as a function of pH and buffer species. Normally, the pH of maximum solubility is not the pH of maximum stability. However, a pH range that is a good compromise between solubility and stability can be selected and maintained with the proper selection of the appropriate buffer component.

In the pH range of 7–12, buffer concentration can have a significant effect on the rate of deamidation indicating general acid–base catalysis. Generally, deamidation is much slower at acidic pH than at neutral or alkaline pH. ACTH deamidation in the pH range of 7–11 is catalysed by increasing buffer concentrations, whereas there is no buffer catalysis

Table 8.2 Buffers used in protein formulations

Buffer system	pKa	pH Range of use
Acetate	4.8	2.5–6.5
Carbonate	6.4	5.0–11.0⁻
Citrate	3.14, 4.8, 5.2	3.0–8.0
Glutamate	2.2, 4.25, 9.67	8.2–10.2
Glycinate	2.4, 9.8	6.5–7.5
Histidine	1.8, 6.0, 9.2	6.2–7.8
Lactate	3.8	3.0–6.0
Maleate	1.92, 6.23	2.5–5.0
Phosphate	7.2 (pKa 2)	3.0–8.0
Succinate	4.2, 5.64	4.8–6.3
Tartrate	2.93, 4.23	3.0–5.0
Tris	6.2 (pKb 7.8)	6.8–7.7

at pH 5–6.5 (Patel, 1993). However, insulin deamidation at the A21 position predominates at acidic pH while deamidation at B3 predominates at neutral pH (Brange *et al.*, 1992b).

Several potential problems are associated with using buffers in parenteral solutions. For example, it should not be expected that in large-scale manufacturing the compounded solution containing the buffer will always result in the exact pH specified. Dilute solutions of strong acids (hydrochloric acid) or bases (sodium hydroxide) are usually required to 'fine-tune' the final solution pH. Excessive use of the pH adjustment solutions may alter the buffer capacity and ionic strength of the buffered solution (Niebergall, 1990). Increasing buffer capacity to control pH better could significantly increase ionic strength which, in turn, may cause increased potential of pain upon injection due to the increase in solution osmolality.

General acid and/or general base buffer catalysis can accelerate the hydrolytic degradation of the protein. An example is shown in Figure 8.3 (Yoshioka *et al.*, 1993) where the inactivation rate of β-galactosidase increased with increasing concentrations of phosphate buffer up to 0.5 M, then decreased, presumably related to higher buffer components causing a reduction in water mobility. Cleland *et al.* (1993) cite several examples where the rate of protein deamidation was markedly dependent on the buffer anion. Capasso *et al.* (1991) compared the deamidation rate of a small peptide using different buffers and found that the peptide was most unstable in a phosphate buffer and most stable in Tris buffer. Wang *et al.* (1996) found that buffer type and concentration affected aggregation of basic fibroblast growth factor depending on pH. At pH 5, aggregation increased as citrate buffer concentration increased. Citrate buffer at pH 3.7 caused aggregation, whereas acetate buffer at pH 3.8 did not. At pH 5.5–5.7, phosphate, acetate and citrate buffers all showed similar aggregation rates.

8.2.2 *Ionic strength*

Ionic strength is a measure of the intensity of the electrical field in a solution. It depends on the total concentration of ions in solution and the valence of each ion. The ionic strength of a 0.1 M solution of sodium chloride is 0.1. The ionic strength of a 0.1 M

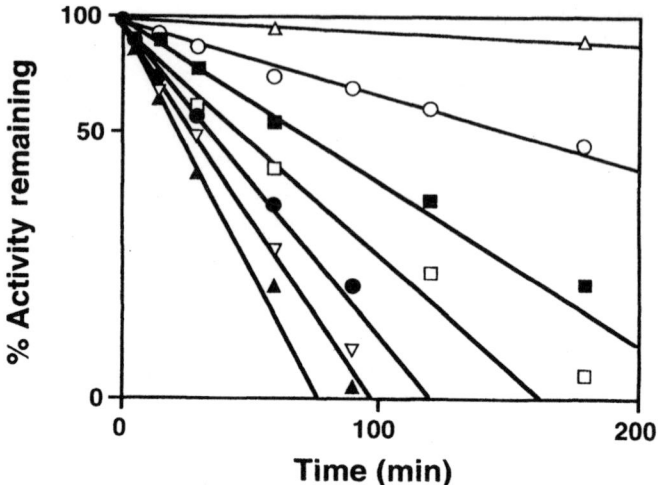

Figure 8.3 Inactivation of β-galactosidase in pH 7.4 phosphate buffer solution at 50°C, as a function of phosphate buffer concentration: (Δ) 10, (O) 50, (□) 100, (∇) 200, (▲) 500, (●) 700, (■) 900 mM. The concentration of β-galactosidase was 0.1 mg/ml. Reprinted with permission from Yoshioka *et al.* (1993); ©Plenum Press.

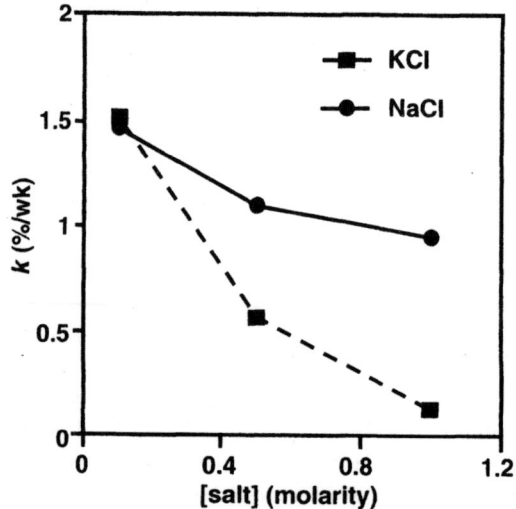

Figure 8.4 Effect of salt concentration on rAAT solution stability. Reprinted with permission from Vemuri *et al.* (1993); ©Plenum Press.

solution of sodium sulphate is 0.3, because sulphate ions have a valence of 2 added to the valence of 1 for the sodium ions. Ionic strength may have an effect on protein stability in solution. The Debye-Huckel theory predicts that increased ionic strength would be expected to decrease the rate of degradation of oppositely charged reactants, and increase the rate of degradation of similarly charged reactants.

Ionic strength will affect the stability of a protein, but in which direction (increase or decrease) differs with different proteins. For example, increasing ionic strength will increase the stability of recombinant alpha$_1$ antitrypsin (Vemuri *et al.*, 1993) (Figure 8.4).

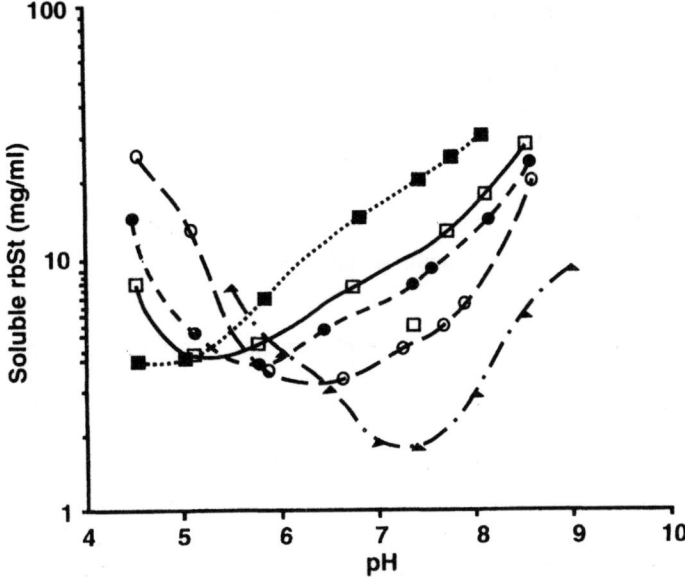

Figure 8.5 Effect of ionic strength on rbSt solubility–pH profiles. Ionic strength μ increased with increasing NaCl: (■) μ = 0.24; (□) μ = 0.12; (●) μ = 0.072; (○) μ = 0.024; (▲) μ = 0.008. Reprinted with permission from Davio and Hageman (1993); ©Plenum Press.

Conversely, increasing ionic strength will increase the rate of deamidation of human growth hormone (Pearlman and Bewley, 1993) and bovine somatotropin (Davio and Hageman, 1993) (Figure 8.5).

8.3 Optimizing oxidative stability

Proteins containing methionine, cysteine, cystine, histidine, tryptophan, and tyrosine may be sensitive to oxidative and/or photolytic degradation depending on the conformation of the protein and resultant exposure of these amino acids to the solvent and environmental conditions such as presence of oxygen, light, high temperature, metal ions, and variours free radical initiators. Oxidation of sulphydryl-containing amino acids (e.g. methionine and cysteine) will lead to disulphide bond formation and loss of biological activity. The free thiol group that is present in a Cys residue of any native biologically active protein not only may oxidize to produce an incorrect disulphide bridge, but also can result in other degradation reactions such as alkylation, addition to double bonds and complexation with heavy metals.

Human growth hormone, chymostrypsin, lysosyme, parathyroid hormone, human granulocyte-colony stimulating factor, insulin-like growth factor I, acidic and basic fibroblast growth factors, relaxin, the monoclonal antibody OKT3, interleukin 1β, and glucagon are examples of proteins that will degrade by this mechanism. Haemoglobin, with its oxygen-carrying properties dependent on the reduced state of ferrous iron, is very sensitive to oxidation and, as a commercial product in the deoxy state, must contain antioxidants to maintain stability of the haem groups (unpublished results). Cleland *et al.*

(1993) list 61 different proteins that can be oxidized with varying degrees of loss of biological activity.

For protection against oxidation, choice of an effective antioxidant is one of several precautions that must be practised in formulation development and final product manufacture. Other factors that contribute to protein stability against oxidative degradation include:

- preparation and storage at low temperatures
- use of chelating agents to eliminate metal catalysis
- increasing ionic strength
- elimination of peroxide and metallic contaminants in formulation additives
- protection from light
- awareness of possible interaction of light exposure and phosphate buffer in forming free radicals (Fransson and Hagman, 1996)
- replacing oxygen with nitrogen or argon during manufacturing
- removing oxygen from the headspace of the final container
- formulation at the lowest pH possible while maintaining desired protein solubility and hydrolytic stability, since there is an inverse correlation between oxidative stability and pH (Akers, 1982)
- use of a container/closure system that allows no oxygen transmission through the package during distribution and storage
- assuring that phenolic or other oxidizing cleaning agent residues are minimal in the production environment, including the freeze-dry chamber (Kirsch *et al.*, 1993).

Fransson (1997) published an excellent paper on methionine oxidation and covalent aggregation in aqueous solution as studied with human insulin-like growth factor-I (hIGF). Oxidation of methionine 59 was catalysed by light and by ferric ions in combination with EDTA. In this example, Fransson suggests that EDTA actually enables ferric ions to be active by stabilizing the transfer of electrons from ferric ions to ferrous ions. Figure 8.6 shows the relationship of EDTA and ferric ion in the oxidation of methionine in hIGF-I in aqueous solution in the presence of light. Methionine in this protein is radicalized by light, then oxidized to methionine sulphoxide. Light may also trigger the generation of hydroxyl radicals by decomposition of water that may oxidize the methionine.

Formulators should be aware of the potential for polysorbate 80 to affect adversely the oxidative stability of proteins. Polysorbate 80 is a commonly used surface active agent in protein formulations to minimize surface aggregation problems. However, it has the tendency to produce peroxides which can oxidize methionine and cysteine residues. This phenomenon was reported in studies involving formulation development of Neupogen® (Herman *et al.*, 1996) and recombinant human ciliary neurotrophic factor (Knepp *et al.*, 1996).

8.3.1 *Antioxidants*

There are several choices of antioxidant that can be used in protein formulations (see Table 8.3). Those which have been used most frequently are ascorbic acid, salts of sulphurous acid (sodium bisulphite, sodium metabisulphite or sodium thiosulphate), and

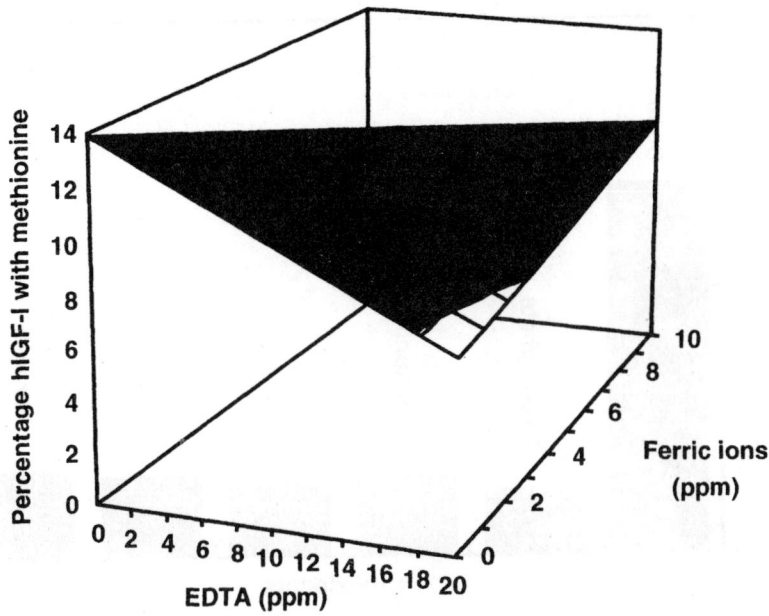

Figure 8.6 Response surface for the oxidation of methionine 59 in hIGF-I in aqueous solution at 25°C. Reprinted with permission from Fransson (1997); ©American Chemical Society and American Pharmaceutical Association.

Table 8.3 Antioxidants and chelating agents for protein formulations

Antioxidants (all water-soluble)	*Normal % used in formulation*
Ascorbic acid (including isoascorbic acid, sodium ascorbate)	0.1–1.0
Sulphurous acid salts (sodium bisulphite, sodium metabisulphite, sodium sulphite)	0.1–0.5
Thioglycerol	0.1–0.5
Thioglycollic acid	0.05–0.2
Cysteine hydrochloride	0.1–0.5
Chelating agents (all water-soluble)	
Ethylenediaminetetraacetic acid (EDTA) (usually the disodium salt)	0.05–0.1
Citric acid/sodium citrate	0.02–1

thiols such as thioglycerol and thioglycolic acid. Dithriothreitol, reduced glutathione, acetylcysteine, mercaptoethanol, and thioethanolamine are thiols which usually oxidize too readily to be of practical use in pharmaceutical formulations requiring long-term storage.

Lam *et al*. (1997b) studied the inhibitory effect of various antioxidants on the oxidation of recombinant monoclonal antibody HER2. As shown in Figure 8.7, sodium thiosulphate, methionine, catalase, or platinum, as antioxidants, were effective in reducing the oxidation

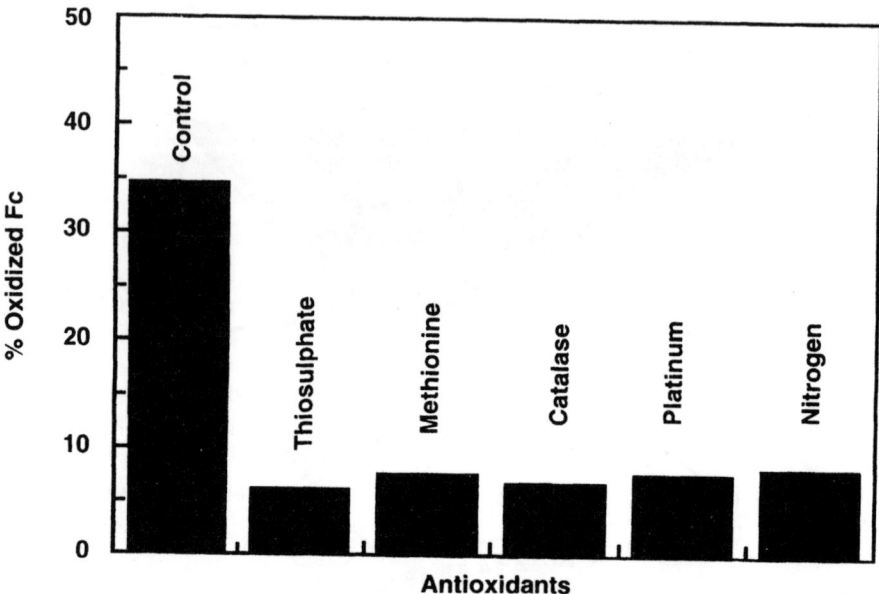

Figure 8.7 Methods to prevent temperature-induced oxidation of rhuMAb HER2 in formulation containing 5 mM sodium acetate, pH 5.0, 147 mM NaCl, and 0.01% polysorbate 20. Reprinted with permission from Lam *et al.* (1997b); ©American Chemical Society and American Pharmaceutical Association.

of the antibody. Figure 8.7 also shows that replacing air with nitrogen in sample vials was also effective in reducing antibody oxidation. The authors proposed that thiosulphate and methionine either inhibit free-radical induced oxidation by terminating the chain reaction or compete with the methionine residues in the antibody for reaction with the free hydroxyl radicals. Catalase and platinum serve as free-radical scavengers.

Precautions must be applied when considering ascorbic acid as an antioxidant in protein formulations. Li *et al.* (1993) found that ascorbate in the presence of Fe^{3+} and oxygen actually induces the oxidation of methionine in small model peptides. Ascorbate is a powerful electron donor in that it is readily oxidized to dehydroascorbate. It also generates highly reactive oxygen species such as hydrogen peroxide and peroxyl radicals. These, in turn, will accelerate the oxidation of methionine. Phosphate buffer compared to other buffer systems (e.g. Tris, HEPES, and MOPS) accelerated the degradation of methionine in the presence of ascorbic acid. The addition of EDTA did not enhance stability even though ferric ion and other transition metals were components in the formulation, either purposely added or as trace components of the buffer and peptide. This prooxidant effect of ascorbate methionine oxidation was concentration-dependent and occurred most readily at pH 6–7.

It is also known that sodium bisulphite can cause stability problems with certain drugs and proteins. For example, bisulphite will rapidly destroy insulin (Asahara *et al.*, 1991). These authors studied the compatibility of human insulin in solutions containing sodium bisulphite since human insulin and sodium bisulphite were being used in some intravenous admixtures in their hospital practice. Bisulphite was found to cleave the interchain disulphide bonds of insulin. The addition of glucose to these solutions stabilized human insulin in the presence of sodium bisulphite, with the stabilization postulated to be the formation of a bisulphite–glucose adduct in solution.

8.3.2   Chelating agents

Chelating agents (see Table 8.3) are used in protein formulations to aid in inhibiting free-radical formation and resultant oxidation of proteins caused by trace metal ions such as copper, iron, calcium, manganese, and zinc. While organic buffer salts, such as sodium citrate, have some capability of binding trace heavy metal contaminants in protein solutions, the major chelating agent used is disodium ethylenediaminetetraacetic acid (DSEDTA). The concentration of DSEDTA is usually very small, e.g. ≤0.04%. DSEDTA tends to dissolve slowly and is usually among the first formulation ingredients to be dissolved during compounding before adding other ingredients, including the protein.

EDTA should not be used in formulations of metalloproteins such as insulin and fibrolase (Pretzer *et al.*, 1993), as the chelating agent will attack the metal that is part of the stable conformation of the protein.

As already discussed, EDTA can also accelerate the oxidative degradation of methionine in hIGF-I solutions (Fransson, 1997). Thus, the formulator must not indiscriminately include EDTA in protein formulations, but must carefully determine that its presence aids in oxidative stabilization of the protein.

8.3.3   Inert gases

Inert gases are frequently used in production of protein dosage forms. The most commonly used inert gas is nitrogen. Other inert gases which can be used, although not very often primarily because of expense, include argon and helium. Argon, however, has been shown to be more efficient in displacing oxygen because it is heavier than air and will more readily stay in the vial compared to nitrogen, which is lighter than air (Harwood *et al.*, 1993). Inert gases are normally used in protein formulation and production in two ways.

1   Added to water and compounding solutions prior to aseptic filtration to saturate the solution and minimize the level of dissolved oxygen. However, oxygen is never completely displaced with an inert gas when the solution is sparged. Many manufacturers use a dual needle which permits filling of a liquid and purge gas at the same time.

2   Introduced into the headspace of a filled vial right before the vial is stoppered with a rubber closure, thereby, theoretically, displacing oxygen in the headspace. Again, a dual needle can be used to fill solution and purge gas into the final container at the same time.

The inert gas must be high-quality grade and must be sterilized, usually with a 0.45 μm hydrophobic membrane filter. The integrity of the gas filter is tested before and after use by diffusion flow methods.

8.3.4   Packaging and oxidation

All the appropriate formulation and processing procedures can be in place for stabilizing protein solutions against oxidation, but if the packaging system is inadequate from an integrity standpoint, the product will readily degrade. Most protein products are packaged in glass vials with rubber closures. The rubber–glass interface and the oxygen transmission coefficient of the rubber closure will dictate the quality of the container/closure system. The integrity of the rubber–glass interface can be tested by a variety of techniques including helium mass spectometry (Kirsch *et al.*, 1997a, 1997b, 1997c). This technique is used to validate the integrity of the specific type of rubber closure with the specific type of glass vial and, then, can be used to check integrity of a representative number of product vials per lot.

Oxygen transmission coefficients are determined for a particular rubber closure formulation by the rubber closure manufacturer. Rubber formulations having the lowest oxygen transmission coefficients are the synthetic butyl and halobutyl types. The formulator should determine from the rubber manufacturer how the halobutyl rubber is cured (shaped, moulded) since common curing agents are zinc oxide, aluminium, and peroxide, which potentially can leach out of the rubber formulation with time and catalyse oxidative degradation (Boyett and Avis, 1976, Milano *et al.*, 1982, Danielson *et al.*, 1984, Liebe, 1995).

8.3.5 *Other chemical stabilizers*

Sugars and polyols, such as ethylene glycol, glycerol, glucose, and dextran, at high concentrations, can inhibit the metal-catalysed oxidation of human relaxin (Li *et al.*, 1996). All but dextran act as chelating agents in complexing transition metal ions whereas dextran, which has a higher binding affinity to metal ions and undergoes depolymerization in a metal-catalysed oxidation, protects relaxin by a radical scavenging mechanism.

Mannitol has been shown to inhibit the iron-catalysed oxidation of Met-containing peptides (Li *et al.*, 1995). Mannitol is the most commonly used excipient in freeze-dried formulations, often serving a dual role as a bulking agent and a stabilizer.

Fibroblast growth factors, both acidic and, basic, possess nearly identical three-dimensional structures of 12 antiparallel β-strands arranged with approximate three-fold internal symmetry (Tsai *et al.*, 1993). Acidic fibroblast growth factor formulation design was studied by Tsai *et al.* (1993), who found that its tendency to aggregate in solution was inhibited by a variety of polyanionic additives such as inositol hexasulphate or sulphated β-cyclodextrin and by a number of commonly used exicipients such as sucrose, dextrose, trehalose, glycerol, and glycine. The polyanionic additives interacted with the polyanion binding site of the protein while the non-specific agents were thought to be preferentially hydrated. In all cases, these interactions between acidic fibroblast growth factor and various excipients resulted in an increase in the protein's T_m, the midpoint of the temperature of the transition from the folded to unfolded protein. Basic fibroblast growth factor formulation design was studied by Shahrokh *et al.* (1994a), who found that its major degradation pathway involves not only aggregation and precipitation, but also a succinimide replacement of aspartate at position 15 of the protein sequence. Adjusting solution pH from 5 to 6.5 and storage at low temperatures will help to avoid this reaction.

A variety of co-solvents can stabilize proteins in solution because the co-solvent is preferentially excluded from surface interaction with the protein (Arakawa *et al.*, 1991). Co-solvents behaving this way include glycerol and sorbital. Polyethylene glycol is also preferentially excluded from the protein, yet will still denature or destabilize proteins in solution.

8.4 *Optimizing physical stability*

Unlike small molecules, where physical instability is rarely encountered except for poorly water-soluble compounds, proteins, because of their unique ability to adopt three-dimensional forms, tend to undergo a number of structural changes, independent of chemical modifications. Physical instability of proteins is sometimes a greater cause for concern

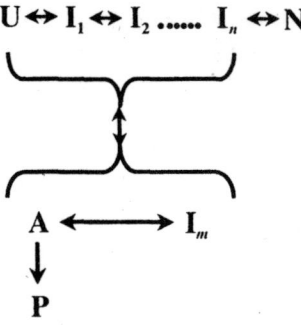

Figure 8.8 Aggregation scheme. The potential folding pathways for an unfolded protein (U) are shown assuming that it is refolded *in vitro* by dilution with a simple buffer (no additives). The unfolded protein will fold to form an intermediate structure (I_1) which has some secondary structure. This intermediate and others in the folding pathway ($I_2 \ldots I_n$) often associate to form soluble aggregates (A). These soluble aggregates can agglomerate to form large irreversible precipitates (P) that must be resolubilized in denaturants. All intermediates as well as the unfolded protein can form misfolded or off-pathway intermediates (I_m) that can reduce the yield of native protein by becoming a kinetic trap for the preceding species. Some intermediates in the later portion of the pathway obtain a native-like conformation (I_n) and eventually assume the native state (N). Reprinted with permission from Cleland (1993); ©American Chemical Society.

and is more difficult to control than chemical instability. All protein structures are hydrophobic to some extent. Many proteins, particularly when exposed to stressful conditions, e.g. extremes in temperature, will unfold such that the hydrophobic portions become exposed to the aqueous environment. Such exposure will promote aggregation or self-association, possibly leading to physical instability and loss of biological activity since the interaction with the receptor site requires folded structures with correct conformation. The relationship of the different pathways of physical destabilization of proteins is shown in Figure 8.8. The physical stability of proteins is also dealt with in Chapter 6.

8.4.1 Denaturation

Denaturation is unique to proteins and occurs when their native quarternary, tertiary and, frequently, secondary structure is disrupted. Denaturation can lead to unfolding and the unfolded polypeptide chain may undergo further reactions. Such inactivation could be association with surfaces and/or interaction with other protein molecules, leading to aggregation and precipitation. Denaturation is of two types: (1) reversible denaturation, caused by temperature or exposure to chaotropic agents (urea, guanidine hydrochloride) where, if the denaturing condition is removed, the protein will regain its native state and maintain its activity, and (2) irreversible denaturation where the protein, once unfolded, will not regain its native form and activity. However, there are several instances where a protein which is 'irreversibly denatured' is returned to its native state by the use of denaturant followed by dialysis. One example is T4 lysozyme, where its lost activity can be restored by renaturation with guanidine hydrochloride (Wetzel *et al.*, 1988). When a protein recovers its activity by the addition of denaturants such as guanidine hydrochloride or urea and by subsequent dialysis, such a process may indicate the influence of events such as aggregation, precipitation, and adsorption. Regardless of the observed phenomenon, the approach is of little use in solution formulation design.

For stabilizing proteins against denaturation in solution, Arakawa and Timasheff (1982, 1984) and Arakawa *et al.* (1993) have shown that reversible denaturation can be decreased by the use of additives such as salts that bind to non-specific binding sites on the proteins. Dahlquist *et al.* (1976) demonstrated increased thermal stability of thermolysin by the binding of ions to specific sites on the protein. Roe *et al.* (1988) showed the ability of Zn $(NO_3)_2$ to increase significantly the thermal stability of superoxide dismutase. Gekko and Timasheff (1981) concluded that the preferential hydration of proteins observed at all conditions in the presence of a glycerol–water mixed solvent system is a prerequisite for stabilizing the native structure of several globular proteins. Pace and Grimsley (1988) found the stability of ribonuclease T_1 to increase in the presence of 0.1M NaCl, $MgCl_2$ and Na_2HPO_4, respectively. Through genetic engineering they were able to introduce appropriate amino acid substitution in ribonuclease T_1 creating specific cation/anion binding sites on the protein. The stability profile of ribonuclease T_1 was enhanced considerably with this approach. Pantoliano *et al.* (1988) successfully introduced negatively charged side-chains such as Asp in the vicinity of the weak Ca^{2+} binding site of subtilisin. Such modifications caused an increase in binding affinity for Ca^{2+}, thereby increasing the thermal stability of subtilisin. These engineering approaches are acceptable provided that protein activity and other quality aspects are not adversely affected.

8.4.2 *Protein aggregation*

Aggregation of peptides and proteins is caused mainly by hydrophobic interactions that eventually lead to denaturation. When the hydrophobic region of a partially or fully unfolded protein is exposed to water, this creates a thermodynamically unfavourable situation due to the fact that the normally buried hydrophobic interior is now exposed to a hydrophilic aqueous environment. Consequently, the decrease in entropy from structuring water molecules around the hydrophobic region forces the denatured protein to aggregate, mainly through the exposed hydrophobic regions. Thus, solubility of the protein may also be compromised. In some cases self-association of protein subunits, either native or misfolded, may occur under certain conditions and this may lead to precipitation and loss in activity (Mitraki and King, 1989; Shahrokh *et al.*, 1994b; Brange *et al.*, 1992b; Brange and Langkjaer, 1993; Silvestri *et al.*, 1993). The protein, antithrombin, aggregates when denatured with guanidine hydrochloride which proceeds through an intermediate partially-unfolded state. It is possible to return antithrombin to its native state by dialysing the partially unfolded form which aggregates slowly. However, once aggregation occurs, the native state cannot be reformed by this approach (Fish *et al.*, 1985). Irreversible aggregation which occurs due to denaturation can be prevented by the use of surfactants, polyols, or sugars. The use of surfactants is elaborated in section 8.4.5.

Factors that affect protein aggregation in solution generally include protein concentration, pH, temperature, other excipients, and mechanical stress. Some factors (e.g. temperature) can be more easily controlled during compounding, manufacturing, storage and use than others (e.g. mechanical stress). Formulation studies will dictate appropriate choice(s) of pH and excipients that will not induce aggregation and/or, in fact, will aid in the prevention of aggregation. Protein concentration is dictated by the required therapeutic dose and, depending on what this concentration is, will determine whether the potential for higher associated states (dimers, tetramers, etc.) exists, which can then lead to aggregation in solution. Careful studies must be done during formulation development to determine what factors influence protein aggregation and then how these factors can be eliminated or controlled.

The desire to identify stable solution preparations of insulin for use in novel delivery systems such as continuous infusion pumps has led to the development of test methodology for assessing the impact of various additives on physical stability. Based on the known factors influencing protein aggregation and the requirements of such applications, physical stability has been evaluated using thermomechanical procedures involving agitation or rotation of insulin solutions at elevated temperature. In some cases, Teflon beads have been added to the solutions to introduce hydrophobic surfaces in order to avoid complications with air headspace (Sluzky *et al.*, 1991). Turbidity resulting from aggregation is usually determined as a function of time by visual inspection or light scattering analysis. Alternatively, reductions in the soluble protein content due to precipitation can be quantitated by HPLC assay as a function of time. Relative stability is defined by the length of time a preparation remains on the test without showing a change in either parameter. It should be noted that the greatest difficulty in applying such testing strategies is in interpreting the experimental data and correlating them in a practical way to 'real life' conditions that the formulation may actually experience. Nevertheless, regulatory agencies may request data from such testing to support dating periods or other product claims. Physical stress testing, however, is more appropriately used as a development screening tool to identify the capability of various additives to prevent aggregation, as we shall discuss. Since the principles involved in physical stress testing studies are fairly similar, we will only summarize some of the methodology and formulation strategies described for insulin.

Brange and Havelund (1983) report the effect of carbohydrate additives on the physical stability of neutral insulin formulations in vials subjected to shaking (amplitude 5 cm, frequency 100 rev/min) at 41°C. While some of the carbohydrates improved physical stability, formulations containing these additives also demonstrated a reduction in both chemical and biological stability. However, the addition of dissociable calcium salts improved physical stability without impacting other quality attributes. A variety of additives to insulin solutions including bacteriostatic agents, non-ionic detergents, anionic detergents, physiologic compounds and extracts, salts, buffers, and alcohols were examined in other work (Lougheed *et al.*, 1983). The physical stability of solutions prepared with the various additives and stored in vials with an air headspace was assessed at defined time intervals after exposure to 37°C and either shaking (130 cpm in the horizontal plane) or rotation (20 cm from the axis of a wheel rotating at 60 rev/min in the vertical plane). The formulation compatibility with certain device materials was also examined by inserting samples directly into the solution being tested. Additives that reduced solvent polarity (e.g. anionic and non-ionic surfactants with long hydrophobic chains or alcohols) were most effective in preventing insulin aggregation. However, the concentrations of these additives necessary to achieve the beneficial effect on physical stability may not be appropriate for pharmaceutical preparations. In similar, but somewhat more sophisticated stress-testing experiments, Sluzky *et al.* (1992) reported on the mechanism of additive stabilization. It was concluded that the presence of surfactants enhanced physical stability by reducing unfavourable interfacial interactions between insulin and hydrophobic surfaces rather than by lowering surface tension. Thurow and Geisen (1984) examined the physical stability of insulin and other protein solutions containing polypropylene glycol/polyethylene glycol block polymers. Samples were filled into ampoules with an air headspace and subjected to rotation (20 cm from the axle, 60 rev/min) at 37°C. The additive, Genapol PF-10, was shown to have a stabilizing effect.

8.4.3 *Adsorption*

Proteins exhibit a certain degree of surface activity, i.e. they adsorb to surfaces due to their innate nature as amphiphilic polyelectrolytes. Consequently biological activity may be either reduced or totally lost if such adsorption occurs during manufacturing, storage, or use of the final product. The process of adsorption in the case of proteins depends on protein–protein interactions, time, temperature, pH and ionic strength of the medium and the nature of the surface (Absolom *et al.*, 1987). Norde (1995) reviewed the general principles underlying protein adsorption from aqueous solution onto a solid surface. Interactions that determine the overall adsorption process between a protein and a surface include redistribution of charged groups in the interfacial layer, changes in the hydration of the sorbent and the protein surface, and structural rearrangements in the protein molecule. Surface denaturation which commonly takes place at the liquid–solid and liquid–air interfaces has been shown by Lenk *et al.* (1989) to involve conformational changes such as loss of α-helices to ß-sheets and certain random structures. These structural changes, which are determined by the nature of the interfaces, are similar to those observed with the aggregation phenomenon caused by heat, high pressure, or chemical denaturants. In the case of proteins, sources such as the polymer of the membrane filter, the administration set, agitation that occurs during the purification process, and the method of manufacture are known or at least suspected to cause surface denaturation. Strategies often used to overcome protein denaturation due to adsorption are:

1   increase protein concentration during filtration and/or using extra volume to saturate the filter with protein solution

2   modify (e.g. siliconize) the surface of the glass containers, providing a resistant barrier to protein–surface interaction

3   decrease the rate of mixing when it is known that shear will affect protein adsorption

4   add excipients such as surfactants that have higher surface activity

5   add macromolecules such as albumin and gelatin (although one must realize the increased concern regarding these natural materials because of their potential for pyrogenic and/or BSE[1] contamination).

The literature is replete with problems encountered while delivering insulin because of its ability to adsorb onto the surfaces of delivery pumps and glass containers and to the inside of the intravenous bags (James *et al.*, 1981; Iwamoto *et al.*, 1982; Mitrano and Newton, 1982; Twardowski *et al.*, 1983a, 1983b, 1983c; Lougheed *et al.* 1983; Sato *et al.*, 1984; Brennan *et al.*, 1985). Insulin adsorption usually is finite once binding sites are covered, and such adsorption is usually not clinically significant.

There are several approaches to minimize or overcome protein adsorption. Adsorption to filters and tubing can often be overcome by first saturating the surfaces with excess protein solution, then discarding the filtrate or wash. This approach will waste valuable protein and so may not be a good choice. In some cases, adsorption can be minimized by using certain additives. For example, Oshima (1989) showed that surface denaturation of chymotrypsin can be prevented by the addition of 0.1 M NaCl as well as by coating

---

[1] BSE is bovine spongiform encephalopathy, also known as 'mad cow disease', a chronic degenerative disease affecting the central nervous system of cattle. First diagnosed in Great Britain in 1986, this transmissible disease could be a contaminant in bovine-sourced pharmaceutical excipients such as gelatin, bovine serum albumin, and polysorbate 80.

the surface of the container with either lecithin or BSA. Adsorption of urokinase to glass can be prevented by the addition of 0.25% gelatin to the container surface (Patel, 1990). Calcitonin adsorption to glass syringes can be prevented by benzalkonium chloride or benzethenium chloride, presumably due to coating of glass silanol groups by these positively charged antimicrobial preservatives (Kakimoto *et al.*, 1985). Johnston (1996) reported that adsorption of recombinant human granulocyte to glass, polyvinyl chloride, and polypropylene surfaces can be minimized by the addition of certain additives such as polysorbate 20, polysorbate 80, or Pluronic F-127.

8.4.4 *Precipitation*

Precipitation of proteins occurs subsequent to denaturation and is a consequence of aggregates combining to form large particles. The mechanism of aggregation leading to precipitation is beyond the scope of this chapter; the reader is referred to Glatz (1992) for more detailed information. Brennan *et al.* (1985) have established the tendency of insulin to precipitate when loaded into a long-term infusion device, making it cumbersome for delivery.

8.4.5 *Surfactants*

Surfactants are surface active agents that can exert their effect at surfaces of solid–solid, solid–liquid, liquid–liquid, and liquid–air because of their chemical composition, containing both hydrophilic and hydrophobic groups. These materials reduce the concentration of proteins in dilute solutions at the air–water and/or water–solid interfaces where proteins can be adsorbed and potentially aggregated. Surfactants can bind to hydrophobic interfaces in protein formulations and packaging. Glass, rubber, or plastic adsorption of proteins is well documented (Christensen *et al.*, 1978; Hirsch *et al.*, 1977, 1981; Anik and Hwang, 1983; Suelter and DeLuca, 1983; Wang and Chien, 1984; Chawla *et al.*, 1985; Dong *et al.*, 1987; Seres, 1990; Johnston, 1996). Proteins on the surface of water will aggregate, particularly when shaken, because of unfolding and subsequent aggregation of the protein monolayer.

Surfactants can denature proteins, but can also stabilize them against surface denaturation. Generally, ionic surfactants can denature proteins. However, non-ionic surfactants usually do not denature proteins even at relatively high concentrations (1% w/v) (Cleland *et al.*, 1993). Most parenterally acceptable non-ionic surfactants come from either the polysorbate (sorbitol–polyethylene oxide polymers) or polyether (polyethylene oxide–polypropylene oxide block co-polymers) groups. Polysorbate 20 and 80 and sodium dodecyl sulphate are effective and acceptable surfactant stabilizers in marketed protein formulations (see Table 8.4). However, other surfactants used in protein formulations for clinical studies and/or found in the patent literature include Pluronic F-68 and other polyoxyethylene ethers (e.g. the 'Brij' class) (Wang and Hanson, 1988).

Surfactants are well known to prevent the denaturation and aggregation of insulin (Lougheed *et al.*, 1983; Sato *et al.*, 1984; Chawla *et al.*, 1985). However, the choice of surfactant and the final concentration optimal for stabilization is quite dependent on a variety of factors including other formulation ingredients, protein concentration, headspace in the container, the type of container, and test methodology.

Table 8.4 Examples of commercial protein solution formulations containing surface active agents

Generic name	Brand name	Manufacturer	Surface active agent in formulation
Aldesleukin	Proleukin	Chiron	Sodium dodecyl sulphate, 0.018% (following reconstitution)
Filgrastim	Neupogen	Amgen	Polysorbate 80, 0.004%
Interferon gamma 1B	Actimmune	Genentech	Polysorbate 20, 0.01%
Muromonab CD3	Orthoclone	Ortho Biotech	Polysorbate 80, 0.1%

Recombinant human growth hormone will aggregate readily under mechanical and thermal stress. Aggregation from mechanical stress can be substantially reduced in the presence of surfactants (Katakam *et al.*, 1995). Mechanical stress may cause proteins to be more exposed to air–water interfaces where denaturation is more likely to occur than in the bulk phase of water. Surfactants will preferentially compete with proteins for accumulation at the air–water interface and keep the protein from undergoing interfacial denaturation resulting from mechanical stress. Pluronic F-68 and Brij 35 will stabilize hGH at their critical micelle concentrations (CMCs) (0.1% and 0.013%, respectively), while stabilization with polysorbate 80 requires a concentration of 0.1%, higher than the CMC value for polysorbate 80 of 0.0013%. The reasons for these differences in stabilizing concentrations are not clear, but simply reflect differences in interactions between different surfactants and proteins. It is interesting to note that these surfactants do not stabilize hGH from aggregation due to high temperature stress.

Further substantiation of the important role of surfactants, particularly polysorbate 80, in protecting proteins against surface-induced denaturation during freezing was reported by Chang *et al.* (1996). They found a strong correlation between freeze denaturation (quick freezing of the protein) and surface denaturation (shaking the protein in solution). Proteins that tend to denature under these conditions are protected by the addition of polysorbate 80 (0.1%). Other surfactants – Brij 35, Lubrol-px, Triton X-10, and even the ionic surfactant sodium dodecyl sulfate – also protected the protein from denaturation although these surfactants have not yet been approved for use in injectable formulations. The authors pointed out that surfactants may be needed to protect proteins from denaturation during the freezing step only, and that other stabilizers, e.g. sucrose, may be needed to protect the protein further during freeze-drying.

Bovine somatotropin (bSt) presents an example where surfactants are not effective in preventing protein aggregation and precipitation in solution at elevated temperature,[2] whereas other stabilizers such as sucrose are effective. Figure 8.9 shows the effect of polysorbate 80 on bSt precipitation at 54°C, where bSt is more stable without the presence of polysorbate 80 and increasing the amount of polysorbate 80 increases the extent of bSt precipitation. Interestingly, for bSt, increasing concentrations of sucrose have a positive effect on stabilizing bSt against aggregation and precipitation.

---

[2] While polysorbate 80 was not effective in stabilizing bSt at elevated temperature, it was effective when the applied stress was agitation. Also, the authors noted that polysorbate 80 destabilization of bSt was not observed at ambient or refrigerated temperatures as other decomposition pathways, e.g., deamidation, became more predominant at lower temperatures.

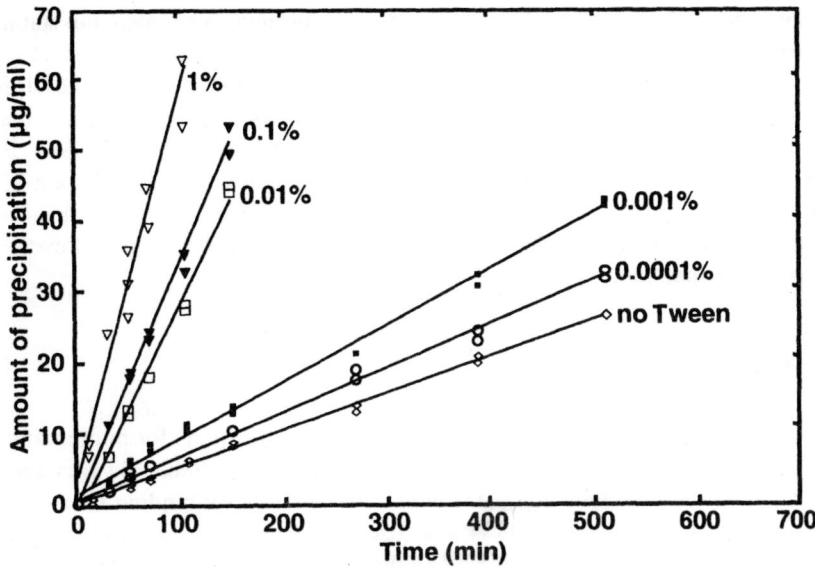

Figure 8.9 Effect of Tween 80 concentration on precipation of rbSt as a function of thermal stress at 54°C. Reprinted with permission from Hageman *et al.* (1993); ©Plenum Press.

Peroxides are known contaminants of non-ionic surfactants. Knepp *et al.* (1996) reported on the peroxide levels of polysorbate 80 obtained from different manufacturers using a colorimetric titration method. Levels ranged from less than 1 mEq/kg to more than 27 mEq/kg. They found that peroxide levels increased upon storage at ambient temperatures, probably due to headspace oxygen and/or the container–closure interface allowing ingress of air. Peroxides in polysorbate can result in oxidative degradation of proteins. Formulators need to screen sources of polysorbate 80 or other polymeric additives used in protein formulations for peroxide contamination and establish peroxide specifications for using the additive. Also, as a precaution, incorporation of an antioxidant can help to overcome the potential for non-ionic surfactants to serve as oxidative catalysts for oxygen-sensitive proteins.

Studies on protein–surfactant interactions are beginning to be published (Bam *et al.*, 1995) where electron paramagnetic resonance (EPR) spectroscopy is used to determine the binding stoichiometry of the surfactant to the protein and, thus, what potentially is the optimal amount of surfactant to use to stabilize the protein against surface denaturation and other physical instability reactions.

8.4.6 *Cyclodextrins*

Cyclodextrins are cyclic (α-1,4)-linked oligosaccharides of α-D-glucopyranose containing a relatively hydrophobic central cavity and hydrophilic outer surface (Loftsson and Brewster, 1996). Cyclodextrins come in a wide variety of structural derivatives, the most common being α-, β-, and γ-cyclodextrins, which consist of six, seven, and eight glucopyranose units, respectively. Two parenteral cyclodextrins are Encapsin™, a hydroxylpropyl-β-cyclodextrin, and Captisol™, a sulphobutyletherβ-cyclodextrin. They have been used widely for increasing the solubility stability and bioavailability of small drug molecules (Thompson, 1997). Peptides and proteins can also be stabilized in cyclodextrin complexes. Irwin *et al.* (1994) used β-cyclodextrins at a 25-fold excess to stabilize leucine enkephalin against enzymatic degradation in sheep nasal mucosa. Hydroxypropyl-β-cyclodextrin at a 1% concentration was shown to enhance the reconstituted solution stability of keratinocyte growth factor (Zhang *et al.*, 1995). Brewster *et al.* (1991) presented several examples (ovine growth hormone, interleukin-2, bovine insulin) where 2-hydroxypropyl-β-cyclodextrin enhanced the solubility and physical stability of proteins in solution. Johnson *et al.* (1994) used chemically modified cyclodextrins to solubilize a tripeptide.

8.4.7 *Albumin*

Serum albumin is a widely used stabilizer in protein formulations for minimizing protein adsorption to glass and other surfaces (Wang and Hanson, 1988, Edwards and Huber, 1992). Albumin preferentially competes with other proteins for binding sites on surfaces, but why this is so is not clear. It is also used as a protectant in several lyophilized formulations and has stabilizing effects on other proteins (Wang and Hanson, 1988), yet the mechanism whereby it is an effective stabilizer is not understood. Examples of commercial protein formulations containing albumin are given in Table 8.5.

Because albumin is a natural protein, concerns have been raised about its potential contamination with human prion protein, which is thought to be the infectious agent in bovine spongiform encephalopathy (BSE) (Pharmaceutical Research and Manufacturers of America BSE Committee, 1998). Epidemiological data collected in the United Kingdom suggest a link between BSE and the feeding to ruminants (e.g. cattle, sheep, goats) of animal feed containing protein derived from transmissible spongiform encephalopathy (TSE)-contaminated tissues. Additionally, the data suggest that exposure to BSE may account for outbreaks of Creutzfeldt-Jakob disease (v-CJD), a CNS disease, in humans in the UK. The formulator should review materials such as enzymes and excipients used in the manufacture of the drug substance and drug product to ensure that there are no concerns with potential TSE contamination. Examples of animal source materials include not only albumin but also gelatin, glycerol and polysorbate 80. Ideally, the use of synthetic versions of these materials would eliminate concerns over potential disease transmission. If animal source materials are used in the manufacture of the product, then assurance must be provided to show what steps are being taken to prevent transmission of BSE. This could include sourcing of material from BSE-free countries or processing at high temperatures and pressure to achieve inactivation of TSE agents. Several regulatory

Table 8.5 Examples of commercial protein solution formulations containing albumin

Generic name	Brand name	Manufacturer	Human serum albumin in product
Alglucerase	Cerdase	Genzyme	1.0%
Erythropoietin	Epogen	Amgen	0.25%
Interferon Alpha-2a	Roferon-A	Roche	0.5%
Interferon Alpha-2b	Intron A	Schering	0.1% (following reconstitution)
Urokinase	Abbokinase	Abbott	5.0% (following reconstitution)

agencies such as the EU, FDA, and TGA (Australia) are critically reviewing the need to establish guidelines in this area (Federal Register, 1997).

8.4.8 *Other physical complexing/stabilizing agents*

Polyethylene glycol (PEG) is a common co-solvent for solubilizing small non-proteinaceous molecules, yet it also has been reported to minimize the aggregation of several peptides and proteins (Lee and Lee, 1987; Arakawa and Timasheff, 1985; Powell *et al.*, 1991; Bhat and Timasheff, 1992). PEG modification of proteins for sustained-release purposes is beyond the scope of this chapter (see Francis *et al.*, 1992). The concentration of PEG needs to be fairly low (<1% w/v) to serve as a stabilizer; at higher concentrations (>10% w/v) it can cause precipitation (Cleland and Randolph, 1992).

Poly(vinylpyrrolidone) (PVP) is like PEG in that at low concentrations it can stabilize proteins while at high concentrations it may help lead to protein aggregation and precipitation. Gombotz *et al.* (1994) reported that PVP at low concentrations (≤2.0%) effectively stabilizes human IgM monoclonal antibody against heat-induced aggregation, while PVP concentrations of ≥5.0% will cause aggregation.

Fibroblast growth factors, acidic and basic, are prone to acid and thermal inactivation and can be stabilized by a number of heparin and heparin-like molecules (Tsai *et al.*, 1993). Human keratinocyte growth factor, also prone to aggregation at high temperature, is stabilized by heparin, sulphated polysaccharides, anionic polymers, and citrate ion (Chen *et al.*, 1994).

8.5 Optimizing microbiological activity: antimicrobial preservatives (APs)

Preservatives are required for parenteral products intended for multiple-dose use. Many protein products, because of their expense and, more importantly, because of their ability to support the growth of microorganisms, are packaged as multiple-dose products, the AP being either formulated with the protein or, more commonly, formulated in a special diluent to be combined with the protein before use. The most common APs used in protein dosage forms are phenol, meta-cresol, benzyl alcohol, methylparaben and propylparaben. Examples of use of these preservatives are listed in Table 8.6.

Table 8.6 Antimicrobial preservative agents for protein products

Type	Concentration	Antimicrobial Activity*				pH	Other comment
		Gram+	Gram−	Fungi	Yeasts		
Benzyl alcohol	0.1–3.0%	+++	+++	−	−	3–6	Not effective pH>7
m-Cresol	0.1–0.3%	++	++	++	++	4–10	Most effective
Methylparaben	0.08–0.1%	++	++	++	++	3–9	Slowly soluble
Propylparaben	0.001–0.02%	+++	+++	+++	+++	3–9	Very slowly soluble
Phenol	0.2–0.5%	++	++	++	++	4–10	Most effective
Thimerosal	0.1–0.4%	++	++	++	++	4–8	Japan won't allow

* +++, most effective;   ++, moderately effective;   −, poor.

Table 8.7 Comparison of USP 24 and EP 2 requirements for preservative efficacy testing

Time after inoculation with microorganisms	Log reduction in microorganism count		
	USP 24	EP 'A' criteria	EP 'B' criteria
6 hours	Not required	3 (bacteria)	Not required
24 hours	Not required	No recovery (bac)	1 (bacteria)
2 days	Not required	No recovery (bac)	Not required
7 days	1 (bacteria)	No recovery (bac) 2 (fungi)	3 (bacteria)
14 days	3 (bacteria) No increase (fungi)	No recovery (bac) No increase (fungi)	1 (fungi)
21 days	No increase	No increase	Not required
28 days	No increase	No recovery (bac) No increase (fungi)	No increase (bac) No increase (fungi)

Antimicrobial agents must pass a preservative efficacy test (PET). Unfortunately, the USP and the British and/or European Pharmacopeial (BP/EP) tests for PET are different in their requirements. Table 8.7 summarizes the differences between the tests. The USP basically requires a bacteriostatic preservative system, while the BP/EP requires a bacteriocidal system. For example, the USP requires a 3 log reduction in the bacterial challenge by the 14th day after inoculation, while criteria A of the BP/EP test requires that same 3 log reduction within 24 hours. This great difference in compendial requirements for preservative efficacy has caused many problems in the formulation of protein dosage forms for various markets. Passing the BP/EP preservative efficacy test requires the use of relatively high amounts of phenol or cresol or other AP, which may have an impact on the stability of the formulation and could result in sorption of the preservative into the rubber closure. The formulator must keep in mind that increasing the concentration of APs may have a negative impact on protein physical stability (precipitation, aggregation, etc.). Increasing AP levels will increase the hydrophobicity of the formulation and could affect the aqueous solubility of the protein. Increasing AP concentrations also increases the potential for toxicological hazards.

It is well known that APs not only protect insulin formulations against inadvertent contamination, but also may have a significant effect on protein stability. For example, phenolic preservatives have a profound effect on the conformation of insulin in solution (Wollmer et al., 1987) and the assembly of the specific type of LysPro insulin hexamer (Birnbaum et al., 1997). Furthermore, phenol and/or m-cresol in insulin solutions will have a tendency to be adsorbed by and permeate rubber closures (Brange, 1987). Therefore, rubber formulations must be designed to minimize these potential problems.

Antimicrobial preservatives are known to interact with proteins and can cause stability problems such as aggregation. For example, phenolic compounds will cause aggregation of human growth hormone (Kirsch et al., 1993; Maa and Hsu, 1996). Phenol will produce a significant decrease in the α-helix content of insulinotropin resulting in aggregation of β-sheet structures (Kim et al., 1994). Benzyl alcohol, above certain concentrations and depending on other formulation factors, will interact with recombinant human interferon-

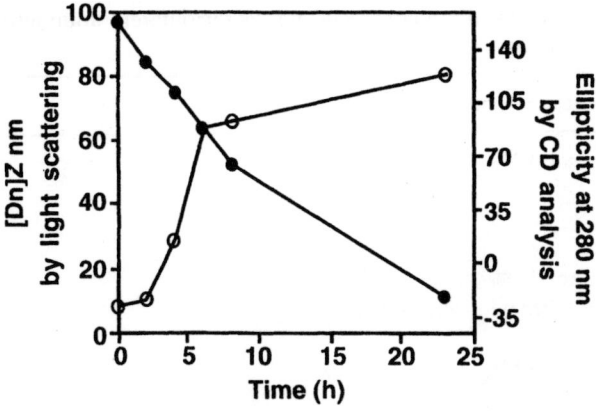

Figure 8.10 Time course of aggregate formation of 1.0 mg/ml rhIFN-γ in 5 mM succinate, pH 5.0 in the presence of 0.9% benzyl alcohol as determined by dynamic light scattering (○) and circular dichroism analysis (●). Reprinted with permission from Lam *et al.* (1997a); ©Plenum Press.

γ, causing aggregation of the protein (Figure 8.10) (Lam *et al.*, 1997a). These examples point out the need for the formulation scientist to understand the importance of potential effects of preservative type and concentration and other formulation additives on the interaction with proteins in solution while balancing the needs for antimicrobial efficacy.

In determining the appropriate AP agent or agents, the model described by Akers *et al.* (1984) might be suitable. The authors used insulin as the protein to be preserved and combined insulin with different types of AP agents, either alone or in combination. These formulations were challenged with the five USP preservative efficacy test organisms, and *D* values[3] were determined. The *D* value determination allows a single quantitative estimate of the AP effectiveness of a certain agent or combination of agents in a specific formulation against a specific microorganism. An example of the *D* value data obtained for insulin formulations with different AP systems against *Staphylococcus aureus* is given in Table 8.8, with the AP systems listed in order of effectivess (e.g. 0.2% phenol + 0.3% m-cresol was the most effective AP system).

There are instances where a manufacturer, because of concerns regarding aseptic processing and sterility assurance of the product throughout its shelf-life, will include an AP agent in the protein formulation even though it is intended only for a single-dose injection. This is a very controversial practice. Regulatory agencies worldwide object to this approach if, in their opinion, APs are used in a single-dose injectable product in order to 'cover up' for inadequate aseptic manufacturing practices and controls.

Many countries require preservative effectiveness tests (PETs) be performed for routine stability protocols and for special stability studies. Also, there may be requests from agencies to do PETs on containers that have been used (i.e. penetrated; partial volume withdrawn) to demonstrate that the product can still kill microorganisms. In mid-1995, the Australian Drug Evaluation Committee (ADEC) passed resolutions that, in light of safety concerns with contamination and cross-contamination, the use of injectable products in multi-dose packages is discouraged. In order to support the use of a multi-dose

---

[3] *D* value = time required for a 1 log reduction in the microbial population due to the effect of the antimicrobial preservative system. The smaller the *D* value, the greater is the effect of the preservative on the microorganism in question.

Table 8.8 *D* values against *Staphylococcus aureus* for different antimicrobial preservative systems in insulin solutions

Antimicrobial preservative system	D value (hours)
Phenol 0.2% + m-cresol 0.3%	0.5
m-Cresol 0.3%	0.6
Phenol 0.5%	0.8
Benzyl alcohol 2.0%	0.8
Phenol 0.2% + m-cresol 0.2%	1.3
Methylparaben 0.2% + benzyl alcohol 1.0%	1.4
Chlorobutanol 0.5%	1.8
Phenol 0.2% + m-cresol 0.1%	2.2
m-Cresol 0.2%	2.2
Methylparaben 0.2% + propylparaben 0.02%	3.0
Benzyl alcohol 1.0%	4.2
Methylparaben 0.1%	9.5
Methylparaben 0.1% + propylparaben 0.01%	12.3
Phenol 0.2%	16.2

Reprinted with permission from Akers *et al.* (1984). Copyright American Pharmaceutical Association.

product and the shelf-life once a package has been reconstituted or opened for use, antimicrobial preservative efficacy data are required for approval.

8.6   Osmolality (tonicity) agents

Salts or non-electrolytes (e.g. glycerine) are added to protein formulations in order to achieve an isotonic solution. Non-electrolytes are often preferred to salts as tonicity adjusters because of the potential problems salts cause in precipitating proteins (Pikal, 1990). Generally, solutions containing proteins administered IV, IM, or SC do not have to be precisely isotonic because of immediate effects from dilution by the blood. Intrathecal and epidural injections into the cerebrospinal fluid require very precise specifications for the product to be isotonic and at physiological pH. This is because extremes in osmolality and/or pH can damage or destroy cells, and cerebrospinal cells cannot be reproduced or replaced (Cradock *et al.*, 1977).

8.7   Packaging

Packaging issues are addressed in section 8.1 and in discussion of protein adsorption. Some additional information and guidance are summarized here. Solution dosage forms of peptides and proteins most commonly are packaged in glass vials sealed with rubber closures and aluminium seals. Other packaging systems used are cartridges, syringes, and plastic vials and bottles. In all these packaging systems, the formulator needs to be very concerned about potential reactivity of the peptide/protein and other ingredients in the formulation (e.g. antimicrobial preservatives) with the packaging components.

Selection of the packaging system not only depends on compatibility with the product formulation and the convenience to the consumer, but also depends on the integrity of the container–closure interface to assure maintenance of sterility throughout the shelf-life of the product. Container–closure integrity testing has received significant attention recently, and usually is an integral part of the regulatory submission and subsequent regulatory GMP inspections. While it is beyond the scope of this chapter to discuss the various aspects of container–closure integrity, it is emphasized that formulation scientists developing the final product, including the final package, must appreciate the need to develop appropriate testing methods to ensure that the selected packaging system indeed has the proper seal integrity to protect the product during its shelf-life from any ingress of microbiological contamination. There are excellent review articles on this subject by Morton *et al.* (1989), Chrai *et al.* (1994), Guazzo (1994) and Kirsch *et al.* (1997a, 1997b, 1997c).

8.8 Processing

Likewise, the basic concerns to be aware of in the manufacturing of proteins and peptides have been addressed in section 8.1. Unit processes involved in the manufacturing of solution sterile dosage forms include compounding and mixing, filtration, filling, terminal sterilization (although almost always not possible for proteins and peptides), closing and sealing, sorting and inspection, labelling, and final packaging for distribution. Since this chapter deals only with solution dosage forms, one of the most complex of processes – lyophilization or freeze-drying – will not be covered.

The pharmaceutical scientist must be aware of the various issues involved in the manufacturing arena that can impact the stability and quality of the protein or peptide formulation. Among the more relevant areas of concern are shear rate and stress during compounding, filtration, and filling (Charm and Wong, 1976; Thurow and Geisen, 1984), adsorption onto process tubing and filter surfaces (Hawker and Hawker, 1975; Brophy and Lambert, 1994; Brose and Waibel, 1996), and the effects of time and temperature during each step of the manufacturing process (Hsu *et al.*, 1988). Formulation scientists and process engineers should work together to design and implement experiments to determine processing effects on protein stability and establish an appropriate control strategy. In most cases, e.g. protein adsorption onto filter surfaces, the potential problems can be avoided or minimized once understood through experimentation by alternative choices of filter material or predicting the amount of solution to be passed through the filter to saturate the binding sites.

The surge of potential heat-labile products from biotechnology and the inability to sterilize these molecules terminally has accelerated the development of barrier/isolator technology. This technology, when perfected, will enable the processing of protein and peptide solutions to occur under a much higher degree of sterility assurance than is now achievable with conventional aseptic processing. The main features of barrier/isolator technology are the ability to sterilize, not just sanitize, the environment under which sterile solution is exposed during filling and stoppering, and the removal of humans from direct contact with the exposed sterile solution.

8.9 Conclusion

The formulation of stable, manufacturable, elegant, and high-quality protein and peptide solution preparations offers significant challenges to the development scientist. However, these challenges can be overcome through sound, rational formulation approaches with manufacturing processes and packaging systems designed to maintain stability and other quality features of the formulation. This chapter has outlined current thinking on formulation science in developing commercially viable protein and peptide solution dosage forms.

References

ABSOLOM, D.R., ZINGG, W. and NEUMANN, A.W., 1987, Protein adsorption to polymer particles: role of surface properties, *J. Biomed. Mater. Res.*, **21**, 161–171.

AKERS, M.J., 1982, Antioxidants in pharmaceutical products, *J. Parenteral Sci. Tech.*, **36**, 222–228.

AKERS, M.J., BOAND, A. and BINKLEY, D., 1984, Preformulation method for parenteral preservative efficacy evaluation, *J. Pharm. Sci.*, **73**, 903–907.

ANIK, S.T. and HWANG, J.Y., 1983, Adsorption of D-Nal(2)6LHRH, a decapeptide, onto glass and other surfaces, *Int. J. Pharm.*, **16**, 181–190.

ARAKAWA, T. and TIMASHEFF, S.N., 1982, Preferential interaction of protein with salts in concentrated solution, *Biochemistry*, **21**, 6545.

1984, Mechanism of protein salting in and salting out by divalent cation salts: balance between hydration and salt binding, *Biochemistry*, **23**, 5913.

1985, Mechanism of polyethylene glycol interactions with proteins, *Biochemistry*, **24**, 6756.

ARAKAWA, T., KITA, Y. and CARPENTER, J.F., 1991, Protein–solvent interactions in pharmaceutical formulations, *Pharm. Res.*, **8**, 285–291.

ARAKAWA, T., PRESTRELSKI, S.J., KENNEY, W.C. and CARPENTER, J.F., 1993, Factors affecting short-term and long-term stabilities of proteins, *Adv. Drug Del. Rev.*, **10**, 1–29.

ASAHARA, K., YAMADA, H. and YOSHIDA, S., 1991, Stability of human insulin in solutions containing sodium bisulfite, *Chem. Pharm. Bull.*, **39**, 2662–2666.

BAM, N.B., RANDOLPH, T.W. and CLELAND, J.L., 1995, Stability of protein formulations: investigation of surfactant effects by a novel EPR spectroscopic technique, *Pharm. Res.*, **12**, 2–11.

BHAT, R. and TIMASHEFF, S.N., 1992, Steric exclusion is the principal source of the preferential hydration of proteins in the presence of polyethylene glycols, *Protein Sci.*, **1**, 1133–1143.

BIRNBAUM, D.T., KILCOMONS, M.A., DEFELIPPIS, M.R. and BEALS, J.M., 1997, Assembly and dissociation of human insulin and Lys Pro-insulin hexamers: a comparison study, *Pharm. Res.*, **14**, 25–36.

BOYETT, J.B. and AVIS, K.E., 1976, Extraction rates of marker compounds from rubber closures for parenteral use II. Mechanism of extraction and evaluation of select extraction parameters, *Bull. Parenter. Drug Assoc.*, **30**, 169–177.

BRANGE, J., 1987, *Galenics of Insulin*, p. 41, New York: Springer-Verlag.

BRANGE J. and HAVELUND, S., 1983, Insulin pumps and insulin quality – requirements and problems, *Acta Med. Scand.*, Suppl., **671**, 135–138.

BRANGE, J. and LANGKJAER, L., 1993, Insulin structure and stability, *Stability and Characterization of Protein and Peptide Drugs: Case Histories*, WANG, Y.J. and PEARLMAN, R. (eds), p. 315, New York: Plenum Press.

BRANGE, J., LANGKJAER, L, HAVELUND, S. and VOLUND, A., 1992a, Chemical stability of insulin. 1. Hydrolytic degradation during storage of pharmaceutical preparations, *Pharm Res.*, **9**, 715–726.

BRANGE, J., HAVELUND, S. and HOUGAARD, P., 1992b, Chemical stability of insulin. 2. Formation of higher molecular weight transformation products during storage of pharmaceutical preparations, *Pharm Res.*, **9**, 727–734.

BRENNAN, J.R., GEBHART, S.S.P. and BALCKARD, W.G., 1985, Pump-induced insulin aggregation, *Diabetes*, **34**, 353.

BREWSTER, M.E., HORA, M.S., SIMPKINS, J.W., and BODOR, N., 1991, Use of 2-hydroxypropyl-β-cyclodextrin as a solubilizing and stabilizing excipient for protein drugs, *Pharm. Res.*, **8**, 792–795.

BROPHY, R.T. and LAMBERT, W.J., 1994, The adsorption of insulinotropin to polymeric sterilizing filters, *J. Parenteral Sci. Tech.*, **48**, 92–94.

BROSE, D.J. and WAIBEL, P., 1996, Adsorption of proteins in commercial microfiltration capsule, *Pharmaceut. Techol.*, **20**, 48–52.

CAPASSO, S., MAZZARELLA, L. and ZAGARI, A., 1991, Deamidation via cyclic imide of asparaginyl peptides: dependence on salts, buffers, and organic solvents, *Pept. Res.*, **4**, 234.

CHANG, B.S., KENDRICK, B.S. and CARPENTER, J.F., 1996, Surface-induced denaturation of proteins during freezing and its inhibition by surfactants, *J. Pharm. Sci.*, **85**, 1325–1330.

CHARM, S.E. and WONG, B.L., 1976, Enzyme inactivation with shearing, *Biotech. Bioeng.*, **12**, 1103–1109.

CHAWLA, A.S., HINBERG, I., BLAIS, P. and JOHNSON, D., 1985, Aggregation of insulin, containing surfactants, in contact with different materials, *Diabetes*, **34**, 420.

CHEN, B., ARAKAWA, T., MORRIS, C.F., KENNEY, W.C., WELLS, C.M. and PITT, C.G., 1994, Aggregation pathway of recombinant human keratinocyte growth factor and its stabilization, *Pharm. Res.*, **11**, 1581–1587.

CHRAI, S., HEFFERNAN, G. and MYERS, T., 1994, Glass vial container-closure integrity testing – an overview, *Pharm. Tech.*, **18**, 162–173.

CHRISTENSEN, P., JOHANSSON, A. and NIELSEN, V., 1978, Quantitation of protein adsorbance to glass and plastics: investigation of a new tube with low adherence, *J. Immunol. Methods*, **23**, 23–28.

CLELAND, J.L., 1993, Impact of protein folding on biotechnology, *Protein Folding In Vivo and In Vitro*, CLELAND, J.L. (ed.), ACS Symposium Series 526, p. 7, Washington, DC: American Chemical Society.

CLELAND, J.L. and RANDOLPH, T.W., 1992, Mechanism of polyethylene glycol interaction with the molten globule folding intermediate of bovine carbonic anhydrase B, *J. Biol. Chem.*, **267**, 3147–3153.

CLELAND, J.L., POWELL, M.F. and SHIRE, S.J., 1993, The development of stable protein formulations: a close look at protein aggregation, deamidation and oxidation, *Crit. Rev. Ther. Drug Carrier Sys.*, **10**, 307–377.

CRADOCK, J.C., KLEINMAN, L.M. and DAVIGNON, J.P., 1977, Intrathecal injections – a review of pharmaceutical factors, *Bull. Parenter. Drug Assoc.*, **31**, 237.

DAHLQUIST, F.W., LONG, J.W. and BIGBEE, W.L., 1976, Role of calcium in thermal stability of thermolysin, *Biochemistry*, **15**, 1103.

DANIELSON, J.W., OXBORROW, G.S. and PLACENCIA, A.M., 1984, Quantitative determination of chemicals leached from rubber stoppers into parenteral solutions, *J. Parenter. Sci. Tech.*, **38**, 90–93.

DAVIO, S.R. and HAGEMAN, M.J., 1993, Characterization and formulation considerations for recombinantly derived bovine somatotropin, *Stability and Characterization of Protein and Peptide Drugs: Case Histories*, WANG, Y.J. and PEARLMAN, R. (eds), pp. 76–80, New York: Plenum Press.

DONG, D.E., ANDRADE, J.D. and COLEMAN, D.L., 1987, Adsorption of low density lipoproteins onto selected biomedical polymers, *J. Biomed. Mater. Res.*, **21**, 683–700.

EDWARDS, R.A. and HUBER, R.E., 1992, Surface denaturation of proteins: the thermal inactivation of β-galactosidase (*Escherichia coli*) on the wall–liquid surfaces, *Biochem. Cell Biol.*, **70**, 63–69.

FEDERAL REGISTER, 1997, Proposed Rule 21CFR Part 589, *Substances Prohibited from Use in Animal Food or Feed, Animal Proteins Prohibited in Ruminant Feed*, 3 January, pp. 551–583.

FISH, W.W., DANIELSSON, A., NORDLING, L., MILLER, S.H., LAM, C.F. and BJORK, I., 1985, Denaturation behavior of antithrombin in guanidinium chloride: irreversibility of unfolding caused by aggregation, *Biochemistry*, **24**, 1510.

FRANCIS, G.E., DELGADO, C. and FISHER, D., 1992, PEG-modified proteins, *Stability of Protein Pharmaceuticals, Part B: In Vivo Pathways of Degradation and Strategies for Protein Stabilization*, AHERN, T.J. and MANNING, M.C. (eds), pp. 235–263, New York: Plenum Press.

FRANSSON, J.R., 1997, Methionine oxidation and covalent aggregation in aqueous solution, *J. Pharm. Sci.*, **86**, 1046–1050.

FRANSSON, J. and HAGMAN, A., 1996, Oxidation of human insulin-like growth factor I in formulation studies 2. Effects of oxygen, visible light, and phosphate on methionine oxidation in aqueous solution and evaluation of possible mechanisms, *Pharm Res.*, **13**, 1476–1481.

GEKKO, K. and TIMASHEFF, S.N., 1981, Mechanism of protein stabilization by glycerol: preferred hydration in glycerol–water mixtures, *Biochemistry*, **20**, 466.

GLATZ, C.E., 1992, Modeling of aggregation–precipitation phenomena, *Stability of Protein Pharmaceuticals, Part A: Chemical and Physical Pathways of Protein Degradation*, AHERN, T.J. and MANNING, M.C. (eds), pp. 135–166, New York: Plenum Press.

GOMBOTZ, W.R., PANKEY, S.C., PHAN, D., DRAGER, R., DONALDSON, K., ANTONSEN, K.P., HOFFMAN, A.S. and RAFF, H.V., 1994, The stabilization of a human IgM monoclonal antibody with poly(vinylpyrrolidone), *Pharm. Res.*, **11**, 624–632.

GUAZZO, D., 1994, Package integrity testing, *Parenteral Quality Control: Sterility, Pyrogen, Particulate Matter and Package Integrity Testing*, AKERS, M.J. (ed.), New York: Marcel Dekker.

HAGEMAN, M.J., TINWALLA, A.Y. and BAUER, J.M., 1993, Kinetics of temperature-induced irreversible aggregation/precipitation of bovine somatotropin as studied by initial rate methods, *Pharm. Res.*, **10**, S-85.

HARWOOD, R.J., PORTNOFF, J.B. and SUNBERY, E.W., 1993, The processing of small volume parenterals and related sterile products, *Pharmaceutical Dosage Forms: Parenteral Medications*, Vol. 2, 2nd edn, AVIS, K.E., LIEBERMAN, H.A. and LACHAMAN, L. (eds), pp. 70–73, New York: Marcel Dekker.

HAWKER, R.J. and HAWKER, L.M., 1975, Protein losses during sterilizing by filtration, *Lab. Pract.*, **24**, 805–807.

HERMAN, A.C., BOONE, T.C. and LU, H.S., 1996, Characterization, formulation, and stability of Neupogen® (Filgrastim), a recombinant human granulocyte-colony stimulating factor, characterization, stability, and formulations of basic fibroblast growth factor, *Formulation, Characterization, and Stability of Protein Drugs: Case Histories*, PEARLMAN, R. and WANG, Y.J. (eds), pp. 324–325, New York: Plenum Press.

HIRSCH, J.I., FRATKIN, M.J., WOOD J.H. and THOMAS, R.B., 1977, Clinical significance of insulin adsorption by polyvinylchloride infusion systems, *Am. J. Hosp. Pharm.*, **34**, 583–588.

HIRSCH, J.I., WOOD, J.H. and THOMAS, R.B., 1981, Insulin adsorption to polyolefin infusion bottles and polyvinylchloride administration sets, *Am. J. Hosp. Pharm.*, **38**, 995–997.

HSU, C.C., PEARLMAN, R. and CURLEY, J.C., 1988, Some factors causing protein denaturation and aggregate formation in pharmaceutical processing, *Pharm. Res.*, Suppl., **5**, S34.

IRWIN, W.J., DWIVEDI, A.K., HOLBROOK, P.A. and DEY, M.J., 1994, The effect of cyclodextrins on the stability of peptides in nasal enzymatic systems, *Pharm. Res.*, **11**, 1698–1703.

IWAMOTO, G.K., VAN WAGENEN, R.A. and ANDRADE, J.D., 1982, Insulin adsorption: intrinsic tyrosine interfacial fluorescence, *J. Colloid Interface Sci.*, **86**, 581.

JAMES, D.E., JENKINS, A.B., KRAEGAN, E.W. and CHISHOLM, D.J., 1981, Insulin precipitation in artificial infusion devices, *Diabetologia*, **21**, 554.

JOHNSON, M.D., HOESTEREY, B.L. and ANDERSON, B.D., 1994, Solubilization of a tripeptide HIV protease inhibitor using a combination of ionization and complexation with chemically modified cyclodextrins, *J. Pharm. Sci.*, **83**, 1142.

JOHNSTON, T.P., 1996, Adsorption of recombinant human granulocyte colony stimulating factor (rGH-CSF) to polyvinyl chloride, polypropylene, and glass: effect of solvent additives, *PDA J. Pharm. Sci. Tech.*, **50**, 238–245.

KAKIMOTO, F., ASAKAWA, N., ISHIBASHI, Y. and MIYAKE, Y., 1985, *Prevention of adsorption of peptide hormones*, Eisai Co. Ltd, Japan Patent JP61221125A2861001.

KATAKAM, M., BELL, L.N. and BANGA, A.K., 1995, Effect of surfactants on the physical stability of recombinant human growth hormone, *J. Pharm. Sci.*, **84**, 713–716.

KIM, Y., ROSE, C.A., LIU, Y., OZAKI, Y., DATTA, G. and UT, A.T., 1994, RT-IR and near-infared FT-Raman studies of the secondary structure of insulinotropin in the solid state: α-helix to β-sheet conversion induced by phenol and/or high shear force, *J. Pharm. Sci.*, **83**, 1175–1180.

KIRSCH, L.E., RIGGIN, R.M., GEARHART, D.A., LEFEBER, D.S. and LYTLE, D.L., 1993, In-process protein degradation by exposure to trace amounts of sanitizing agents, *J. Parenter. Sci. Technol.*, **47**, 155.

KIRSCH, L.E., NGUYEN, L. and MOECKLY, C.S., 1997a, Pharmaceutical container/closure integrity I. Mass spectrometry-based helium leak rate detection for rubber-stoppered glass vials, *PDA J. Pharm. Sci. Tech.*, **51**, 187–194.

KIRSCH, L.E., NGUYEN, L., MOECKLY, C.S. and GERTH, R., 1997b, Pharmaceutical container/closure integrity II. The relationship between microbial ingress and helium leak rates in rubber-closured glass vials, *PDA J. Pharm. Sci. Tech.*, **51**, 195–202.

KIRSCH, L.E., NGUYEN, L. and GERTH, R., 1997c, Pharmaceutical container/closure integrity III. Validation of the helium leak rate method for rigid pharmaceutical containers, *PDA J. Pharm. Sci. Tech.*, **51**, 203–207.

KNEPP, V.M., WHATLEY, J.L., MUCHNIK, A. and CALDERWOOD, T.S., 1996, Identification of antioxidants for prevention of peroxide-mediated oxidation of recombinant human ciliary neurotrophic factor and recombinant human nerve growth factor, *PDA J. Pharm. Sci. Tech.*, **50**, 163.

LAM, X.M., PATAPOFF, T.W. and NGUYEN, T.H., 1997a, The effect of benzyl alcohol on recombinant human interferon-γ, *Pharm. Res.*, **14**, 725–729.

LAM, X.M., YANG, J.Y. and CLELAND, J.L., 1997b, Antioxidants for prevention of methionine oxidation in recombinant monoclonal antibody HER2, *J. Pharm. Sci.*, **86**, 1250–1255.

LEE, J.C. and LEE, L.L.Y., 1987, Thermal stability of proteins in the presence of polyethylene glycols, *Biochemistry*, **26**, 7813–7819.

LENK, J.R., RATNER, B.D., GENDREAU, R.M. and CHITTUR, K.K., 1989, IR spectral changes of bovine serum albumin upon surface adsorption, *J. Biomed. Mater. Res.*, **23**, 549–569.

LI, S., SCHONEICH, C., WILSON, G.S. and BORCHARDT, R.T., 1993, Chemical pathways of peptide degradation. V. Ascorbic acid promotes rather than inhibits the oxidation of methionine to methionine sulfoxide in small model peptides, *Pharm. Res.*, **10**, 1572.

LI, S., SCHONEICH, C. and BORCHARDT, R.T., 1995, Chemical pathways of peptide degradation. VIII. Oxidation of methionine in small model peptides by prooxidant/transition metal ion systems: influence of selective scavengers for reactive oxygen intermediates, *Pharm. Res.*, **12**, 348–355.

LI, W., PATAPOFF, T.W., NGUYEN, T.H. and BORCHARDT, R.T., 1996, Inhibitory effect of sugars and polyols on the metal-catalyzed oxidation of human relaxin, *J. Pharm. Sci.*, **85**, 868–872.

LIEBE, D.C., 1995, Pharmaceutical packaging, *Encyclopedia of Pharmaceutical Technology*, Vol. 12, SWARBRICK, J. and BOYLAN, J.C. (eds), pp. 16–28. New York: Marcel Dekker.

LOFTSSON, T. and BREWSTER, M.E., 1996, Pharmaceutical applications of cyclodextrins. 1. Drug solubilization and stabilization, *J. Pharm. Sci.*, **85**, 1017–1025.

LOUGHEED, W.D., ALBISSER, A.M., MARTINDALE, H.M., CHOW, J.C. and CLEMENT, J.R., 1983, Physical stability of insulin formulations, *Diabetes*, **32**, 424.

MAA, Y.F. and HSU, C.C., 1996, Aggregation of recombinant human growth hormone induced by phenolic compounds, *Int. J. Pharm.*, **140**, 155–168.

MANNING, M.C., PATEL, K. and BORCHARDT, R.T., 1989, Stability of protein pharmaceuticals, *Pharm. Res.*, **6**, 903–918.

MILANO, E.A., WARASZKIEWICZ, S.M. and DIRUBIO, R., 1982, Extraction of soluble aluminum from chlorobutyl rubber closures, *J. Parenter. Sci. Tech.*, **36**, 116–120.

MITRAKI, A. and KING, J., 1989, Protein folding intermediates and inclusion body formation, *Biotechnology*, **7**, 690.

MITRANO, F.P. and NEWTON, D.W., 1982, Factors affecting insulin adherence to Type I glass bottles, *Am. J. Hosp. Pharm.*, **39**, 1491–1495.

MORTON, D.K., LORDI, N.G., TROUTMAN, L.H. and AMBRIOSIO, T.J., 1989, Quantitative and mechanistic measurements of container/closure integrity. Bubble, liquid, and microbial leakage tests, *J. Parenter. Sci. Tech.*, **43**, 104–108.

NAIL, S.L. and AKERS, M.J. (eds), 2000, *Development and Manufacture of Protein Pharmaceuticals*, New York: Plenum Press.

NIEBERGALL, P.J., 1990, Ionic solutions and electrolytic equilibria, *Remington's Pharmaceutical Sciences*, 18th edn, p. 243, Philadelphia: Mack Publishing Co.

NORDE, W., 1995, Adsorption of proteins at solid–liquid interfaces, *Cells Mater.*, 5, 97–112.

OLIYAI, C. and BORCHARDT, R.T., 1994, Chemical pathways of peptide degradation. VI. Effect of the primary sequence on the pathways of degradation of aspartyl residues in model hexapeptides, *Pharm Res.*, 11, 751–758.

OSHIMA, G., 1989, Solid surface-catalyzed inactivation of bovine alpha-chymotrypsin in dilute solution, *Int. J. Biol. Macromol.*, 11, 43.

PACE, C.N. and GRIMSLEY, G.R., 1988, Ribonuclease T1 is stabilized by cation and anion binding, *Biochemistry*, 27, 8311.

PANTOLIANO, M.W., WHITLOW, M., WOOD, J.F., ROLLENCE, M.L., FINZEL, B.C., GILLIAND, G.L., POULOS, G. and BRYANT, P.N., 1988, The engineering of binding affinity at metal ion binding sites for the stabilization of proteins: subtilisin as a test case, *Biochemistry*, 27, 8311.

PATEL, J.P., 1990, Urokinase: stability studies in solution and lyophilized formulations, *Drug Dev. Ind. Pharm.*, 16, 2613–2626.

PATEL, K., 1993, Stability of adrenocorticotropic hormone (ACTH) and pathways of deamidation of asparaginyl residue in hexapeptide segments, *Stability and Characterization of Protein and Peptide Drugs: Case Histories*, WANG, Y.J. and PEARLMAN, R. (eds), pp. 207–212, New York: Plenum Press.

PATEL, K. and BORCHARDT, R.T., 1990, Chemical pathways of peptide degradation II. Kinetics of deamidation of an asparaginyl residue in a model hexapeptide, *Pharm. Res.*, 7, 787–793.

PEARLMAN, R. and BEWLEY, T.A., 1993, Stability and characterization of human growth hormone, *Stability and Characterization of Protein and Peptide Drugs: Case Histories*, WANG, Y.J. and PEARLMAN, R. (eds), p. 44, New York: Plenum Press.

PHARMACEUTICAL RESEARCH AND MANUFACTURERS OF AMERICA BSE COMMITTEE, 1998, Assessment of risk of bovine spongiform encephalopathy in pharmaceutical products, part 1, *Biopharm*, 11, 20–31.

PIKAL, M.J., 1990, Freeze drying of proteins II: formulation selection, *Biopharm*, 3, 26–30.

POWELL, M.F., SANDERS, L.M., ROGERSON, A. and SI, V., 1991, Parenteral peptide formulations: chemical and physical properties of native luteinizing hormone-releasing hormone (LHRH) and hydrophobic analogues in aqueous solution, *Pharm. Res.*, 8, 1258–1263.

PRETZER, D., SCHULTEIS, B.S., SMITH, C.D., VANDERVELDE, D.G., MITCHELL, J.W. and MANNING, M.C., 1993, *Stability and Characterization of Protein and Peptide Drugs: Case Histories*, WANG, Y.J. and PEARLMAN, R. (eds), pp. 305–309, New York: Plenum Press.

ROE, J.A., BUTLER, A., SCHALLER, D.M. and VALENTINE, J.S., 1988, Differential scanning calorimetry of Cu, Zn-superoxide dismutase, the apoprotein, and its Zn-substituted derivatives, *Biochemistry*, 27, 950.

SATO, S., EBERT, C.D. and KIM, S.W., 1984, Prevention of insulin self-association and surface adsorption, *J. Pharm. Sci.*, 72, 228.

SENDEROFF, R.I., WOOTTON, S.C., BOCTOR, A.M., CHEN, T.M., GIORDANI, A.B., JULIAN, T.N. and RADEBAUGH, G.W., 1994, Aqueous stability of human epidermal growth factor 1–48, *Pharm. Res.*, 11, 1712–1720.

SERES, D.S., 1990, Insulin adsorption to parenteral infusion systems; case report and review of the literature, *Nutr. Clin. Pract.*, 5, 111–117.

SHAHROKH, Z., EBERLEIN, G., BUCKLEY, D., PARANANDI, M.V., ASWAD, D.W., STRATTON, P., MISCHAK, R. and WANG, Y.J., 1994a, Major degradation products of basic fibroblast growth factor: detection of succinimide and iso-aspartate in place of aspartate, *Pharm. Res.*, 11, 936–944.

SHAHROKH, Z., EBERLEIN, G. and WANG, Y.J., 1994b, Probing the conformation of protein (bFGF) precipitates by fluorescence spectroscopy, *J. Pharm. Biomed. Anal.*, 12, 1035–1041.

SILVESTRI, S., LU, M.Y. and JOHNSON, H., 1993, Kinetics and mechanism of peptide aggregation I: aggregation of a cholecystokinin analog, *J Pharm. Sci.*, 92, 689–693.

SLUZKY, V., TAMADA, J.A., KLIBANOV, A.M. and LANGER, R., 1991, Kinetics of insulin aggregation in aqueous solutions upon agitation in the presence of hydrophobic surfaces, *Proc. Natl Acad. Sci. USA*, 88, 9377–9381.

SLUZKY, V., KLIBANOV, A.M. and LANGER, R., 1992, Mechanism of insulin aggregation and stabilization in agitated aqueous solutions, *Biotechnol. Bioeng.*, **40**, 895–903.

SUELTER, C.H. and DELUCA, P.P., 1983, How to prevent losses of protein by adsorption to glass and plastic, *Anal. Biochem.*, **135**, 112.

THOMPSON, D.O., 1997, Cyclodextrins–enabling excipients: their present and future use in pharmaceuticals, *Crit. Rev. Ther. Drug Carrier Sys.*, **14**, 1–104.

THUROW, H. and GEISEN, K., 1984, The stabilisation of dissolved proteins against denaturation at hydrophobic interfaces, *Diabetologica*, **27**, 212.

TSAI, P.K., VOLKIN, D.B., DABORA, J.M., THOMPSON, K.C., BRUNER, M.W., GRESS, J.O., MATUSZEWSKA, B., KEOGAN, M., BONDI, J.V. and MIDDAUGH, C.R., 1993, Formulation design of acidic fibroblast growth factor, *Pharm. Res.*, **10**, 649–659.

TWARDOWSKI, Z.J., NOLPH, K.D., MCGARY, T.J., MOORE, H.L., COLLIN, P., AUSMAN, R.K. and SLIMACK, W.S., 1983a, Insulin binding to plastic bags: a methodologic study, *Am. J. Hosp. Pharm.*, **40**, 575.

TWARDOWSKI, Z.J., NOLPH, K.D., MCGARY, T.J. and MOORE, H.L., 1983b, Nature of insulin binding to plastic bags, *Am. J. Hosp. Pharm.*, **40**, 579.

1983c, Influence of temperature and time on insulin adsorption to plastic bags, *Am. J. Hosp. Pharm.*, **40**, 583.

VEMURI, S., YU, C.T. and ROOSDORP, N., 1993, Formulation and stability of recombinant αl-antitrypsin, *Stability and Characterization of Protein and Peptide Drugs: Case Histories*, WANG, Y.J. and PEARLMAN, R. (eds), pp. 269–270, New York: Plenum Press.

WANG, Y.J. and CHIEN, Y.W., 1984, *Sterile pharmaceutical packaging: compatibility and stability*, Technical Report 5, Parenteral Drug Association, Philadelphia.

WANG, Y.J. and HANSON, M.A., 1988, Parenteral formulations of proteins and peptides: stability and stabilizers, *J. Parenter. Sci. Tech.*, Technical Report 10, **42**, Suppl.

WANG, Y.J., SHAHROKH, Z., VEMURI, S., EBERLEIN, G., BEYLIN, I. and BUSCH, M., 1996, Characterization, stability, and formulations of basic fibroblast growth factor, *Formulation, Characterization, and Stability of Protein Drugs: Case Histories*, PEARLMAN, R. and WANG, Y.J. (eds), pp. 164–165, New York: Plenum Press.

WETZEL, R., PERRY, L.J., BAASE, W.A. and BECKTEL, W.J., 1988, Disulfide bonds and thermal stability of T4 lysozyme, *Proc. Natl Acad. Sci. USA*, **85**, 401–405.

WOLLMER, A., RANNEFELD, B., JOHNASEN, B.R., HEJNAES, K.R., BALSCHMIDT, P. and HANSEN, F.B., 1987, Phenol-promoted structural transformation of insulin in solution, *Biol. Chem. Hoppe-Seyler*, **368**, 903–911.

YOSHIOKA, S., ASO, Y., IZUTSU, K. and TERAO, T., 1993, Stability of β-galactosidase, a model protein drug, is related to water mobility as measured by ^{17}O nuclear magnetic resonance (NMR), *Pharm. Res.*, **10**, 103–108.

ZHANG, M.Z., WEN, J., ARAKAWA, T. and PRESTRELSKI, S.J., 1995, A new strategy for enhancing the stability of lyophilized protein: the effect of the reconstitution medium on keratinocyte growth factor, *Pharm. Res.*, **12**, 1447–1452.

9

Roles of Protein Conformation and Glassy State in the Storage Stability of Dried Protein Formulations

JOHN F. CARPENTER,[1,2] LOTTE KREILGAARD,[1,2] S. DEAN ALLISON[1,2] AND THEODORE W. RANDOLPH[1,3]

[1]University of Colorado Center for Pharmaceutical Biotechnology, Denver, Colorado, USA
[2]Department of Pharmaceutical Sciences, University of Colorado Health Sciences Center, Denver, Colorado, USA
[3]Department of Chemical Engineering, University of Colorado, Boulder, Colorado, USA

9.1   Introduction

The medical and financial successes of a protein therapeutic depend on designing a formulation that has adequate stability during shipping and long-term storage. There is often intense pressure on the research staff to accomplish this task in a short period of time. At least in part, this is usually because substantial clinical testing has been completed by the time the decision has been made to make a full-scale attempt at final formulation development. Then the project is deemed 'fast track'. Of course, it would be advantageous to have sufficient time to explore numerous possible formulations and to optimize protein stability completely. Furthermore, if formulation development were accomplished early in the investigation of a new protein therapeutic, there would be less risk of delays in marketing of the new drug and the associated absence of income from the product. But, in reality, there is rarely the luxury of time in formulation development. Therefore, it is critical that the researchers involved in these projects take a rational approach to the problem. This approach is derived from a clear understanding of the fundamental mechanisms governing stability of protein therapeutics during processing, shipping and storage (Manning *et al.*, 1989; Arakawa *et al.*, 1993; Pikal, 1994; Carpenter and Chang, 1996; Carpenter *et al.*, 1997). As we will explain in this chapter, if the essential properties of a successful formulation are known, and can be tested for immediately after sample preparation, a formulation that is sufficient to meet stability requirements can be designed fairly rapidly. Conversely, if a range of formulations are merely screened, without attention to rational development, there is a risk of failure to obtain a marketable product, at least within the time constraints that are present in industry.

The purpose of the current chapter is to describe the recent evidence that critical characteristics for stable dried protein products are: (1) using excipients that minimize protein unfolding during processing, and (2) obtaining a formulation in which the protein is in a fraction with amorphous excipients wherein the glass transition temperature is greater than the targeted storage temperature. These factors should play the same role in products prepared by lyophilization, spray-drying or air-drying. However, here we will limit our comments to lyophilized proteins, for which the bulk of published data are available. Other factors that dictate protein stability in the solid state include residual moisture and specific chemical degradative reactions (Manning *et al.*, 1989; Hageman, 1992; Arakawa *et al.*, 1993; Pikal, 1994; Carpenter and Chang, 1996). These aspects will be described briefly in the context of how they interplay with protein stability and glassy state.

Let us first consider the rationale behind preparing a dried formulation. An aqueous liquid formulation is the easiest and most economical to handle during manufacturing, and is the most convenient for the end-user. However, many proteins are susceptible to chemical (e.g. deamidation or oxidation) and/or physical degradation (e.g. aggregation and precipitation) in liquid formulations (Manning *et al.*, 1989; Arakawa *et al.*, 1993; Cleland *et al.*, 1993). An aqueous formulation that slows protein degradation adequately under controlled storage conditions (i.e. constant temperature and minimal agitation) may fail during shipping. During shipping, products can be subjected to numerous stresses that denature proteins, including agitation, high and low temperatures, and freezing (Arakawa *et al.*, 1993). Furthermore, although a formulation and shipping system might be designed to avoid damage from these stresses, it still may not be possible to inhibit damage sufficiently during long-term storage. For example, there are cases where conditions that minimize chemical degradation foster physical damage, and vice versa (Manning *et al.*, 1989; Arakawa *et al.*, 1993). Then, a compromise affording the requisite long-term stability cannot be found.

All these difficulties theoretically can be avoided with a properly-prepared dried formulation. In the dried solid, degradative reactions can be avoided or slowed sufficiently, and the protein product can be stable for months or years at ambient temperatures (Franks *et al.*, 1991; Pikal, 1994; Carpenter and Chang, 1996; Carpenter *et al.*, 1997). Furthermore, short-term excursions in temperature control during shipping are usually not damaging to a lyophilized protein (e.g. Chang *et al.*, 1996). Even in cases where different conditions are needed to minimize two or more degradative pathways, the reduced reaction rates in a dried product can allow for long-term stability.

Without appropriate stabilizing excipient(s), most protein preparations are at least partially denatured by the freezing and dehydration stresses encountered during lyophilization (Carpenter *et al.*, 1993; Prestrelski *et al.*, 1993a, 1993b, 1995; Dong *et al.*, 1995; Strambini and Gabellieri, 1996; Allison *et al.*, 1996, 1998; Kreilgaard *et al.*, 1998, 1999; Chang *et al.*, 1996). The result is often irreversible aggregation of a fraction of the protein population, either immediately after processing or after storage (e.g. Prestrelski *et al.*, 1993a, 1993b; Dong *et al.*, 1995; Allison *et al.*, 1996, 1998; Kreilgaard *et al.*, 1998, 1999). Because most protein drugs are delivered parenterally, the presence of only a few per cent of aggregated protein can render the product unacceptable.

Designing a formulation that allows the protein to survive the lyophilization process does not necessarily assure stability during long-term storage in the dried solid (e.g. Prestrelski *et al.*, 1995; Chang *et al.*, 1996). A poorly formulated lyophilized product, in which the protein is sufficiently reactive to require storage at sub-zero temperature, should not be considered a success. However, as will be seen below, for basic mechanistic studies it is important to understand why such formulations fail and others succeed.

Finally, there are issues beyond protein stability that dictate success of a lyophilized product and its rational design. These include: (1) configuration needed for product use (e.g. dose, volume and container/delivery system); (2) formulation tonicity; (3) collapse temperature, and ease and economy of processing; and (4) structure and strength of the dried cake. These factors have been discussed in detail elsewhere (Pikal, 1994; Carpenter and Chang, 1996; Carpenter *et al.*, 1997), and will not be considered further here.

9.2 Infrared spectroscopy to study protein secondary structure

Elucidation of the importance of protein structure in storage stability of dried formulations has only been possible because of the use of infrared spectroscopy to compare the secondary structure of the native aqueous protein to that for the dried solid. Prior to 1993, researchers were judging the effects of processing and excipients on protein structure solely by results obtained after rehydration, which sometimes misled them into thinking that a relatively poor formulation was stabilizing a protein adequately. We now know that many proteins that are unfolded during lyophilization can readily refold upon rehydration (e.g. Prestrelski *et al.*, 1993b; Dong *et al.*, 1995; Allison *et al.*, 1996). But, as will be described below, even these easily refolded proteins often have poor storage stability in the dried solid.

The assumption that data obtained after rehydration were sufficient to judge formulation quality was a common mistake fostered, in part, by reports in the protein chemistry literature that dehydration did not alter a protein's conformation (reviewed in Dong *et al.*, 1995). Such a claim was clearly counter to the known contributions of water to the formation of the native, folded protein (Kuntz and Kauzman, 1974; Edsall and McKenzie, 1983). Also, it was difficult to reconcile the finding that proteins could be irreversibly inactivated and aggregated after rehydration with the contention that protein structure was not perturbed by dehydration.

Reconciliation of this apparent dilemma was provided by application of modern infrared spectroscopic methods to the study of structure of therapeutic and other proteins in the dried state. Infrared spectroscopy has long been used for studying stress-induced alterations in protein conformation and for quantitation of protein secondary structure (e.g. Byler and Susi, 1986; Susi and Byler, 1986; Surewicz and Mantsch, 1988; Krimm and Bandekar, 1986; Dong and Caughey, 1994; Mantsch and Chapman, 1997; Dong *et al.*, 1995). Structural information is obtained by analysis of the conformationally-sensitive amide I band, which is located between 1600 and 1700 cm^{-1}. This band is due to the in-plane C=O stretching vibration, weakly coupled with C–N stretching and in-plane N–H bending (Krimm and Bandekar, 1986; Mantsch and Chapman, 1997). Each type of secondary structure (i.e. α-helix, β-sheet, β-turn and disordered) gives rise to different C=O stretching frequencies (Byler and Susi, 1986; Susi and Byler, 1986; Surewicz and Mantsch, 1988; Krimm and Bandekar, 1986; Dong and Caughey, 1994; Mantsch and Chapman, 1997), and, hence, results in characteristic band positions, which are designated by wavenumber, cm^{-1}. Band positions are used to determine the secondary structural types present in a protein. The relative band areas (determined by curve fitting) can then be used to quantitate the relative amount of each structural component. Therefore, an analysis of the infrared bands in the amide I region can provide quantitative as well as qualitative information about protein secondary structure (Byler and Susi, 1986; Susi and Byler, 1986; Surewicz and Mantsch, 1988; Krimm and Bandekar, 1986; Dong and Caughey, 1994; Mantsch and Chapman, 1997; Dong *et al.*, 1995).

To obtain detailed structural information, it is necessary to enhance the resolution of the protein amide I band, which usually appears as a single broad absorbance contour. The widths of the overlapping component bands are often greater than the separation between the absorbance maxima of neighbouring bands. Because the band overlapping is beyond instrumental resolution, several mathematical band-narrowing methods (i.e. resolution enhancement methods) have been developed to overcome this problem. For studies of lyophilization-induced structural changes and assessment of effects of excipients on these alterations, calculation of the second derivative spectrum is recommended (Dong *et al.*, 1995; Carpenter *et al.*, 1998). This method is objective, and alterations in component band widths, heights and positions, which are due to protein unfolding, are preserved in the second derivative spectrum. For most unprotected proteins (i.e. lyophilized in the presence of only buffer) the second derivative spectra for the dried solid are greatly altered, relative to the respective spectra for the native proteins in aqueous solutions (Carpenter *et al.*, 1993; Prestrelski *et al.*, 1993a, 1993b, 1995; Dong *et al.*, 1995; Allison *et al.*, 1996, 1998; Kreilgaard *et al.*, 1998, 1999; Chang *et al.*, 1996). Two different behaviours of proteins unfolded in the dried solid are displayed during rehydration (reviewed in Carpenter *et al.*, 1998): (1) the protein regains the native conformation upon rehydration (reversible unfolding), as observed for α-lactalbumin, lysozyme, chymotrypsinogen, ribonuclease, β-lactoglobulins A and B, α-chymotrypsin and subtilisin; (2) a significant fraction of the protein molecules aggregate upon rehydration (irreversible unfolding), as noted for lactate dehydrogenase, phosphofructokinase, interferon-gamma, basic fibroblast growth factor and interleukin-2.

It has been documented with several proteins in the latter class that prevention of aggregation and recovery of activity after rehydration correlate directly with retention of the native structure in the dried solid (reviewed in Carpenter and Chang, 1996; Carpenter *et al.*, 1998). Thus, the mechanism by which stabilizing additives (e.g. disaccharides) minimize loss of activity and aggregation during lyophilization and rehydration is to prevent unfolding during freezing and drying. Also, unfolding of proteins that refold if immediately rehydrated can be inhibited by stabilizing additives (Prestrelski *et al.*, 1993b; Dong *et al.*, 1995; Allison *et al.*, 1996, 1998; Kreilgaard *et al.*, 1998, 1999; Chang *et al.*, 1996).

9.3 Physical factors affecting storage stability of dried protein formulations

Before discussing the mechanistic studies that have addressed these issues, it is important to consider the routes by which proteins may degrade and how this may impinge on development of stable lyophilized protein formulations. Several review papers have been published to describe the physical (e.g. aggregation and precipitation) and chemical (i.e. covalent alterations) degradation pathways of proteins during storage in aqueous solution (Wang and Hanson, 1988; Manning *et al.*, 1989; Wang and Pearlman, 1993; Cleland *et al.*, 1993). Very little published information is available about the effects of lyophilization on the rates of individual degradation pathways in protein formulations. However, it is important to understand the major degradation pathways arising during storage in the dried solid and to develop the appropriate analytical methods. This is especially important if the protein undergoes chemical degradation (e.g. deamidation), for which very specific adjustments (e.g. appropriate pH or removal of oxygen from the vial headspace) are needed to prevent protein damage. To develop the analytical methods needed and to identify the types of degradation products which may arise, degradation can be accelerated. The protein is lyophilized in the absence of stabilizers and stored at relatively high temperatures (e.g. >50°C). The protein is then rehydrated and analysed for changes, such as aggregation and specific chemical alterations. Since this analysis is performed in solution, the approaches outlined in reviews on solution stability can be used. Detailed accounts can be found in Manning and Ahern (1992), Manning *et al.* (1989) and Cleland *et al.* (1993).

Given that there can be highly specific routes of degradation for individual proteins, it would seem impossible to discern general rules that would govern the storage stability of most proteins in dried solids. However, it appears that in formulation development what allows the most rational and rapid progress to a successful formulation is a combined approach. First, the formulation is designed to inhibit protein unfolding during the lyophilization process itself. This can be accomplished for most proteins by including a stabilizing additive such as sucrose in the formulation. Also, for some proteins (e.g. Prestrelski *et al.*, 1995) it has been found that altering the initial pH can help to minimize lyophilization-induced protein unfolding. Secondly, the formulation and storage conditions will need to be adjusted to minimize specific routes of chemical degradation that have been identified for the given protein. Finally, combined with these two approaches, optimization of the formulation for the critical physical factors (i.e. protein native structure and glassy state) should generally be needed to obtain optimal storage stability of any protein. The roles of these physical factors in storage stability will now be described.

In the earlier days of mechanistic studies of protein solid-state stability, it was proposed that all that was needed to assure the long-term stability of dried proteins was to maintain the formulation below its glass transition temperature, T_g (Franks, 1990; Franks *et al.*, 1991). In the dried powder, the protein is a component of a glassy phase that includes amorphous excipients and the residual water. If this phase is held below its characteristic glass transition temperature (T_g), which can be determined with DSC or other thermal scanning methods (Nail and Gatlin, 1993; Chang and Randall, 1992; Crowe *et al.*, 1996), the rate of diffusion-controlled reactions, including protein unfolding and chemical degradative processes, should be greatly reduced relative to rates noted above the transition temperature (Roy *et al.*, 1989; Franks, 1990; Franks *et al.*, 1991). Theoretically, degradative reactions will be limited by sample mobility and, hence, the further below the glass transition temperature that a sample is stored, the more stable the protein will be. Storage above the glass transition temperature should greatly accelerate reaction rates to a much greater level than expected, based on Arrhenius kinetics, due to the greatly increased mobility.

It is important to realize that water is a potent plasticizer for glasses. Thus, in addition to playing a direct role in degradation reactions, water in the dried formulation can greatly reduce T_g. For example, increasing the residual moisture of pure sucrose from 1% to about 3–4% (g H_2O/100 g dried powder) is sufficient to reduce T_g to below room temperature (Crowe *et al.*, 1998). Clearly, it is critical to achieve a sufficiently low residual moisture for a given formulation such that T_g exceeds the planned storage temperature.

More recently, it has been proposed that to obtain long-term storage stability of a dried protein formulation it is also necessary to minimize protein unfolding during lyophilization. To date there have been published studies on five different proteins that test the relative role of protein structure and formulation glassy state in protein storage stability. These will described in turn.

The first published data came from a study by Prestrelski *et al.* (1995) with interleukin-2, in which they found that when lyophilized from a solution of pH 7, the protein was unfolded in the dried solid and unstable during storage at 45°C. The dominant degradation product was aggregated protein, which was quantified after storage and rehydration. In contrast, lyophilization from a pH 4 solution led to a native protein and storage stability. This storage temperature was most likely below the sample T_g, because characteristically dried proteins have T_g values greater than 100°C (Angell, 1995). Using protein in a pH 7 solution, they also compared protection afforded during lyophilization and storage stability conferred by carbohydrates of increasing molecular weight (Prestrelski *et al.*, 1995). As molecular weight increased, the inhibition of unfolding during processing decreased, but the formulation glass transition temperature increased. The direct relationship between carbohydrate molecular weight and T_g is well established, and appears to be generally applicable to most individual carbohydrates (Levine and Slade, 1988a, 1988b, 1992). With interleukin-2, the optimum stability during storage at 45°C was noted in samples stabilized with the tetrasaccharide, stachyose. In these samples structure of the dried protein was native and T_g was greater than 45°C. The protein degraded in samples with lower molecular weight carbohydrates, which had T_g values less than 45°C, even though native protein structure was initially present in the dried solid. Thus, even with a native protein, stability is dependent on maintaining the temperature below T_g. With the highest molecular weight dextrans, for which $T_g > 45°C$, the protein was initially unfolded. However, no further damage was noted during storage, which supports the stabilizing role of the glassy state.

In the next published study, Chang and colleagues (1996) investigated stability of lyophilized interleukin-1 receptor antagonist. In support of the glass transition theory, they found that for any given formulation (among more than 15 tested), storage above T_g greatly accelerated degradation, which was due to deamidation and aggregation (measured after storage and rehydration). However, in a series of formulations (100 mg/ml protein) with varying initial sucrose concentrations ranging from 0 to 10% (w/v), all of which had a T_g of $66 \pm 2°C$, those with sucrose concentrations <5% degraded rapidly during storage below T_g at 50°C. In contrast, formulations with higher sucrose concentrations had less than 2% deamidation and no detectable aggregation after 14 months at 50°C. Infrared spectroscopy indicated that these stable formulations contained native protein in the dried solid, whereas the protein was unfolded in the unstable formulations. Thus, it appears that storage of a dried formulation below T_g is necessary but not sufficient for stability. It is also necessary to obtain a native protein during the lyophilization cycle.

More recently Kreilgaard *et al.* (1998) used various formulations to test the role of protein structure and glassy state on the stability of recombinant Factor XIII during storage at 40 and 60°C. First, they found that formulations prepared with excipients that crystallized during lyophilization, mannitol and polyethylene glycol, did not inhibit protein unfolding during lyophilization and also did not confer stability during storage. Thus, excipients that do not remain amorphous do not protect during processing or storage. Next they studied dextran, which does remain amorphous during lyophilization and storage, but does not inhibit lyophilization-induced unfolding (Prestrelski *et al.*, 1995). However, T_g values for dextran formulations usually greatly exceed the highest storage temperature tested (e.g. Prestrelski *et al.*, 1995). So if all that is needed for storage stability is to maintain a formulation below its T_g, then dextran should be excellent at conferring storage stability. With Factor XIII, the protein's structure in the dried solid containing dextran was even more perturbed than that noted in the sample prepared without excipients. This formulation underwent extensive degradation during storage at either temperature, which was manifested as protein aggregates and activated zymogen in the rehydrated samples. Thus, it appears that the failure of dextran to provide storage stability could be due to the poor retention of native protein structure noted immediately after lyophilization.

Optimal storage stability was found with a trehalose formulation, which had a T_g of 110°C and provided partial inhibition of lyophilization-induced unfolding. Similar results were noted with a sucrose formulation stored at 40°C. However, when this sample was stored at 60°C, there was rapid degradation of the dried protein. This was due to crystallization of sucrose during storage. This effect was most likely fostered by moisture transfer from the stopper to the dried formulation, and the resulting reduction of the formulation T_g and crystallization temperature to less than 60°C. The accelerated damage to the protein was attributed to the fact that sucrose forms an anhydrous crystal. As a result, the effective residual moisture of the remaining amorphous phase containing the protein would be increased upon crystallization of sucrose, leading to a decrease in T_g.

This effect could be dramatic. For example, let's consider a formulation containing 2 mg of dried protein and 50 mg of sucrose, with an initial moisture content of 1%. This means that the total amorphous fraction containing protein and sucrose has 0.52 mg of water. For simplicity let's assume that there was no transfer of moisture from the vial stopper or other source to the dried formulation. If 80% of the sucrose crystallized during storage, the 0.52 mg of water would be distributed in the 12 mg of material remaining amorphous. Thus, the moisture content of this fraction would increase from 1% to 4.3%. This increase would be more than sufficient to lower the formulation T_g to below room temperature (Crowe et al., 1998). In practice, the risk of sucrose crystallization and its disastrous consequences for protein stability can be minimized by using vial stoppers that are treated to reduce moisture transfer to the product. Also, avoiding storage temperatures approaching the T_g of the formulation reduces the risk of sucrose crystallization.

It should be noted that in addition to having a higher T_g for a given moisture content than sucrose, trehalose is also different in that it forms a dihydrate crystal (Crowe et al., 1998). Thus, if trehalose did crystallize due to storage conditions (e.g. relatively high storage temperature and moisture transfer into product), the remaining amorphous phase would actually have a lower moisture content and a higher T_g. Thus, trehalose crystallization might be expected to increase the storage stability of the protein, assuming that no other adverse effects were induced. Such effects could include perturbation of protein structure or reduction of the mass of amorphous phase sufficiently to eliminate effective 'dilution' of protein molecules in the dried solid.

It is also important to realize that disaccharide stabilizers are not the only formulation components for which there is a risk of crystallization and potential adverse effects on protein stability. Mannitol, which is often employed as a 'crystalline' bulking agent, is an example which has often been encountered in industry, but rarely reported in the literature. Often what has been seen is that product stored at a relatively high test temperature of about 40°C in the laboratory is stable for several months. During the winter the product retains stability during shipping, and there are no reports of problems from the consumers. However, occasionally during summer, after shipping excessive degradation is found after only a few weeks or months of storage at room temperature. Examination of the initial dried powder with differential scanning calorimetry (DSC) can provide the explanation. It is found that mannitol in the formulation is not completely crystallized, but rather forms a metastable glass with a T_g of greater than 40°C (e.g. about 45°C). When this temperature is exceeded, even for relatively short periods of time during shipping in the summer, the mannitol can crystallize. Since mannitol is an anhydrous crystal, as is the case with sucrose, the water originally 'associated' with mannitol stays in the remaining amorphous phase, which lowers its T_g. This problem can be avoided by using DSC to design an appropriate annealing protocol to complete mannitol crystallization, which can be implemented prior to primary drying and/or under vacuum after secondary drying (Carpenter and Chang, 1996). Alternatively, one may be able to adjust mannitol concentration or residual water down, to prevent crystallization even at temperatures around 45°C.

In a study with a lipase, which has a much more hydrophobic surface than typical globular proteins, Kreilgaard *et al.* (1999) obtained results similar to those noted with Factor XIII. Similar relative stabilities were noted with mannitol, polyethylene glycol, dextran and trehalose formulations, both immediately after lyophilization and during long-term storage at 40 and 60°C. The optimal stability was noted with trehalose, which greatly inhibited lyophilization-induced unfolding and provided a T_g exceeding 60°C. Also, in this study the initial sucrose–protein ratio was varied to determine whether this could affect propensity of sucrose to crystallize during storage, and, hence, protein storage stability. It was found that with an initial sucrose concentration of 50 mM (with 2 mg/ml protein), the protein had storage stability at either 40 or 60°C that was equal to that seen with trehalose. Under these conditions the sucrose did not crystallize. With an initial sucrose concentration of 300 mM, there was also excellent protein stability at 40°C, but rapid degradation at 60°C. Again, the degradation was linked to the crystallization of sucrose during storage, which was fostered by moisture transfer from the stopper to the dried formulation. Sarciaux and Hageman (1997) found that somatotropin was able to increase the crystallization temperature of sucrose as the mass fraction of sucrose–protein was decreased. These results show that if sucrose is the excipient, it is important to use a sufficient amount to protect the protein during lyophilization, but not to have an excessive amount that is prone to crystallization during storage.

Finally, Allison *et al.* (1998) have compared the capacities of dextran, sucrose and trehalose alone, and disaccharide-dextran mixtures to stabilize actin during lyophilization and storage in the dried solid. They found that dextran alone failed to confer optimum stability, again because protein unfolding was not inhibited during lyophilization. Optimal stability was found with mixtures of sucrose or trehalose and dextran. In these systems, the disaccharide protected protein structure during lyophilization. The presence of dextran increased the T_g of the dried powder relative to that measured for formulations lyophilized with just the disaccharides. In addition, dextran provided an amorphous 'bulking agent', which fostered the formation of strong, elegant, dried cakes.

Dextran is not always acceptable for parenteral products. However, other polymers have the potential to offer the same benefits as dextran to the physical properties of a lyophilized product. For example, hydroxyethyl starch also has a high T_g and generally is more acceptable than dextran for parenteral administration. It is hoped that the rational use of polymers as 'T_g modifiers' will make formulations more robust and much easier to lyophilize rapidly.

Finally, it is of interest to consider why non-native proteins undergo degradation during storage at temperatures well below T_g. It has been shown recently that mobility, which occurs on the time-scale of months and is of relevance to pharmaceutical products, is not completely eliminated unless the product is stored at at least 50 K below its T_g (Hancock *et al.*, 1995). Thus, many degradative reactions, especially those involving small reactive species and/or small-scale motions, may not be tightly coupled to the formulation glassy state (Pikal, 1994; Duddu *et al.*, 1997; Duddu and Dal Monte, 1997; Yoshioka *et al.*, 1997, 1998). For example, damage due to the presence of oxygen in the headspace (probably due to methionine oxidation) was noted with lyophilized factor VIII formulations at temperatures that should be well below the T_g of the formulation (Osterberg *et al.*, 1997). For a protein with an internal reactive methionine, damage may be avoided if the native conformation is retained during lyophilization. However, for a protein with reactive groups on the surface, even the native structure could be susceptible to degradation. Simply storing such proteins below T_g may also not be sufficient. In these cases more specific efforts to minimize degradation (e.g. remove oxygen from the vial headspace) may also be needed.

9.4   Summary and conclusions

For optimal long-term stability of dried protein formulations it is essential to inhibit protein unfolding during lyophilization and to achieve a formulation T_g that exceeds the product storage temperature. To date, it appears that the disaccharides sucrose and trehalose are most effective at meeting these criteria. Sucrose has been used in a number of protein products approved for parenteral administration. Recently, a product containing trehalose has been approved.

For rational formulation development it is critical that protein structure in the dried solid be assessed with infrared spectroscopy and that some thermal method be used to determine formulation T_g. Optimization of these parameters is the minimum requirement for development of stable lyophilized formulations. It may also be necessary to take other actions to reduce specific routes of chemical degradation.

Acknowledgements

JFC and TWR gratefully acknowledge support by grants from the National Science Foundation (BES9816975), the Whitaker Foundation, Boehringer Mannheim Therapeutics, Genentech, Inc., Genetics Institute, Inc., Genencor International, Amgen, Inc., and Zymogenetics, Inc. We also thank the National Science Foundation, the American Foundation of Pharmaceutical Education, the American Pharmaceutical Manufacturing Association and the Colorado Institute for Research in Biotechnology for providing predoctoral fellowships to our graduate students.

References

ALLISON, S.D., DONG, A. and CARPENTER, J.F., 1996, Counteracting effects of thiocyanate and sucrose on chymotrypsinogen secondary structure and aggregation during freezing, drying and rehydration. *Biophys. J.*, **71**, 2022–2032.

ALLISON, S.D., MANNING, M.C., RANDOLPH, T.W., MIDDLETON, K., DAVID, A. and CARPENTER, J.F., 1998, Effects of drying methods and additives on structure and function of action: mechanisms of dehydration-induced damage and its inhibition. *Arch. Biochem. Biophys.*, **358**, 171–181.

ANGELL, C.A., 1995, Formation of glasses from liquids and biopolymers, *Science*, **267**, 1924–1935.

ARAKAWA, T., PRESTRELSKI, S.J., KINNEY, W. and CARPENTER, J.F., 1993, Factors affecting short-term and long-term stabilities of proteins, *Adv. Drug Del. Rev.*, **10**, 1–28.

BYLER, D.M. and SUSI, H., 1986, Examination of the secondary structure of proteins by deconvoluted FTIR spectra, *Biopolymers*, **25**, 469–487.

CARPENTER, J.F. and CHANG, B.S., 1996, Lyophilization of protein pharmaceuticals, *Biotechnology and Biopharmaceutical Manufacturing, Processing, and Preservation*, K.E. AVIS and V.L. WU (eds), pp. 199–264, Buffalo Grove, IL: Interpharm Press.

CARPENTER, J.F., PRESTRELSKI, S.J. and ARAKAWA, T., 1993, Separation of freezing- and drying-induced denaturation of lyophilized proteins by stress-specific stabilization: I. Enzyme activity and calorimetric studies, *Arch. Biochem. Biophys.*, **303**, 456–464.

CARPENTER, J.F., PIKAL, M.J., CHANG, B.S. and RANDOLPH, T.W., 1997, Rational design of stable lyophilized protein formulations: some practical advice, *Pharm. Res.*, 969–975.

CARPENTER, J.F., PRESTRELSKI, S.J. and DONG, A., 1998, Application of infrared spectroscopy to the development of stable lyophilized protein formulations, *Eur. J. Pharm. Biopharm.*, **45**, 231–238.

CHANG, B.S. and RANDALL, C.S., 1992, Use of subambient thermal analysis to optimize protein lyophilization, *Cryobiology*, **29**, 632–656.

CHANG, B.S., BEAUVAIS, R.M., DONG, A. and CARPENTER, J.F., 1996, Physical factors affecting the storage stability of freeze-dried interleukin-1 receptor antagonist: glass transition and protein conformation, *Arch. Biochem. Biophys.*, **331**, 249–258.

CLELAND, J.L., POWELL, M.F. and SHIRE, S.J., 1993, The development of stable protein formulations: a close look at protein aggregation, deamidation, and oxidation, *Crit. Rev. Ther. Drug Carrier Sys.*, **10**(4), 307–377.

CROWE, L.M., REID, D.S. and CROWE, J.H., 1996, Is trehalose special for preserving dry biomaterials? *Biophys. J.*, **71**, 2087–2093.

CROWE, J.H., CARPENTER, J.F. and CROWE, L.M., 1998, Role of vitrification in anhydrobiosis, *Ann. Rev. Physiol.*, **60**, 73–103.

DONG, A. and CAUGHEY, W.S., 1994, Infrared methods for study of hemoglobin reactions and structures, *Methods Enzymol.*, **232**, 139–175.

DONG, A., PRESTRELSKI, S.J., ALLISON, S.D. and CARPENTER, J.F., 1995, Infrared spectroscopic studies of lyophilization- and temperature-induced protein aggregation, *J. Pharm. Sci.*, **84**, 415–424.

DUDDU, S.P. and DAL MONTE, P.R., 1997, Effect of glass transition temperature on the stability of lyophilized formulations containing a chimeric therapeutic monoclonal antibody, *Pharm. Res.*, **14**, 591–595.

DUDDU, S.P., ZHANG, G. and DAL MONTE, P.R., 1997, The relationship between protein aggregation and molecular mobility below the glass transition temperature of lyophilized formulations containing a monoclonal antibody, *Pharm. Res.*, **14**, 596–600.

EDSALL, J.T. and McKENZIE, H.A., 1983, Water and proteins II. The location and dynamics of water in protein systems and its relation to their stability and properties, *Adv. Biophys.*, **16**, 53–183.

FRANKS, F., 1990, Freeze drying: from empiricism to predictability, *Cryo-Letters*, **11**, 93–110.

FRANKS, F., HATLEY, R.H.M. and MATHIAS, S.F., 1991, Material science and the production of shelf stable biologicals, *BioPharm*, **4**(9), 38–55.

HAGEMAN, M., 1992, Water sorption and solid state stability of proteins, *Stability of Protein Pharmaceuticals. Part A. Chemical and Physical Pathways of Protein Degradation*, T. AHERN and M.C. MANNING (eds), pp. 273–309, New York: Plenum Press.

HANCOCK, B.C., SHAMBLIN, S.L. and ZOGRAFI, G., 1995, Molecular mobility of amorphous pharmaceutical solids below their glass transition temperature, *Pharm. Res.*, **12**, 799–806.

KREILGAARD, L., FROKJAER, S., FLINK, J.M., RANDOLPH, T.W. and CARPENTER, J.F., 1998, Effects of additives on the stability of recombinant human Factor XIII during freeze-drying and storage in the dried solid, *Arch. Biochem. Biophys.*, **360**, 121–134.

1999, Effects of additives on the stability of *Humicola lanuginosa* lipase during freeze-drying and storage in the dried solid, *J. Pharm. Sci.*, **88**, 281–290.

KRIMM, S. and BANDEKAR, J., 1986, Vibrational spectroscopy and conformation of peptides, polypeptides and proteins. *Adv. Protein Chem.*, **38**, 181–364.

KUNTZ, I.D. and KAUZMAN, W., 1974, Hydration of proteins and polypeptides, *Adv. Protein Chem.*, **28**, 239–345.

LEVINE, H. and SLADE, L., 1988a, Principles of 'cryostabilization' technology from structure/property relationships of carbohydrate/water systems – a review, *Cryo-Letters.*, **9**, 21–63.

LEVINE, H. and SLADE, L., 1988b, Thermomechanical properties of small-carbohydrate–water glasses and 'rubbers': kinetically metastable systems at sub-zero temperatures, *J. Chem. Soc. Faraday Trans.*, **184**, 2619–2633.

1992, Glass transitions in foods, *Physical Chemistry of Foods*, H.S. SHARTZBERG and R.W. HARTEL (eds), pp. 83–221, New York: Dekker.

MANNING, M.C. and AHERN, T.J., 1992, *Stability of Protein Pharmaceuticals: Part A. Chemical and Physical Pathways of Protein Degradation*, New York: Plenum Press.

MANNING, M.C., PATEL, K. and BORCHARDT, R.T., 1989, Stability of protein pharmaceuticals, *Pharm. Res.*, **6**, 903–918.

MANTSCH, H.H. and CHAPMAN, D., 1997, *Infrared Spectroscopy of Biomolecules*, New York: Wiley-Liss.

NAIL, S.L. and GATLIN, L.A., 1993, Freeze-drying: principles and practice. *Pharmaceutical Dosage Forms: Parenteral Medications*, Vol. 2, K.E. AVIS, H.A. LIEBERMAN and L. LACHMAN (eds), pp. 163–233, New York: Marcel Dekker.

OSTERBERG, T., FATOUROS, A. and MIKAELSSON, M., 1997, Development of a freeze-dried albumin-free formulation of recombinant Factor VIII SQ, *Pharm. Res.*, **14**, 892–898.

PIKAL, M.J., 1994, Freeze-drying of proteins, *Formulation and Delivery of Proteins and Peptides*, J.L. CLELAND and R. LANGER (eds), ACS Symposium Series, Vol. 567, pp. 120–133.

PRESTRELSKI, S.J., ARAKAWA, T. and CARPENTER, J.F., 1993a, Separation of freezing- and drying-induced denaturation of lyophilized proteins by stress-specific stabilization: II. Structural studies using infrared spectroscopy, *Arch. Biochem. Biophys.*, **303**, 465–473.

PRESTRELSKI, S.J., TEDESCHI, N., ARAKAWA, T. and CARPENTER, J.F., 1993b, Dehydration-induced conformational changes in proteins and their inhibition by stabilizers, *Biophys. J.*, **65**, 661–671.

PRESTRELSKI, S.J., PIKAL, K.A. and ARAKAWA, T., 1995, Optimization of lyophilization conditions for recombinant human interleukin-2 by dried-state conformational analysis using Fourier transform infrared spectroscopy, *Pharm. Res.*, **12**, 1250–1259.

ROY, M.L., PIKAL, M.J., RICKARD, E.C. and MALONEY, A.M., 1989, The effects of formulation and moisture on the stability of a freeze-dried monoclonal antibody–vinca conjugate: a test of the WLF glass transition theory, *Dev. Biol. Standard.*, **74**, 323–340.

SARCIAUX, J.-M.E. and HAGEMAN, M.J., 1997, Effects of bovine somatotropin (rbSt) concentration at different moisture levels on the physical stability of sucrose in freeze-dried rbSt/sucrose mixtures, *J. Pharm. Sci.*, **86**, 365–371.

STRAMBINI, G.B. and GABELLIERI, E., 1996, Proteins in frozen solutions: evidence of ice-induced partial unfolding, *Biophys. J.*, **70**, 971–976.

SUREWICZ, W.K. and MANTSCH, H.H., 1988, New insights into protein conformation from infrared spectroscopy, *Biochim. Biophys. Acta*, **953**, 115–130.

SUSI, H. and BYLER, D.M., 1986, Resolution enhanced Fourier transform infrared spectroscopy of enzymes, *Methods Enzymol.*, **130**, 290–311.

WANG, Y.J. and HANSON, M.A., 1988, Parenteral formulations of proteins and peptides: stability and stabilizers, *J. Parenter. Sci. Technol.*, **42**, 2–26.

WANG, Y.J. and PEARLMAN, R., 1993, *Stability and Characterization of Protein and Peptide Drugs: Case Histories*, Pharmaceutical Biotechnology, Volume 5, New York: Plenum Press.

YOSHIOKA, S., ASO, Y. and KOJIMA, S., 1997, Softening temperature of lyophilized bovine serum albumin and gamma-globulin as measured by spin-spin relaxation time of protein protons, *J. Pharm. Sci.*, **86**, 470–474.

YOSHIOKA, S., ASO, Y., NAKAI, Y. and KOJIMA, S., 1998, Effect of high molecular mobility of poly(vinyl alcohol) on protein stability of lyophilized gamma-globulin formulations. *J. Pharm. Sci.*, **87**, 147–151.

Peptide and Protein Drug Delivery Systems for Non-parenteral Routes of Administration

METTE INGEMANN, SVEN FROKJAER, LARS HOVGAARD AND
HELLE BRØNDSTED

Department of Pharmaceutics, The Royal Danish School of Pharmacy, Copenhagen, Denmark

10.1 Introduction

Proteins for systemic effect are most commonly given by the parenteral route. The preferred oral administration has been considered one of the greatest challenges in drug delivery of peptide and protein drugs, and is extensively described in a number of reviews (Langguth *et al.*, 1997; Lee, 1991; Wearley, 1991). However, the inherent barrier function of the intestine towards the absorption of larger peptides and proteins in therapeutic concentrations has prompted scientists to explore alternative routes. In general, it is the physicochemical properties of peptide and protein drugs that limit their bioavailability, including high molecular weight, susceptibility to enzymatic hydrolysis and high hydrophilicity.

10.2 Non-parenteral routes of delivery for peptides and proteins

Delivery of peptide and protein drugs by non-parenteral routes is obviously useful for local treatment. However, for systemic delivery of peptide and protein drugs, parenteral administration is still the most commonly used route. Numerous reviews on potential non-parenteral routes of administration for peptide and protein drugs can be found in the

Table 10.1 Non-parenteral routes of administration: absorption area and proteolytic barrier

Route	Absorption area*	Proteolytic activity level[†]
Oral	200 m^2	+++++++
Rectal	$200–400 \text{ cm}^2$	+++
Buccal	100 cm^2	++
Nasal	150 cm^2	++
Transdermal	2 m^2	+
Pulmonary	75 m^2	+
Vaginal		++
Ocular	1 cm^2	

* From Lee (1991), Cumming (1980) and Wearley (1991);
[†] From Zhou (1994)

literature (Eppstein and Longenecker, 1988; Wearley, 1991; Zhou and Wan Po, 1991; Reddy and Banga, 1993). Among the alternative routes of administration are the oral, rectal, buccal, nasal, transdermal, pulmonary, vaginal and ocular routes (Table 10.1). In general, oral delivery is by far the most favoured route of administration. However, drugs absorbed from the intestine are subjected to the hepatic first-pass metabolism before entering the systemic circulation. In comparison to oral delivery, the other alternative mucosal routes have a number of advantages in common: bypass of or reduced hepatic first-pass metabolism, sufficient absorption areas, and a reduced presence of proteolytic enzymes. The different anatomical and physical characteristics of the various mucosae, together with the characteristics of the drug molecule, set the limitations for the absorption of a certain peptide or protein drug. Therefore, the choice of administration site for a specific disease relies on integrated considerations of drug characteristics, membrane properties and level of treatment. The rectal route has, in spite of not being well accepted by patients, obvious advantages for peptide and protein drug delivery. The rectal environment is quite constant with respect to the amount and viscosity of the rectal fluid (Wearley, 1991). Moreover, the proteolytic activity is significantly lower than in the upper intestine. The nasal mucosa is highly vascularized, and the absorptive epithelium consists of only a single cellular barrier. Therefore, the possibility of a rapid delivery of the peptide or protein drug exists, despite the existence of an often limited and variable residence time (Verhoef *et al.*, 1990). The buccal route is much less efficient with respect to transport due to a higher epithelial thickness of the multilayer, presumably due to keratinization. Potential advantages of transdermal peptide or protein delivery are the ease of discontinuing drug administration, the possibility of long-term application, and excellent patient compliance. Although the skin is extremely impermeable, the transdermal administration of peptide and protein drugs is expected to be important in the future, e.g. by the use of iontophoresis or other electrostimulus-induced absorption processes (Merino *et al.*, 1997). Until recently, the pulmonary route has not been explored in detail as a possible route of administration of peptide and protein drugs for systemic effect. The potential of this route relies on the large area available for absorption and the thin epithelial barrier. The pharmaceutical challenges of pulmonary delivery lie in the inefficient deposition of aerosol particles in the alveoli (Verhoef *et al.*, 1990). Vaginal application of peptide and protein drugs gained some attention 10–15 years ago. However, in spite of a rich blood supply, good permeability and a relatively large surface area, the cyclic changes and aesthetic concerns make the route unrealistic for widespread use in drug delivery (Richardson and Illum, 1992). The potential use of the ocular route for systemic delivery of peptides and proteins is suggested by the fact that more than 90% of an applied dose to the eye can be absorbed into the systemic circulation, bypassing the nasal mucosa due to the tear drainage through the nasolacrimal duct (Reddy and Banga, 1993).

10.2.1 Barriers to non-parenteral administration of peptides and proteins

There are many obstacles to non-parenteral administration of peptides and proteins. The barriers may broadly be categorized as (a) enzymatic barriers that exist at the site of administration or as a part of the transport pathway, and (b) physical barriers to the transport of drug substances through the epithelium.

Enzymatic barriers

The enzymatic barrier is often underrated and is probably the most aggressive barrier. On the other hand, it is the easiest barrier to overcome by the use of enzyme inhibitors. The enzymatic barrier involves enzymatic degradation either by hydrolysis of peptide bonds by exopeptidases or endopeptidases or by chemical modification such as oxidation, phosphorylation or deamidation (Lee, 1991). The hydrolytic cleavage by peptidases is by far the dominant degradation pathway for peptides and proteins. This is an irreversible reaction which cleaves peptides and proteins into amino acids and smaller peptide segments. The peptides and proteins are theoretically accessible for degradation at several linkages within the amino acid chain, by peptidases that act synergistically (Woodley, 1997). In general, larger peptides and proteins are initially degraded by enzymes not related to the surface of an epithelium, followed by the action of membrane-bound brush border enzymes on the ensuing smaller peptides.

Both exopeptidases and endopeptidases are present in each mucosal tissue, and what distinguishes the enzymatic activity of one mucosa from another is probably the relative proportion of these peptidases as well as their subcellular distribution (Lee, 1991). Comparing the proteolytic barrier level for possible non-parenteral routes by listing the content of soluble proteins in the various tissue homogenates (Table 10.1), Zhou (1994) has shown that the oral route presents the most intensive enzymatic barrier to the absorption of peptide and protein drugs and the skin the least intensive barrier. In addition to the massive amount of peptidase enzyme activity secreted daily by the pancreas, the hydrochloride content of the stomach probably denatures the peptide or protein drug.

Absorption barriers

Crossing of the epithelial cell layer is basically possible via five routes: by passive paracellular diffusion through the hydrophilic tight junctions, by passive or facilitated transcellular diffusion through the lipophilic absorptive cells, by active carrier-mediated transport systems, or by transcytosis (Figure 10.1). Most larger peptides and proteins, if they are absorbed at all, are absorbed by passive paracellular diffusion due to their macromolecular dimensions and hydrophilic nature. However, there are exceptions, e.g. immunoglobulin G and epidermal growth factor, which are absorbed by transcytosis (Lee, 1991). Certain smaller peptides such as dipeptides and tripeptides have been shown

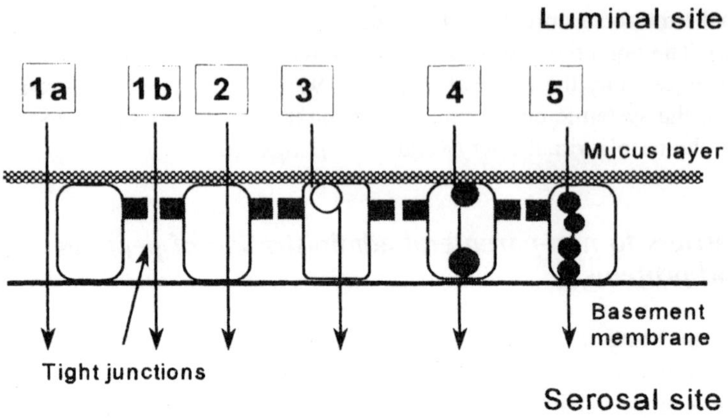

Luminal site

Figure 10.1 boxes: 1a | 1b | 2 | 3 | 4 | 5

Mucus layer

Basement membrane

Tight junctions

Serosal site

Figure 10.1 Routes for the transport of drugs across a mucous cell barrier: (1a) passive paracellular diffusion; (1b) passive paracellular diffusion with existing tight junctions; (2) passive transcellular diffusion; (3) facilitated transcellular diffusion; (4) active carrier-mediated transport; (5) transcytosis.

to have affinity for an active nutrient transporter and to be absorbed by this transport mechanism in the intestine (Bai and Amidon, 1992; Taub *et al.*, 1997). In addition to the passage of the absorptive cells, the physical absorption barrier is often complicated by the existence of the highly viscous and water-containing mucin layer consisting of a negatively charged glycoprotein structure.

10.2.2 *General approaches to bypass enzymatic and absorption barriers*

In order to bypass or minimize the enzymatic and absorption barriers for the purpose of increasing the bioavailability of peptide and protein drugs, several pharmaceutical approaches are available: (1) inhibition of enzymatic degradation by the co-administration of protease inhibitors; (2) increasing permeability across the membrane by the co-administration of absorption enhancers; (3) chemical modification of the drug substance to improve resistance to degradation and transport across the barrier; and (4) designing special pharmaceutical formulations which protect against enzymatic degradation and, in some cases, improve transport across the absorption barrier as well. In the following, methods to overcome the enzymatic and absorption barriers with a focus on the pharmaceutical formulation approach are described. Chemical modification of peptide and protein drugs is described in Chapter 11.

Protease inhibitors

Co-administration of inhibitors of proteolytic enzymes provides a viable means to circumvent the enzymatic barrier in achieving the delivery of peptide and protein drugs. A number of inhibitors including aprotinin (trypsin/chymotrypsin inhibitor), amastatin, bestatin, boroleucine and puromycin (aminopeptidase inhibitors) have been reported for that purpose. One example of the usefulness of protease inhibitors of proteolytic enzymes is the increase in insulin absorption after oral administration, shown by a sharp drop in blood glucose levels in healthy volunteers, when co-administered with a chymotrypsin inhibitor, FK-448 (Shinomiya *et al.*, 1985). Drawbacks of using protease inhibitors are

Table 10.2 Classes of absorption enhancers

Class	Examples	Proposed modes of action
Chelators	EDTA Citric acid Salicylates	Complexation with Ca^{2+} Membrane protein leaching
Synthetic surfactants	Sodium lauryl sulphate Polyoxyethylene alkyl ethers Polyoxyethylene alkyl esters	Micelle formation Membrane lipid solubilization
Natural and semisynthetic surfactants	Bile salts such as: sodium glycocholate sodium taurocholate sodium deoxycholate	Micelle formation Membrane lipid solubilization
Fatty acids	Oleic acid Sodium caprylate Sodium caprate Sodium laurate	Membrane lipid leaching Mucus barrier reduction

that the inhibition may also affect the absorption of other peptides or proteins that normally would be degraded. Moreover, high doses of inhibitor are needed, and the bioavailability of the drug is still limited by the physical barrier of the cells.

Absorption enhancers

In the 1960s, the first use of compounds which could temporarily diminish the integrity of physical barriers, such as the skin and the gastrointestinal wall, was reported. Since then, absorption enhancers that are less tissue-damaging and less toxic have been found, arising from both natural substances and synthesized products. Ideal absorption enhancers should show no toxicity, high efficacy and high specificity. In general, absorption enhancers can be grouped in four classes: (a) chelators; (b) synthetic surfactants; (c) natural and semisynthetic surfactants; and (d) fatty acids and their derivatives. Examples are listed in Table 10.2. The mechanism of absorption enhancement is often complex and not always known in detail. However, much work has dealt with clarifying the mechanisms of action and evaluating the toxicities of absorption enhancers (Anderberg *et al.*, 1992; Hovgaard *et al.*, 1995). Possible mechanisms include an increase in membrane fluidity, a decrease in mucous viscosity, solubilization of the drug, leaching of proteins from the membrane, and opening of the tight junctions (Reddy and Banga, 1993).

The chelators are believed to exert their action by complexation with calcium ions and by facilitating leaching of proteins from the membrane. The result is a contraction of the cell followed by an increase in paracellular transport due to opening of the tight junctions and an increase in membrane fluidity. The absorption enhancement by bile salts and surfactants is always related to the micelle formation and is brought about in several ways. It has been suggested that the delivery of the peptide or protein drug to the absorptive surface of the mucosa is increased by micellar protection or that the membrane integrity is disrupted due to solubilization of membrane phospholipids (Aungst *et al.*, 1996). Fatty acids are believed to have a more direct action on the organization of the membrane lipid bilayer, causing changes in the membrane fluidity (Muranishi, 1990). An effect from the reduction of the thickness of the mucus layer observed in rats has also been reported as a possible enhancing mechanism for fatty acids (Lee and Yamamoto, 1990).

Formulation approach

The general strategy for improving peptide or protein delivery is to incorporate the drug substance in a delivery system designed to protect the peptide or protein drug from contact with the proteolytic enzymes present at the site of administration, and, further, to release the drug substance only upon reaching an area favourable for its absorption. A combination of the formulation approach and the use of enzyme inhibitors or absorption enhancers is often used. In the following, different drug delivery systems used for non-parenteral administration of peptide and protein drugs in order to overcome enzymatic and absorption barriers will be described.

10.3   Formulation principles for peptides and proteins

A prerequisite for a successful drug delivery system, which should resist the enzymatic barrier at the site of administration and improve the transport across the absorption barrier, is a maintenance of the biological activity of the peptide or protein drug. There-fore, a drug delivery system has no relevance if the preparation of the system, the storage, or handling afterwards results in an inactivation of the drug. Stabilization techniques developed can be divided into four groups: adsorption to a carrier, covalent binding to a carrier, entrapment in a matrix, and encapsulation in a semipermeable membrane. The most frequently used formulation principles for the purpose of non-parenteral peptide or pro-tein drug delivery, i.e. entrapment, encapsulation and covalent binding, will be discussed.

10.3.1   Entrapment and encapsulation

Entrapment and encapsulation are the most widely used pharmaceutical techniques for designing drug delivery systems for peptide or protein drugs to optimize delivery properties including protection against enzymatic degradation. Entrapment or encapsula-tion could be in the form of polymeric drug delivery systems such as hydrogels and nanocapsules/microspheres, and lipid drug delivery systems such as liposomes and micro-emulsions. The proposed effects of the drug delivery systems are compared in Table 10.3.

Table 10.3   Expected effect on the enzymatic and absorption barrier of the presented drug delivery systems for peptide and protein drug delivery

Effect on	Hydrogel	Nanocapsules/ microspheres	Liposomes	Microemulsions
Enzymatic barrier	+++	++	++	+
Absorption barrier	+	++	++	+++

+++, major effect; ++, of importance; +, minor effect.

Hydrogels

The use of hydrogels has become increasingly important in the pharmaceutical and medical areas. Hydrogels have especially been used for wound-healing purposes, in non-thrombogenic coating of surgery materials, as contact lenses, as implants and in vaginal, oral or rectal controlled release systems to deliver drugs or bioactive macromolecules (Peppas, 1987). Hydrogels are by definition, crosslinked hydrophilic polymers of natural or synthetic origin and have the ability to swell in an aqueous environment without dissolving. The hydrophilic hydrogels exhibit a range of potential physical, chemical and biological properties which make them attractive as drug delivery systems, e.g. high biocompatibility and release control.

Preparation of hydrogels Hydrogel entrapment is a technique by which monomers form a crosslinked polymeric network around the material to be entrapped. The surrounding polymer matrix permits entrapment of large quantities of peptide or protein drugs. The polymerization reaction can be carried out in bulk, in solution or in suspension. Most pharmaceutical hydrogels are synthesized in solution by free radical polymerization initiated by the generation of free radicals by thermal, ionization, radiation or redox means (Peppas and Khare, 1993). This is usually performed by mixing of monomers and cross-linking agent in an organic solvent or buffered solution followed by the addition of a catalyst system, which initiates the polymerization process (Roorda *et al.*, 1986). The peptide or protein drug to be entrapped can be introduced into the hydrogel matrix during or after synthesis (Figure 10.2). Entrapment after synthesis is done by soaking of pre-washed hydrogels in a concentrated drug solution (Gyselinck *et al.*, 1983). Due to the fact that peptides or proteins may crosslink to the polymer network if introduced before crosslinking, the post-crosslinking soaking is usually the method of choice.

The polymerized hydrogel may be produced in a wide variety of geometric shapes including films, membranes, rods and particles. Common examples of synthetic and natural polymers which are used for hydrogels are acrylics, vinyl alcohols, ethylene oxides, cellulose ethers and starch, albumin or dextran (Mishra *et al.*, 1996). In addition, hydrogels with enzyme digestible crosslinkers or polymer backbones are used as biodegradable drug delivery systems (Park, 1988). The use of biodegradable polymers is an interesting principle for delivering peptides and proteins to the colon with a possibility of improving systemic bioavailability.

Nanocapsules and microspheres

Encapsulation has been used to avoid the involvement of administered peptides and proteins in immunological reactions, as well as to protect the drug against degradation and to control its release from the site of administration. Nanocapsules, nanospheres, microcapsules, and microspheres differ by their size (below and above 1 µm, respectively) and the process used for preparation. Nanocapsules and microcapsules are vesicular systems in which the drug is confined to a cavity surrounded by a polymer membrane, whereas nanospheres or microspheres are matrix systems in which the drug is dispersed throughout the particle.

Preparation of nanocapsules and microcapsules The preparation methods commonly used are divided into two main categories according to whether the formation requires an *in situ* polymerization reaction or whether it is achieved directly from a preformed polymer or a natural macromolecule.

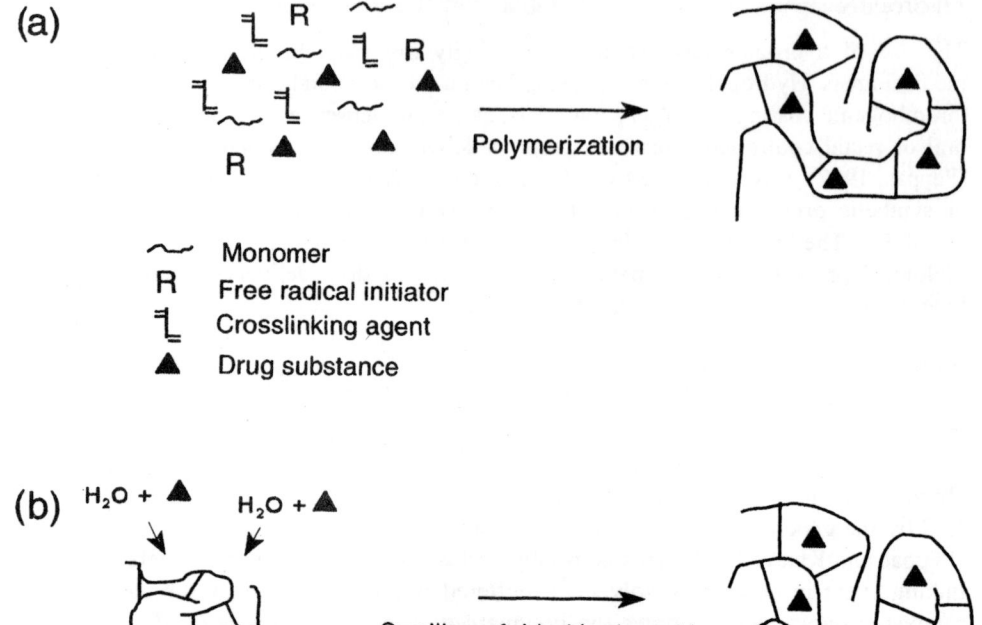

Figure 10.2   Introduction of drug: (a) before crosslinking in free radical hydrogel synthesis; (b) after hydrogel synthesis by soaking of dried hydrogel in a concentrated drug solution.

In general, preparation via a polymerization reaction takes place in two steps initiated by the introduction of the monomer into a dispersed phase of an emulsion or an inverse microemulsion, or dissolved in a non-solvent of the polymer. Subsequently, a nucleation phase is followed by a growth phase. Earlier, acrylic monomers were frequently used, and recently much work has been published on cyanoacrylate nanocapsules which are easier to form. However, *in situ* polymerization is often a harsh technique by which peptides and proteins, due to their chemical reactivity, can participate in the polymerization reaction in an uncontrolled manner.

Starting from preformed polymers, nanocapsules or microcapsules are formed by the precipitation of synthetic polymers or by denaturation of natural macromolecules (Couvreur *et al.*, 1995). Tedious purification to remove residual monomers can be avoided by using preformed polymers for the preparation.

Preparation of nanospheres and microspheres   The processes commonly used to make nanospheres and microspheres include spray-drying, solvent-evaporation, and phase-separation techniques.

Spray-drying consists of the transformation of a liquid into a solid, where the drug is dispersed or dissolved in an organic solution of the polymer to be sprayed (Giunchedi and Conte, 1995). Both the presence of organic solvents and increased temperatures may result in chemical and physical stability problems during preparation.

Solvent evaporation involves an initial solubilization of the polymer in a suitable water immiscible solvent, followed by an emulsification in an aqueous continuous phase to form discrete droplets. The organic phase will then evaporate at the water–air interface. The resulting microspheres are isolated by filtration or centrifugation, washed and dried (O'Donnell and McGinity, 1997). Preparation of microspheres may result in degradation of the peptides or proteins due to hostile processing conditions such as high shear forces and organic solvents.

For preparing microspheres by a phase-separation technique, i.e. simple or complex coacervation, a typical procedure involves preparation of a water/oil emulsion, with the drug and a polymer dissolved in the water phase. Once the emulsion is formed, the polymer is precipitated by reducing the aqueous solubility by one of several techniques, e.g. gelation by cooling, thermal denaturation or precipitation by adding salts or organic solvents to the water phase. Again, the resulting microspheres can be further stabilized by crosslinking and isolated by filtration or centrifugation, washed and dried. This is a rather simple and mild technique in which the drug often will remain intact during the preparation process. Preparation of nanospheres by phase separation can be conducted by the same procedure by reducing the size of the primary emulsion droplets or by a controlled polymer precipitation from an aqueous solution (Rajaonarivony *et al.*, 1993).

For all types of nanospheres and microspheres, drug incorporation can be performed during preparation, resulting in encapsulated drug, or after preparation, resulting in adsorption to or partitioning into the nanospheres or microspheres.

Liposomes

Liposome formulations are most frequently considered for parenteral administration of drugs, but may also be a potential formulation principle for alternative routes, e.g. topical, nasal, and pulmonary administration. Liposomes are vesicles in which an aqueous core is enclosed by phospholipid bilayers. Liposomes are broadly classified by size and lamellarity as small unilamellar vesicles (SUVs; size 25–50 nm), large unilamellar vesicles (LUVs; 100 nm) and multilamellar vesicles (MLVs; 50–10 000 nm) (Vemuri and Rhodes, 1995). Liposomes present unique therapeutic opportunities because both hydrophilic and hydrophobic drug molecules can be incorporated, they are biodegradable, they can be made in a variety of sizes, and they can be used for controlled release and targeted delivery (Kulkarni *et al.*, 1995). Major disadvantages include drug leakage during storage, *in vivo* instability and production scale-up problems.

Preparation of liposomes Liposomes are spontaneously formed when phospholipids are dispersed in excess water, arranging themselves in bilayers with the enclosure of an aqueous core. Most methods of making liposomes involve a lipid hydration step, yielding SUVs or MLVs, whereas solvent dispersion methods are used for making LUVs (New, 1990; Lichtenberg, 1988). The drug incorporation in liposomes occurs at different stages during the preparation process, depending on the lipophilicity of the drug. Incorporated drugs are localized in the aqueous core and/or in the bilayer, depending on the hydrophobicity and/or the electrostaticity of the drug (Talsma and Crommelin, 1992). In addition to the most widely used phospholipid, phosphatidylcholine, charge-inducing agents or cholesterol are often incorporated into the bilayer structure to facilitate liposome formation or stabilize against fusion–aggregation or drug leakage.

Important parameters to characterize for liposome drug delivery systems are the drug encapsulation efficiency, the average size and size distribution of the liposomes, and the chemical composition of the liposomes after manufacture. With regard to peptide and protein formulation, considerations concerning liposomal membrane association are of special importance due to the amphiphilic character of most peptides and proteins. This means, for instance, that peptide and protein localization and binding to the bilayer structure are very sensitive to changes in the membrane composition, for instance caused by chemical degradation of phospholipids.

Microemulsions

Another formulation strategy for peptide or protein drug delivery is to use microemulsions. Emulsions are two immiscible phases dispersed in one another; microemulsions are more complex quarternary or ternary systems, containing a surfactant/co-surfactant blend added to a two-phase hydrophilic/lipophilic mixture. Microemulsions are spontaneously formed at room temperature, appear transparent, and consist of micro-droplets of a size less than 100 nm (Sarciaux *et al.*, 1995).

Preparation of microemulsions   The microemulsions are obtained by mixing of three or four components: a water phase, an oil phase, a co-surfactant and a surfactant. Only specific concentrations and ratios of the components yield microemulsions, which require the use of phase diagrams for the components involved. Lipophilic drug substances are incorporated into the microemulsions by solubilization in the oil or oil/surfactant blend before mixing, whereas hydrophilic drug substances are preferably solubilized in the aqueous phase. A major concern is the high concentration of surfactants needed, with a potential risk of toxicity. An increasing number of reports in the literature suggest that microemulsions can be used to enhance the oral bioavailability of peptides (Ritschel, 1991).

10.3.2   *Covalent binding*

Conjugation of poly(ethylene) glycol (PEG) to proteins, referred to as 'PEGylation' or 'PEGnology', is a widely used technique to stabilize proteins (Harris, 1997; Abuchowski *et al.*, 1977). In the majority of reported studies PEG–protein conjugation is intended for parenteral injection, where the most dramatic property of the PEG–protein conjugates is their increased *in vivo* circulation time compared to the unconjugated protein. However, this principle may also have great potential for non-parenteral formulations as a method for improving stability against enzymatic degradation at the site of administration.

Preparation of PEG conjugates

The PEG molecules require an activation of the hydroxyl groups before the conjugation reaction. In the majority of cases of conjugation of PEG to peptide or proteins, electrophilic groups are introduced to the polymer and coupled to the nucleophilic ε-amino atoms of lysine residues or the N-terminal amino group on the peptides or proteins. The coupling moiety is either incorporated as a part of the PEG–protein conjugate or lost. Overviews of commonly used PEG-coupling methods have been presented by several authors (Zalipsky and Lee, 1992; Francis *et al.*, 1992; Mumtaz and Bachhawat, 1991). The four most frequently used methods for the preparation of PEG–protein conjugates in development of pharmaceutical preparations are presented here (Figure 10.3).

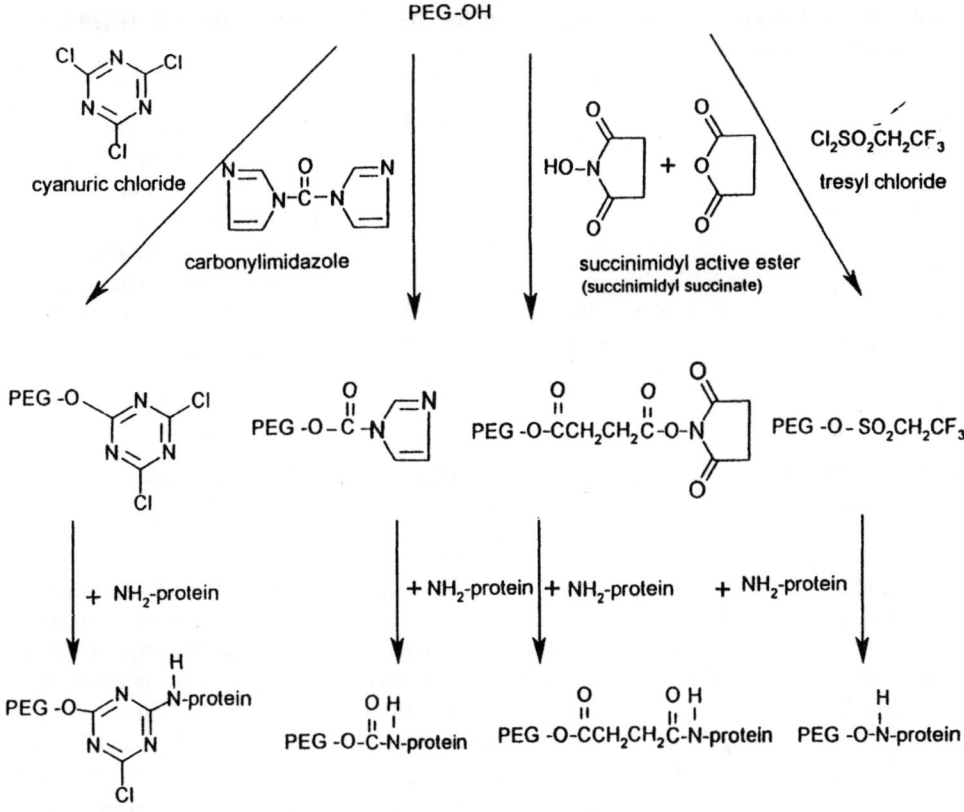

Figure 10.3 PEG-activation and PEG-coupling reactions to proteins for cyanuric chloride, carbonylimidazole, succinimidyl active ester and tresyl chloride (Francis *et al.*, 1992).

- *Cyanuric chloride* has been the most widely used reagent for activation of PEG. However, this method uses harsh conditions and crosslinked products can be formed due to the presence of two reactive chlorines in the triazine ring. A rather large coupling moiety, a triazine ring, is incorporated in the PEG–protein conjugate.

- PEG activated by *carbonylimidazole* requires very long reaction times but leaves only a carbonylgroup–peptide-like conjugation after binding.

- The *succinimidyl active ester* method involves a generation of a carboxylic acid at the PEG hydroxyl end by addition of succinic anhydride followed by activation with N-hydroxysuccinimide. This way of PEG-coupling to proteins does not inactivate SH-dependent proteins, like cyanuric chloride-activated PEG, but the method is limited by the hydrolysis of the labile ester linkage.

- Using *tresyl chloride* as an activation reagent implies that the reaction can be performed fast under very mild conditions, with no incorporation of the coupling moiety (Francis *et al.*, 1992).

The best method for activation and conjugation will depend on the actual peptide or protein drug and will vary from case to case. More than 40 proteins have been conjugated to PEG and a number of these have already been in clinical trials (Fuertges and Abuchowski, 1990).

10.4   Immobilized proteins intended for local effect in the GI tract – a case study

A variety of diseases can be treated by the administration of therapeutic proteins. Examples are enzyme substitution by local-acting pancreatic enzymes in the treatment of digestive organ diseases, enzyme replacement by local or systemic nutrient or metabolite depletion (e.g. in cancer therapy by antitumour enzymes), use of fibrinolytic and clotting proteins to inhibit or promote blood clotting or the treatment of genetically derived enzyme deficiencies such as lysosomal storage diseases (Torchillin, 1987). The following section of this chapter illustrates the design and development of different drug delivery systems for a model enzyme intended for a local effect in the intestinal tract which should maintain a sufficient degree of enzymatic activity and protect against degradation.

10.4.1   *Oral enzyme supplementation therapy – phenylalanine ammonia-lyase*

Phenylalanine ammonia-lyase (PAL) (EC 4.3.1.5) for the treatment of phenylketonuria (PKU) was chosen as a model enzyme. PKU, an inborn error of phenylalanine metabolism, is the result of a genetic deficiency characterized by severe irreversible mental retardation if untreated due to the accumulation of aromatic metabolites from phenylalanine. The essence of treatment involves substrate restriction therapy – reduction of blood phenylalanine – and is continued throughout childhood and in some cases even longer (Newton, 1996). It was shown that oral administration of PAL to a PKU patient resulted in lowering of the blood phenylalanine concentration (Hoskins *et al.*, 1980), in spite of the high proteolytic lability of PAL (Wieder *et al.*, 1979). Ideally, removal of phenylalanine from the intestine after oral administration of PAL would prevent phenylalanine from reaching the blood circulation due to the conversion of phenylanine to cinnamic acid. Using Caco-2 cell monolayers as a model for the intestinal epithelium, it has been shown that the kinetics favours the transport of cinnamic acid across the cell monolayer relative to the phenylalanine in the presence of PAL (Hovgaard *et al.*, 1998).

Drug delivery systems investigated were entrapments in acrylamide hydrogels and PEG-rich hydrogels and covalent binding to linear PEG molecules. The drug delivery systems were characterized and evaluated with respect to maintenance of enzymatic activity after synthesis and degree of stabilization against proteolytic enzymes.

Entrapment of PAL in hydrogels

PAL-containing acrylic hydrogels based on various amounts of acrylamide, acrylic acid, tert. butylacrylate and 2-(N,N-dimethylamino)-ethylacrylate crosslinked with N,N′-methylenebisacrylamide were synthesized by free radical copolymerization in solution (7 wt%) using a redox initiation process as described by Ingemann *et al.* (1996). The PAL-containing PEG-rich hydrogels were prepared by UV-polymerization of a mixture of the macromonomers poly(ethylene glycol) dimethacrylate (PEGDMA, m_w 1000) and poly(ethylene glycol) methacrylate (PEGMA, m_w 400) in various ratios in solution (50 wt%) to which was added a photoinitiator. Both kinds of hydrogels were synthesized as hydrogel films. The hydrogels were characterized by common methods, i.e. equilibrium degree of swelling and mechanical strength. Further, enzymatic activity and conformational

Table 10.4 Remaining enzymatic activity after hydrogel entrapment and after acrylic monomer contact for PAL

% Remaining enzymatic activity (n = 1)		% Remaining enzymatic activity (n = 3)	
Acrylamide hydrogel*	PEG hydrogel[†]	Acrylic monomers absent	Acrylic monomers present[‡]
Wash: <2%	Wash: <3%	95.3 ± 2.1	10.0 ± 5.1
Gel: <2%	Gel: <3%		

* Total monomer concentration 7 wt%, acrylamide:tert. butylacrylate 90:10, crosslinking agent 2 mol%.
[†] Total monomer concentration 50 wt%, PEGDMA:PEGMA 4:1.
[‡] Removal of acrylamide, tert. butylacrylate and crosslinking agent (concentrations as in *) by size exclusion chromatography after 16 hours of contact with the enzyme. $n = 3 \pm$ S.D.

changes of PAL after synthesis were measured in a spectrophotometric assay and estimated by circular dichroism (CD) and by SDS-PAGE, respectively. The degree of PAL immobilization was determined from evaluation of the enzymatic activity of the hydrogel films and the washing water in relation to the added amount of enzymatic activity.

The results given in Table 10.4 indicate that PAL is inactivated when entrapped in both types of hydrogels. No enzymatic activity of PAL is detectable inside the hydrogel or in the washing water. The time for completion of the polymerization reaction was longer for the acrylamide hydrogels than for the PEG-rich hydrogels. Further, the equilibrium degree of swelling was much higher for acrylamide hydrogels than for PEG-rich hydrogels. These parameters could be of importance for the retention of the enzymatic activity and the loss of the entrapped PAL during the washing process, respectively.

It was investigated whether the PAL inactivation was caused by the actual physical entrapment or by chemical inactivation during the polymerization process. For the acrylamide-containing hydrogels, the remaining enzymatic activity was measured after contact with acrylic monomers used for the synthesis. As seen in Table 10.4, PAL was inactivated even before polymerization, which excludes the possibility that the inactivation could have been caused solely by the physical entrapment. As also seen with glucose 6-phosphate dehydrogenase, acrylamide can be highly toxic, thus the contact time with the monomer before polymerization should be as short as possible to retain maximum amount of enzyme activity (Harrison, 1974). The PEG-rich hydrogels prepared by UV-polymerization with an almost instantaneous completion of the polymerization reaction could have overcome the stability problems associated with the contact to the acrylic functional groups, but failed to do so.

Inactive PAL samples were compared to active PAL samples with respect to SDS-PAGE and CD analysis to deduce whether a size or conformational change could have contributed to the enzyme activity loss. It was observed by SDS-PAGE or CD analysis that neither size nor conformational change had occurred, as the protein bands and the CD spectra were identical. The UV radiation included in the synthesis of the PEG-rich hydrogels did not influence the enzymatic activity of PAL, suggesting that the acrylic functional groups in some way contribute to the activity loss of PAL.

This study illustrates that by changing the monomers and the polymerization process, it is possible to influence the properties of the resulting hydrogels. Knowledge of the relationship between chemical structure of hydrogels and the retention of enzymatic

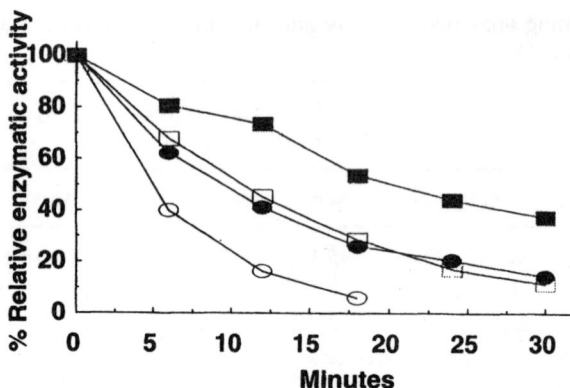

Figure 10.4   PAL degradation (170 μg/ml) in solutions of trypsin (40 μg/ml in Tris buffer pH 8.5) (□, ■) and chymotrypsin (2.4 μg/ml in Tris buffer pH 8.5) (○, ●) for free PAL (open symbols) or SPA-PEG-conjugated PAL (solid symbols).

activity of various enzymes will allow us to design drug delivery systems for the enzymes on a more rational basis.

PEGylation of PAL

Due to the lability of PAL under synthesis conditions of hydrogels, a fundamentally different stabilization principle was examined. PAL was conjugated with linear PEG molecules by mixing the enzyme in an aqueous solution with activated succinimidyl propionate PEG (SPA-PEG, m_w 5000). The enzymatic activity was maintained after the coupling reaction, and the conjugation was qualitatively proved by an increase in molecular weight as seen after SDS-PAGE. The conjugation of PAL to SPA-PEG resulted in a stabilization against degradation by trypsin and chymotrypsin, as illustrated in Figure 10.4. The degradation followed first-order kinetics, and the half-life increased by a factor of 2 after conjugation with PEG. These results indicate that PEGylation may be a promising drug delivery strategy for PAL. However, a number of points still need to be addressed before *in vivo* studies, e.g. effects of the molecular weight of the PEG molecule and the activation moiety of the PEG molecule on the activity and stability of the modified enzyme, and *in vitro* transport studies to illustrate the suitability of the PEGylated PAL in removing phenylalanine from a medium and preventing its absorption.

10.5   Future perspectives

The development of a pharmaceutically acceptable drug delivery system for a peptide or protein is generally more complicated and requires more development time than for traditional drug compounds. This is primarily due to a greater structural complexity of the peptides and proteins. During the past few decades, much effort has been put into research and much knowledge regarding non-parenteral administration of peptides and proteins has been achieved. The problems with peptide and protein delivery are not trivial and will not be overcome by trivial solutions. The use of peptidase inhibitors, absorption enhancers and specialized delivery systems may improve peptide and protein delivery into the systemic circulation. It is unlikely that any one of those approaches would by itself enable the controlled delivery of peptide or protein drugs, or that any single system is going to be universally applicable. Formulations will have to be developed of the needs and requirements as well as the properties of each individual peptide or protein drug. The potential usefulness of the formulations depends on whether or not the enzymatic and absorption barriers existing for the non-parenteral route of delivery can be surmounted. It is important, however, that the natural barriers remain and are altered only transiently. Although only a small amount of a peptide or protein drug needs to be absorbed to get an effect due to its high potency, it seems that the formidable enzymatic barrier present in the intestine means that routes other than the oral one, such as nasal, transdermal or pulmonal administration, may be more promising.

10.6 Summary

Non-parenteral delivery of peptides and proteins requires the overcoming of enzymatic and absorptive barriers present at the site of administration. Co-administration of protease inhibitors or absorption enhancers is a possible way of solving the problems. The formulation approach, by shielding the peptide or protein drug in a polymer or lipid drug delivery system, is another possibility. The preparation and characteristics of hydrogels, nanocapsules, microspheres, liposomes, microemulsions and PEG conjugates for the delivery of peptide and protein drugs are described. The best choice of formulation principle will vary from drug to drug, with a prerequisite of maintenance of bioactivity after synthesis, storage and handling.

References

ABUCHOWSKI, A., VAN ES, T., PALCZUK, N.C. and DAVIS, F.F., 1977, Alteration of immunological properties of bovine serum albumin by covalent attachment of polyethylene glycol, *J. Biol. Chem.*, **252**, 3578–3581.

ANDERBERG, E.K., NYSTRÖM, C. and ARTURSSON, P., 1992, Epithelial transport of drugs in cell culture. VII: Effects of pharmaceutical surfactant excipients and bile acids on transepithelial permeability in monolayers of human intestinal epithelial (Caco-2) cells, *J. Pharm. Sci.*, **81**, 879–887.

AUNGST, B.J., SAITOH, H., BURCHAM, D.L., HUANG, S.-M., MOUSA, S.A. and HUSSAIN, M.A., 1996, Enhancement of the intestinal absorption of peptides and non-peptides, *J. Control. Rel.*, **41**, 19–31.

BAI, J.P.F. and AMIDON, G.L., 1992, Structural specificity of mucosal-cell transport and metabolism of peptide drugs: implication for oral peptide drug delivery, *Pharm. Res.*, **9**, 969–978.

COUVREUR, P., DUBERNET, C. and PUISIEUX, F., 1995, Controlled drug delivery with nanoparticles: current possibilities and future trends, *Eur. J. Pharm. Biopharm.*, **41**, 2–13.

CUMMING, J.S., 1980, Relevant anatomy and physiology of the eye, in ROBINSON, J.R. (ed.), *Opthalmic Drug Delivery Systems*, pp. 1–27, Washington, DC: American Pharmaceutical Association.

EPPSTEIN, D.A. and LONGENECKER, J.P., 1988, Alternative delivery systems for peptides and proteins as drugs, *CRS Crit. Rev. Ther. Drug Car. Sys.*, **5**, 99–139.

FRANCIS, G.E., DELGADO, C. and FISHER, D., 1992, PEG-modified proteins, in AHERN T.J. and MANNING, M.C. (eds), *Stability of Protein Pharmaceuticals. Part B. In Vivo Pathways of Degradation and Strategies for Protein Stabilization*, Pharmaceutical Biotechnology, Vol. 3. pp. 235–261, New York: Plenum.

FUERTGES, F. and ABUCHOWSKI, A., 1990, The clinical efficacy of poly(ethylene glycol)-modified proteins, *J. Control. Rel.*, **11**, 139–148.

GIUNCHEDI, P. and CONTE, U., 1995, Spray-drying as a preparation method of microparticulate drug delivery systems: an overview, *STP Pharm. Sci.*, **5**, 276–290.

GYSELINCK, P., SCHACHT, E., SEVEREN, R.V. and BRAECKMAN, P., 1983, Preparation and characterization of therapeutic hydrogels as oral dosage forms, *Acta Pharm. Technol.*, **29**, 9–12.

HARRIS, J.M., 1997, Introduction to biomedical and biotechnical applications of polyethylene glycol, *Polym. Preprint*, **38**, 520–521.

HARRISON, R.A.P., 1974, The detection of hexokinase, glucosephosphate isomerase and phosphoglucomutase activities in polyacrylamide gels after electrophoresis: a novel method using immobilized glucose 6-phosphate dehydrogenase, *Anal. Biochem.*, **61**, 500–507.

HOSKINS, J.A., JACK, G., PEIRIS, R.J.D., STARR, D.J.T., WADE, H.E., WRIGHT, E.C. and STERN, J., 1980, Enzymatic control of phenylalanine intake in phenylketonuria, *Lancet*, **23**, 392–394.

HOVGAARD, L., BRØNDSTED, H. and NIELSEN, H.M., 1995, Drug delivery studies in Caco-2 monolayers. II. Absorption enhancer effects of lysophosphatidylcholines, *Int. J. Pharm.*, **114**, 141–149.

HOVGAARD, L., JORGENSEN, S.W., EIGTVED, P., FROKJAER, S. and BRØNDSTED, H., 1998, Drug delivery studies in Caco-2 monolayers. VI. Studies of enzyme substitution therapy for phenylketonuria – a new application of Caco-2 monolayers, *Int. J. Pharm.*, **161**, 109–114.

INGEMANN, M., BRØNDSTED, H., HOVGAARD, L. and FROKJAER, S., 1996, Stability of enzymes formulated for oral drug delivery, *Pharm. Res.*, suppl., **13**, s-87.

KULKARNI, S.B., BETAGERI, G.V. and SINGH, M., 1995, Factors affecting microencapsulation of drugs in liposomes, *J. Microencap.*, **12**, 229–246.

LANGGUTH, P., BOHNER, V., HEIZMANN, J., MERKLE, H.P., WOLFFRAM, S., AMIDON, G.L. and YAMASHITA, S., 1997, The challenge of proteolytic enzymes in intestinal peptide delivery, *J. Control. Rel.*, **46**, 39–57.

LEE, V.H.L. (ed.), 1991, *Peptide and Protein Drug Delivery*, New York: Marcel Dekker.

LEE, V.H.L. and YAMAMOTO, A., 1990, Penetration and enzymatic barriers to peptide and protein absorption, *Adv. Drug Deliv. Rev.*, **4**, 171–207.

LICHTENBERG, D., 1988, Liposomes: preparation, characterization and preservation, *Methods Biochem. Anal.*, **33**, 337–462.

MERINO, V., KALIA, Y.N. and GUY, R.H., 1997, Transdermal therapy and diagnosis by iontophoresis, *Trends Biotechnol.*, **15**, 288–290.

MISHRA, P.R., NAMDEO, A., JAIN,.S. and JAIN, N.K., 1996, Hydrogels as drug delivery system, *Indian Drugs*, **33**, 181–186.

MUMTAZ, S. and BACHHAWAT, B.K., 1991, Conjugation of proteins and enzymes with hydrophilic polymers and their applications., *Ind. J. Biochem. Biophys.*, **28**, 346–351.

MURANISHI, S., 1990, Absorption enhancers, *CRS Crit. Rev. Ther. Drug Car. Sys.*, **7**, 1–33.

NEW, R.R.C. (ed.), 1990, Preparation of liposomes, *Liposomes: a Practical Approach*, pp. 33–104, Oxford: IRL Press.

NEWTON, G.D., 1996, Understanding phenylketonuria, *US Pharmacist*, **21**, 56–71.

O'DONNELL, P.B. and MCGINITY, J.W., 1997, Preparation of microspheres by the solvent evaporation technique, *Adv. Drug Del. Rev.*, **28**, 25–42.

PARK, K., 1988, Enzyme-digestible swelling hydrogels as platforms for long-term oral drug delivery: synthesis and characterization, *Biomaterials*, **9**, 435–441.

PEPPAS, N.A. (ed.), 1987, in *Hydrogels in Medicine and Pharmacy, Vol. III, Properties and Applications*, Boca Raton: CRC Press.

PEPPAS, N.A. and KHARE, A.R., 1993, Preparation, structure and diffusional behavior of hydrogels in controlled release, *Adv. Drug Del. Rev.*, **11**, 1–35.

RAJAONARIVONY, M., VAUTHIER, C., COUARRAZE, G., PUISIEUX, F. and COUVREUR, P., 1993, Development of a new drug carrier made from alginate, *J. Pharm. Sci.*, **82**, 912–917.

REDDY, I.K. and BANGA, A.K., 1993, Biotechnology drug delivery: oral vs. alternate routes, *Pharm. Times*, Nov., 92–98.

RICHARDSON, J.L. and ILLUM, L., 1992, The vaginal route of peptide and protein drug delivery, *Adv. Drug Del. Rev.*, **8**, 341–366.

RITSCHEL, W.A., 1991, Microemulsions for improved peptide absorption from the gastrointestinal tract, *Meth. Find. Exp. Clin. Pharmacol.*, **13**, 205–220.

ROORDA, W.E., BODDÉ, H.E., DE BOER, A.G. and JUNGINGER, H.E., 1986, Synthetic hydrogels as drug delivery systems, *Pharm. Weekbl. Sci.*, **8**, 165–189.

SARCIAUX, J.M., ACAR, L. and SADO, P.A., 1995, Using microemulsion formulations for oral drug delivery of therapeutic peptides, *Int. J. Pharm.*, **120**, 127–136.

SHINOMIYA, M., SHIRAI, K., SAITO, Y., YOSHIDA, S. and MATSUOKA, N., 1985, Effect of new chymotrypsin inhibitor (FK-448) on intestinal absorption of insulin, *Lancet*, **1**, 1092–1093.

TALSMA, H. and CROMMELIN, D.J.A., 1992, Liposomes as drug delivery systems, part I: Preparation, *Pharm. Technol.*, **16**, 96–106.

TAUB, M.E., LARSEN, B.D., STEFFANSEN, B. and FROKJAER, S., 1997, β-Carboxylic acid esterified D-Asp-Ala retains a high affinity for the oligopeptide transporter in Caco-2 monolayers, *Int. J. Pharm.*, **146**, 205–212.

TORCHILLIN, V.P., 1987, Immobilised enzymes as drugs, *Adv. Drug Deliv. Rev.*, **1**, 41–86.

VEMURI, S. and RHODES, C.T., 1995, Preparation and characterization of liposomes as therapeutic delivery systems: a review, *Pharm. Acta Helv.*, **70**, 95–111.

VERHOEF, J.C., BODDÉ, H.E., DE BOER, A.G., BOUWSTRA, J.A., JUNGINGER, H.E., MERKUS, F.W.H.M. and BREIMER, D.D., 1990, Transport of peptide and protein drugs across biological membranes, *Eur. J. Drug Metab. Pharmacokinet.*, **15**, 83–93.

WEARLEY, L.L., 1991, Recent progress in protein and peptide delivery by noninvasive routes, *CRS Crit. Rev. Ther. Drug Car. Sys.*, **8**, 331–394.

WIEDER, K.J., PALCZUK, N.C., VAN ES, T. and DAVIS, F.F., 1979, Some properties of polyethylene glycol: phenylalanine ammonia-lyase adducts, *J. Biol. Chem.*, **254**, 12579–12587.

WOODLEY, J.F., 1997, Enzymatic barriers to oral peptide and protein delivery, *Proc. Int. Symp. Control Rel. Bioact. Mater.*, **24**, 99–100.

ZALIPSKY, S. and LEE, C., 1992, Use of functionalized poly(ethylene glycol)s for modification of polypeptides, in Harris, J.M. (ed.), *Poly(ethylene glycol) Chemistry: Biotechnical and Biomedical Applications*, pp. 347–370, New York: Plenum.

ZHOU, X.H., 1994, Overcoming enzymatic and absorption barriers to nonparenterally administered protein and peptide drugs, *J. Control. Rel.*, **29**, 239–252.

ZHOU, X.H. and WAN PO, A.L., 1991, Peptide and protein drugs: II. Non-parenteral routes of delivery, *Int. J. Pharm.*, **75**, 117–130.

Peptide and Protein Derivatives

GITTE JUEL FRIIS

Department of Analytical and Pharmaceutical Chemistry, The Royal Danish School of Pharmacy, Copenhagen, Denmark.

11.1   Introduction

Application of peptides and proteins as clinically useful drugs represents a major challenge. This is due to their poor delivery characteristics caused by their metabolic instability and general non-lipophilic character resulting in poor biomembrane passage. This typically leads to bioavailabilities less than 1–2%. Once within the systemic circulation, short biological half-lives are seen due to rapid metabolism and clearance from the body (Humphrey and Ringrose, 1986; Lee and Yamamoto, 1990; Pauletti *et al.*, 1996a). A possible approach to solve these delivery problems could be chemical derivatization or modification. Two approaches may be possible: either a permanent chemical change in the drug molecule (i.e. the analogue approach) or bioreversible derivatization of the bioactive peptide or protein (i.e. the prodrug approach).

A prodrug is by definition a pharmacological inactive derivative of a drug molecule that is capable of releasing the parent molecule quantitatively, either spontaneously or enzymatically, in the body. The chemical group used for derivatization of the drug molecule, called the pro-group or pro-moiety, should be non-toxic. In contrast to prodrugs, analogues are not bioreversible (Figure 11.1). Both the prodrug and the analogue should possess improved absorption and/or stability characteristics over the parent drug molecule. The analogue should also have high receptor selectivity and affinity (Bundgaard, 1992; Luthman and Hacksell, 1996).

In recent years several types of analogues of various bioactive peptides and proteins have been explored. Possible strategies used in the development of analogs include *N*- and *C*-terminal modifications (e.g. conversion of the *C*-terminal carboxylic acid residue

not bioreversible
Peptide/protein analogue - - - - - - ➤

Bioreversible
Prodrug ⟶ parent peptide/protein + promoiety

Figure 11.1 Illustration of the analogue and prodrug approaches.

to an amide); amino acid manipulations (e.g. systematic replacement of L-amino acids with D-amino acids); peptide backbone modifications, where the use of amide isosters is common (e.g. *N*-methylation of the peptide amide bond); and replacement of larger structural moieties in a compound with dipeptide or tripeptide analogue structures or analogues of the secondary structure (Hruby, 1993; Kahn, 1993; Luthman and Hacksell, 1996; Pauletti *et al.*, 1996a; Sawyer, 1995; Spatola, 1983). In the literature different terms such as analogues and peptidomimetics find use to describe permanent chemical changes in the parent molecule. There seems to be no consistency in the usage, but in general the chemical modified drug molecule is defined by the degree of peptide backbone structure left after derivatization. Thus, analogues have more peptide-like structure left after chemical change than peptidomimetics (Sawyer, 1995).

The prodrug approach was introduced by Albert in 1958, but it was not until the early 1970s that attention was directed to the area. Research in design of prodrugs of various chemical functional groups and of well-known drug substances as well as types of usable pro-groups intensified, and today several drugs are used clinically as prodrug derivatives of a parent drug molecule (for reviews see Christrup *et al.*, 1996; Friis and Bundgaard, 1996; Taylor, 1996). In the late 1980s attention to the use of the prodrug approach to improve the delivery of peptides and proteins increased (for reviews see Bundgaard, 1992; Gangwar *et al.*, 1997b; Møss, 1995; Oliyai, 1996; Oliyai and Stella, 1993).

The objective of the following discussion is to give some examples of the use of the prodrug and analogue approach. It is not intended to give a complete presentation of all literature in the area. The focus of the discussion will be on prodrug derivatization and the use of both the prodrug and analogue approaches at the same time. A peptide is here defined as having a chain length of 2–50 amino acids, whereas proteins include more than 50 amino acids.

11.2 4-Imidazolidinone prodrugs

A potentially useful prodrug type for the α-aminoamide moiety that occurs in a large number of peptides is 4-imidazolidinones. Such derivatives are readily formed by condensing compounds containing an α-aminoamide moiety such as peptides with a free *N*-terminal amino group with aldehydes or ketones. The 4-imidazolidinone derivatives can be seen as cyclic *N*-mannich bases in which the amide and amino functions are placed in the same molecule. The derivatives are cleaved spontaneously to the parent peptide in quantitative amounts (Rasmussen and Bundgaard, 1991a, 1991b).

Shown in Figure 11.2 is the structure of a 4-imidazolidinone derivative of Leu-enkephalin (H-L-Tyr-Gly-Gly-L-Phe-L-Leu-OH). Also shown are the most important sites for enzymatic degradation of Leu-enkephalin. As shown, it is degraded by amino-peptidases, dipeptidyl carboxypeptidases (primarily endopeptidase 24.11 in the intestine and

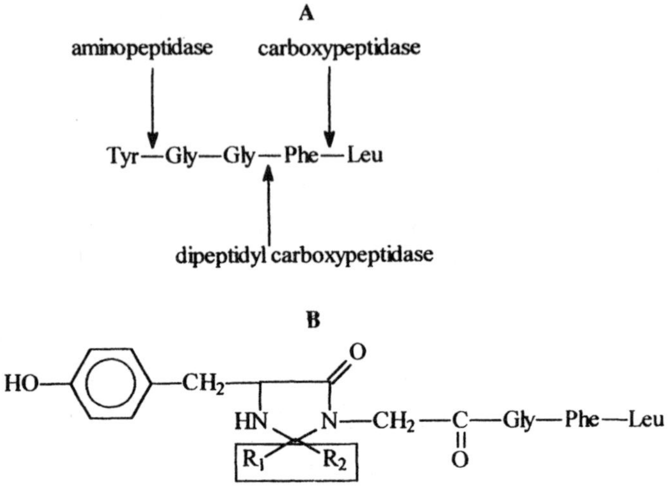

A

aminopeptidase   carboxypeptidase

$\downarrow$ $\downarrow$

Tyr—Gly—Gly—Phe—Leu

$\uparrow$

dipeptidyl carboxypeptidase

B

HO—⟨○⟩—CH$_2$—

HN   N—CH$_2$—C—Gly—Phe—Leu

$\boxed{R_1}$ $\boxed{R_2}$ $\overset{\parallel}{O}$

O (above carbonyl)

Figure 11.2 (A) Enzymatic degradation of Leu-enkephalin; (B) example of the structure of a 4-imidazolidinone prodrug of Leu-enkephalin (H-L-Tyr-Gly-Gly-L-Phe-L-Leu-OH) – the position of the promoiety is highlighted.

peptidylpeptidase A (ACE, angiotensin converting enzyme) in the brain and plasma) and carboxypeptidase A (Geary *et al.*, 1982; Rasmussen and Bundgaard, 1991b; Weinberger and Martinez, 1988). The most important route of enzymatic degradation of Leu-enkephalin seems to be degradation of the Tyr^1-Gly^2. This is seen irrespective of the tissue, preparation or species in question (Friedman and Amidon, 1991; Geary *et al.*, 1982; Kashi and Lee, 1986; Palmieri *et al.*, 1989; Thompson and Audus, 1994; Weinberger and Martinez, 1988).

It has previously been shown that 4-imidazolidinone derivatization can stabilize the enkephalins towards degradation by purified leucine aminopeptidase (Rasmussen and Bundgaard, 1991b). The consequences of the derivatization are that the primary amino group is transformed into a secondary amino group and the vulnerable Tyr^1-Gly^2 bond becomes alkylated. These modifications make the parent compound a poor substrate for aminopeptidases, as substrate specificity requires a primary amino group at the *N*-terminal and an unmodified *N*-terminal peptide bond (Delange and Smith, 1971). The derivatives were also stable in human plasma. In contrast, the derivatives were degraded easily by the pancreatic enzyme carboxypeptidase A, which makes the prodrugs unsuitable for oral delivery (Rasmussen and Bundgaard, 1991b). In another study, different 4-imidazolidinone derivatives of Leu-enkephalin were tested against purified samples of peptidylpeptidase A (ACE) and aminopeptidase N, the two major enzymes responsible for the degradation of Leu-enkephalin in the brain (Thompson and Audus, 1994) and in human plasma (Rasmussen and Bundgaard, 1991b; Bak *et al.*, 1999). The derivatization also stabilized against degradation by these enzymes (Bak *et al.*, 1999). Stabilization was also seen in homogenates of BMEC cells, a cell-model of the blood–brain barrier. On the intact bovine BMEC cells the derivatives were also shown to be stable. A cyclohexanone derivative of Leu-enkephalin was shown to increase the permeation over BMEC monolayer by 61% in relation to the parent peptide (Lund *et al.*, 1998). Thus, formation of 4-imidazolidinone prodrugs helps Leu-enkephalin to overcome the metabolic barrier associated with delivery to the brain. An improved transport is also seen, as described above.

In considering 4-imidazolidinones as a prodrug candidate for a compound, the large decrease seen in the basicity of the *N*-terminal amino group should be appreciated. The 4-imidazolidinones are much weaker bases than the parent peptides. Such depression of amino protonation brings about an increase in lipophilicity of the *N*-terminal amino acid part at physiological pH. This is confirmed by partition experiments in octanol–aqueous buffer, where an increase in the lipophilicity is seen (Rasmussen and Bundgaard, 1991b). The increased lipophilicity will obviously be further influenced by the lipophilicity of the aldehyde or ketone used for the synthesis, i.e. the pro-moiety.

11.3 Prodrugs of TRH

Thyrotropin-releasing hormone (TRH, pGlu-L-His-L-Pro-NH$_2$) is the hypothalamic peptide that regulates the synthesis and secretion of thyrotropin from the anterior pituitary gland. It has been suggested that TRH could have potential as a drug in the management of various neurologic and neuropsychiatric disorders, brain injury, acute spinal trauma, Alzheimer's disease and schizophrenia (Griffiths, 1987; Horita *et al.*, 1986; Jackson, 1982; Metcalf, 1982).

The clinical utilization is, however, greatly hampered due to its rapid metabolism and clearance as well as its poor access to the brain (Griffiths, 1987; Hichens, 1983; Metcalf, 1982). After parenteral administration TRH has a plasma half-life of only 6–8 minutes (Bassiri and Utiger, 1973; Iversen, 1988). This is due to rapid enzymatic degradation of TRH in the blood, mainly by the TRH-specific pyroglutamyl aminopeptidase serum enzyme called PAPase II (Bauer, 1988; Bauer and Novak, 1979; Møss and Bundgaard, 1990a). TRH is also cleaved by the less specific pyroglutamyl aminopeptidase called PAPase I that occurs in many different tissues such as the liver, kidney and brain but not in blood (Bauer, 1988; Szewczuk and Kwiatkowska, 1970; Wilk, 1989). These enzymes cleave the pGlu-L-His binding in TRH. TRH is also degraded by the enzyme prolyl endopeptidase that cleaves TRH at the L-His-L-Pro-NH$_2$ bond (Figure 11.3) (Møss *et al.*,

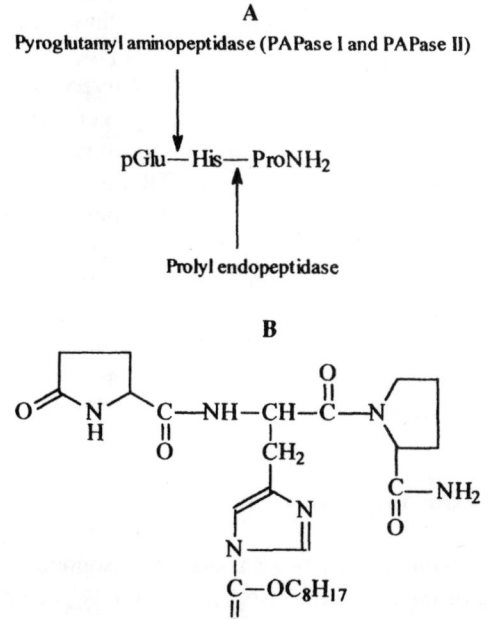

Figure 11.3 (A) Enzymatic degradation of TRH (pGlu-His-Pro-NH$_2$); (B) structure of the N-octyloxycarbonyl TRH prodrug. TRH is derivatized at the imidazole group in the histidine residue.

1990). The lipophilicity of TRH is very low (Bundgaard and Møss, 1989), and this may be the primary reason for the limited ability of the peptide to penetrate the blood–brain barrier (Banks and Kastin, 1985).

To improve the enzymatic stability and the lipophilicity of TRH, various *N*-alkoxycarbonyl prodrugs of the imidazole group of the histidin residue have been prepared (Bundgaard and Møss, 1990). The structure of the *N*-octyloxycarbonyl derivative of TRH is shown in Figure 11.3. The derivatives are more stable towards degradation by the less specific enzyme PAPase I than the parent peptide TRH, but the half-lives are relatively short (PAPase I in phosphate buffer pH 7.4 and 37°C: half-life for TRH is 4 minutes and half-life for *N*-octyloxycarbonyl–TRH is 24 minutes) (Bundgaard and Møss, 1990). The derivatives are stabilized towards degradation by the TRH-specific enzyme PAPase II. Thus, longer half-lives are seen in 80% human plasma at 37°C for the derivatives than for TRH (Bundgaard and Møss, 1990). An improvement of the lipophilicity is also seen, expressed by higher partition coefficients between octanol and buffer for the prodrugs compared to TRH. The log D is −2.46 for TRH and 1.88 for the *N*-octyloxycarbonyl–TRH (octanol–0.02 M phosphate buffer pH 7.4 (21°C)) (Bundgaard and Møss, 1990). An improved penetration across biomembranes of the prodrug derivatives compared to TRH is therefore to be expected. This improvement is not seen in penetration studies done in Ussing chambers using rabbit and rat intestinal tissues. This is explained by enzymatic hydrolysis by prolyl endopeptidase (cleaves off the *C*-terminal proline amide group) and possibly by degradation by non-specific esterases. Thus, a prodrug derivative should also effect protection against this degradation in order to be a successful approach for oral delivery (Møss *et al.*, 1990). In another study, penetration across Caco-2 cells of TRH and the *N*-octyloxycarbonyl–TRH prodrug was explored. No intact TRH prodrug was transported across the cells. Instead TRH released from the prodrug was absorbed at rates comparable to that for TRH alone. No metabolites of TRH could be detected. It was concluded that the increased lipophilicity of the TRH prodrug was without effect on the transport characteristics (Lundin *et al.*, 1991a). In contrast to the failure in providing intestinal penetration of TRH, the *N*-alkoxycarbonyl prodrug derivatives may be useful to improve the transdermal delivery of TRH. Thus, it has been shown that the *N*-octyloxycarbonyl–TRH prodrug can deliver TRH very efficiently though human skin. By applying the prodrug derivative in aqueous solution (pH 6.0) at a concentration of 5%, a steady-state flux of 0.045 μmol TRH h^{-1} cm^{-2} corresponding to 16 μg TRH h^{-1} cm^{-2} was observed. Almost all of the prodrug applied to the skin was converted to the parent TRH during diffusion through the human skin samples. Using propylene glycol as a vehicle and a prodrug concentration of 5%, a flux of 3.6 μg TRH h^{-1} cm^{-2} was obtained. It should be pointed out that the fluxes obtained in the study did not represent the maximally obtainable values, since the prodrug was not applied as a saturated solution. No optimization studies concerning the type of vehicle, use of penetration enhancers, etc. were done (Møss and Bundgaard, 1990b).

11.4   Derivatives of desmopressin

Desmopressin [1-(mercaptopropanoic acid)-8-ᴅ-argininevasopressin; dDAVP] (Figure 11.4) is a synthetic analogue of the hormone vasopressin with a specific antidiuretic effect. In desmopressin the *N*-terminal has been deamidated and ʟ-arginine changed to ᴅ-arginine. It is used in the treatment of central diabetes insipidus and nocturnal enuresis and as a haemostatic agent (Vilhardt, 1990).

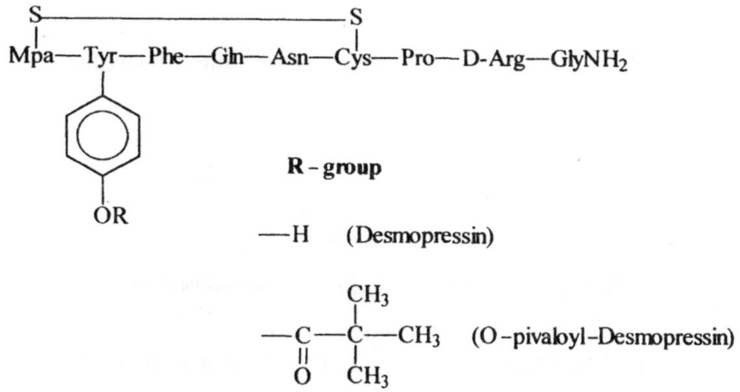

```
S————————————————S
|                 |
Mpa—Tyr—Phe—Gln—Asn—Cys—Pro—D-Arg—GlyNH₂
```

R – group

—H   (Desmopressin)

—C—C—CH₃   (O–pivaloyl–Desmopressin)

OR

Figure 11.4 Structure of desmopressin and the O-pivaloyl ester prodrug of desmopressin.

Desmopressin is usually administered perorally or intranasally but the bioavailability is only 1% and 2–10%, respectively (Harris *et al.*, 1986; Köhler and Harris, 1988). The poor bioavailability is explained by low lipophilicity of desmopressin (Kahns *et al.*, 1993; Lundin and Artursson, 1990; Lundin *et al.*, 1991b) and by enzymatic degradation at the site of administration (Lundin *et al.*, 1989; Morimoto *et al.*, 1991; Saffran *et al.*, 1988). When administered perorally desmopressin is degraded by the proteolytic enzyme α-chymotrypsin (Fredholt *et al.*, 1999; Kahns *et al.*, 1993). The cleavage apparently occurs between the Tyr-Phe bond and the Phe-Gln bonds (Fredholt *et al.*, 1999), which is in accordance with the substrate specificity of the enzyme (Hess, 1971).

Previously it has been shown that ester derivatization of the tyrosine phenolic group in *N*-α-acylated tyrosine amides can protect against degradation by α-chymotrypsin. It was shown that aliphatic carboxylic acid and carbonate esters with a short or branched side-chain were most resistant towards enzymatic degradation (Kahns and Bundgaard, 1991). Such derivatization has also been done at the tyrosine phenolic group in desmopressin. In accordance with previous results, an O-pivaloyl ester (Figure 11.4) showed the highest degree of protection against degradation by the enzyme α-chymotrypsin (Kahns *et al.*, 1993). The permeability relative to the parent peptide has also been studied across Caco-2 cell monolayers. An improved permeation was seen. The highest improvement of transport, by a factor of 3, was seen for the O-pivaloyl ester. The permeability coefficient (P_{app}) was 5.15×10^{-7} cm s^{-1} for desmopressin and 1.51×10^{-6} cm s^{-1} for O-pivaloyl de-smopressin (Kahns *et al.*, 1993). An increase of penetration was also seen for the same ester compared to desmopressin in a rat perfusion model (Lepist *et al.*, 1999).

In another study, analogues of desmopressin have been prepared. Desmopressin had different glycosylated serine amino acids in position 4 instead of glutamine, and the D-form of tyrosine in position 2 was incorporated. The Tyr²-Phe³ was therefore stabilized against degradation by α-chymotrypsin. The glycosylated analogues had significantly higher bioavailabilities than desmopressin when administered intraintestinally in rats. In contrast, the analogues displayed a very low activity at the vasopressin V₂-receptor. Conformational studies with ¹H NMR spectroscopy did not reveal any major change in the conformation of the peptide backbone after glycosylation. The lack of receptor binding is probably explained by steric repulsion between the carbohydrate moiety and the vasopressin receptor (Kihlberg *et al.*, 1995).

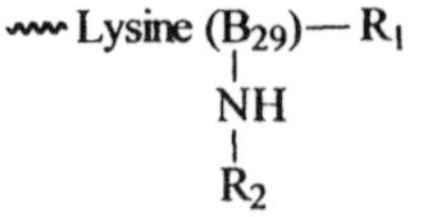

~~~ Lysine (B_{29}) — R_1
$\quad\quad\quad\quad\quad |$
$\quad\quad\quad\quad NH \quad\quad\quad$ (part of structure of insulin)
$\quad\quad\quad\quad\quad |$
$\quad\quad\quad\quad R_2$

$R_1 =$ Thr—OH $R_2 =$ H   human insulin

$R_1 =$ Thr—OH $R_2 = CH_3(CH_2)_8CO$— $N(B_{29})$ decanoylinsulin

$R_1 = $—OH $R_2 = CH_3(CH_2)_{10}CO$— $N(B_{29})$ dodecanoyl desThr (B_{30}) insulin

$R_1 = $—OH $R_2 = CH_3(CH_2)_{12}CO$— $N(B_{29})$ tetradecanoyl desThr (B_{30}) insulin

Figure 11.5 Part of the structure of human insulin acylated with fatty acids at the ε-amino group of Lys^{B29}.

11.5 Derivatives of insulin

Human insulin contains 51 amino acid residues distributed on two polypeptide chains called the A-chain and the B-chain. The A-chain consists of 21 residues and has a disulphide loop between A6 and A11. The longer B-chain has 30 amino acids linked to the A-chain by two disulphide bonds between A7 and B7 and A20 and B19 (Brange and Langkjaer, 1993). Insulin exists as a monomer at low concentrations (< 0.1 μM). It dimerizes at higher concentrations relevant for pharmaceutical formulation. At pH 4–8 in the presence of zinc ions or other divalent metal ions, three dimers form a hexamer. At concentrations ≥ 2 mM the hexamer is formed at neutral pH without the presence of zinc ions (Hansen, 1991).

In a recent study, a Co^{3+}–insulin hexamer was made. The actual structure of the complex was more precisely $(insulin)_6$–$(Co^{3+})_2Ca^{2+}$. No detectable binding of Co^{3+}–insulin complex to the insulin receptor was seen, indicating less than 0.05% receptor affinity compared to that of human insulin. However, Co^{3+}–insulin has a blood glucose lowering effect following subcutaneous incubation, suggesting the active monomer is released from the modified complex *in vivo*. The blood glucose profiles suggest that the complex provides a slow release of the insulin. The complex is possibly activated by bioreduction $(Co^{3+} \rightarrow Co^{2+})$. The compound possesses several advantages that make it a candidate for insulin preparation. Firstly, formation of the non-dissociating Co^{3+}–insulin hexamer might result in physical stabilization, since formation of insoluble fibrils is initiated by partial unfolding of the insulin monomer (Brange *et al.*, 1986). Secondly, the complex is injected as an aqueous solution. The Co^{3+}–insulin complex acts as a prodrug and can be used as a soluble prolonged-acting insulin preparation (Kurtzhals *et al.*, 1995).

Recently a number of insulin analogues acylated with fatty acids at the ε-amino group of Lys^{B29} have been prepared (Figure 11.5). These compounds are prolonged-acting due to albumin binding. The affinity of the acylated insulins for albumin varies considerably among species. Thus, the degree of protraction after subcutaneous injection into rabbits and pigs varies. There is a 20-fold higher affinity of the insulin analogues for the albumin in rabbit serum than in pig serum. This explains the more protracted effect seen in rabbits compared to pigs. The binding affinities for the Lys^{B29}-acylated insulins for albumin in pig and human serum differ by less than a factor of 2, suggesting that the degree of protraction of the insulin derivatives is likely to be similar in humans and in pigs. The authors suggest that derivatization with albumin-binding ligands, e.g. fatty acids, may provide a general approach to prolong the effect of subcutaneously injected peptides. The species differences in ligand binding should be kept in mind, since they can be important in the preclinical evaluation of highly albumin-binding peptide derivatives (Kurtzhals *et al.*, 1996).

11.6 Cyclic prodrugs

In recent years various studies on cyclization of the peptide backbone from the *N*-terminal to the *C*-terminal using different chemical linkers have been reported (Gangwar *et al.*, 1997a; Wang *et al.*, 1996a, 1996b, 1997). Cyclization enhances the extent of intramolecular hydrogen bonding and reduces the potential for intermolecular bonding to aqueous solvents. In addition, cyclization of the backbone blocks the *N*-terminal and the *C*-terminal, thereby protecting the peptide from aminopeptidases and carboxypeptidases. Other advantages could be reduction of the overall charge of the molecule and reduction of size of the peptide (Gangwar *et al.*, 1997b; Okumu *et al.*, 1997). Three different linkers have been reported, namely an acyloxyalkoxy promoiety (Gangwar *et al.*, 1997a), a 3-(2'-hydroxy-4',6'-dimethylphenyl)-3,3-dimethylpropionic acid promoiety (Wang *et al.*, 1997) and a coumaric acid promoiety (Wang *et al.*, 1996a, 1996b). All three prodrug types were designed to be cleaved initially in a slow step by esterases followed by various chemical reactions resulting in release of the parent peptide. The proposed pathway for the conversion of the acyloxyalkoxy-based cyclic prodrug to the parent peptide in esterase media is shown in Figure 11.6. Also shown are the chemical structures of the propylpropionic acid-based prodrug and the coumarin-based prodrug.

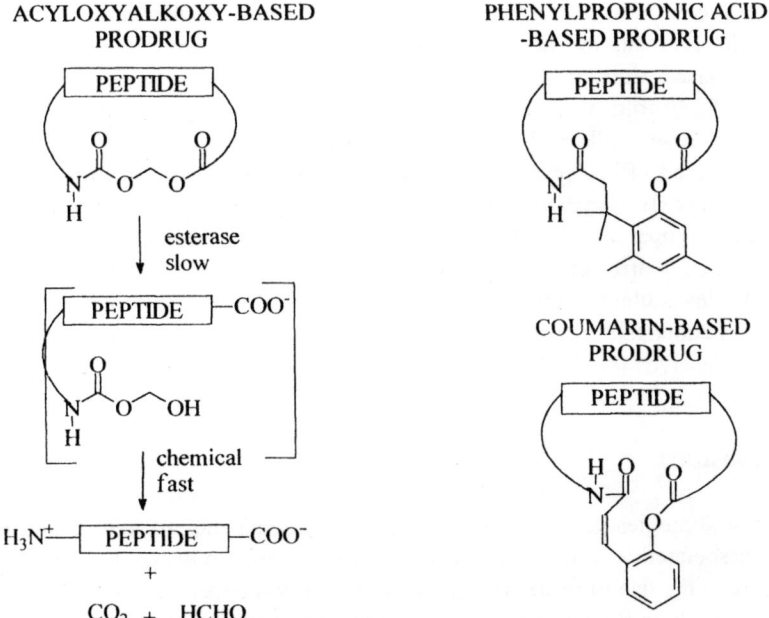

Figure 11.6 Proposed pathway for the conversion of the acyloxyalkoxy-based cyclic prodrug to the parent peptide in esterase media: also shown are the chemical structures of the phenylpropionic acid-based prodrug and the coumarin-based prodrug.

Two cyclic prodrugs have been made of the model hexapeptide H-L-Trp-L-Ala-Gly-Gly-L-Asp-L-Ala-OH, namely the acyloxyalkoxy cyclic prodrug (Pauletti *et al.*, 1996b) and the phenylpropionic acid cyclic prodrug (Pauletti *et al.*, 1997a). It was shown that both prodrugs were converted to the parent peptide more rapidly in human blood than in physiological buffer, supporting the involvement of esterase degradation in their bioconversion. Transport studies conducted using Caco-2 cell monolayers showed for both cyclic prodrugs an increase in permeation at around 70 compared to the parent peptide. The enhanced permeation seen for the acyloxyalkoxy prodrug could in part be explained by stabilization of the peptide against metabolism. Also important for the improved permeation seen is formation of unique solution structures of the prodrug which reduce the hydrogen-bonding potential of the molecule by creation of intramolecular hydrogen bonds (Gangwar *et al.*, 1996).

In another study the coumaric acid-based and the acyloxyalkoxy-based cyclic prodrugs of the opioid peptides Leu-enkephalin (H-L-Tyr-Gly-Gly-L-Phe-L-Leu-OH) and the metabolically stable analogue DADLE (H-L-Tyr-D-Ala-Gly-L-Phe-D-Leu-OH) were prepared. It was also seen here that the prodrugs were degraded significantly faster in 90% human plasma than in physiological buffer (half-lives for prodrugs in Hanks' balanced salt solution, pH 7.4 and 37°C, were around 299–444 minutes; half-lives for prodrugs in 90% human plasma, 37°C, were around 78–215 minutes). The prodrugs were degraded quantitatively to the parent peptides. Thus, during degradation of the prodrugs a simultaneous formation of the parent peptide occurred. Transport studies showed a small permeation of Leu-enkephalin and DADLE across Caco-2 cells (P_{app} value for Leu-enkephalin was estimated to be 0.31×10^{-8} cm s^{-1}; P_{app} for DADLE was $7.84 \times 10^{-8} \pm 0.7$ cm s^{-1}) (Gudmundsson *et al.*, 1997). The P_{app} value for DADLE is very low but in the typical range for pentapeptides or hexapeptides (Pauletti *et al.*, 1997b). The permeability coefficients (P_{app}) from the apical (AP) to the basolateral (BL) side of Caco-2 cells of the coumaric acid-based prodrugs of Leu-enkephalin and DADLE were determined to be 665 and 31 times greater than those for the parent peptides Leu-enkephalin and DADLE, respectively. In contrast, the permeability coefficients (P_{app}) from AP to BL for the acyloxyalkoxy-based prodrugs of Leu-enkephalin and DADLE were shown to be much lower (114- and 130-fold, respectively). An explanation for the low AP-to-BL permeation seen for the acyloxyalkoxy-based prodrugs is that these prodrugs are substrates for an apically polarized efflux system which today is recognized as a possible additional barrier to absorption (Burton *et al.*, 1997). Thus, the BL-to-AP P_{app} values for the prodrugs are much higher compared to the AP-to-BL direction. The different transport characteristics seen for the coumaric acid-based prodrugs and the acyloxyalkoxy-based prodrugs of the opioid peptides could be explained by differences in solution conformations that these cyclic prodrugs adopt rather than the structural features of the chemical linkers, i.e. the promoieties themselves (Gudmundsson *et al.*, 1997).

11.7 Conclusions

Many different strategies have been employed to overcome the enzymatic and penetration barriers associated with the delivery of peptide and protein drugs. Examples of such strategies include the use of protease inhibitors, absorption enhancers, alternative administration routes such as the nasal route, and development of new dosage forms (Lee and Yamamoto, 1990; Sayani and Chien, 1996; Verhoef *et al.*, 1990). Among these approaches are chemical derivatization of peptides and proteins. As illustrated in this chapter the chemical approach, i.e. formation of analogues and prodrugs, is a highly useful means to improve the delivery of peptides and proteins. It is possible to obtain derivatives with increased absorption characteristics and/or metabolic stability. The prodrug approach has until now mainly been used on smaller peptides containing around 10 amino acids. As illustrated here with insulin, the technique can be used successfully on larger molecules. The same is the case with the analogue approach, also illustrated here with insulin. Also described here is the use of the prodrug and analogue approaches together. Chemical derivatization may find greater use in combination with other techniques used to improve the delivery of peptides and proteins. In the future, studies demonstrating the promising results achieved *in vitro* need to be performed to a greater extent *in vivo*. This is especially the case for the prodrug technique. It is hoped that the same encouraging results will be seen.

References

ALBERT, A., 1958, Chemical aspects of selective toxicity, *Nature*, **182**, 421–423.

BAK, A., FICH, M., LARSEN, B.D., FROKJAER, S. and FRIIS, G.J., 1999, 4-Imidazolidinone prodrugs of the N-terminal in Leu-enkephalin: synthesis, chemical and enzymatic stability studies, *European Journal of Pharmaceutical Sciences*, **7**, 317–323.

BANKS, W.A. and KASTIN, A.J., 1985, Peptides and the blood–brain barrier: lipophilicity as a predictor of permeability, *Brain Research Bulletin*, **15**, 287–292.

BASSIRI, R.M. and UTIGER, R.D., 1973, Metabolism and excretion of exogeneous thyrotropin-releasing hormone in humans, *Journal of Clinical Investigation*, **52**, 1616–1619.

BAUER, K., 1988, Degradation and biological inactivation of thyrotropin releasing hormone (TRH): regulation of the membrane-bound TRH-degrading enzyme from rat anterior pituitary by estrogens and thyroid hormones, *Biochimie*, **70**, 69–74.

BAUER, K. and NOVAK, P., 1979, Characterization of a thyroliberin-degrading serum enzyme catalyzing the hydrolysis of thyroliberin at the pyroglutamyl–histidin bond, *European Journal of Biochemistry*, **99**, 239–246.

BRANGE, J. and LANGKJAER, L., 1993, Insulin structure and stability, *Stability and Characterization of Proteins and Peptide Drugs: case histories*, edited by Y.J. WANG and R. PEARLMAN, pp. 315–350, New York: Plenum Press.

BRANGE, J., HAVELUND, S., HOMMEL, E., SØRENSEN, E. and KÜHL, C., 1986, Neutral insulin solutions physically stabilized by addition of Zn^{2+}, *Diabetic Medicine*, **3**, 532–536.

BUNDGAARD, H., 1992, Means to enhance penetration. Prodrugs as a means to improve the delivery of peptide drugs, *Advanced Drug Delivery Reviews*, **8**, 1–38.

BUNDGAARD, H. and MØSS, J., 1989, Prodrug derivatives of thyrotropin-releasing hormone and other peptides, *Biochemical Society Transactions*, **17**, 947–949.

BUNDGAARD, H. and MØSS, J., 1990, Prodrugs of peptides. 6. Bioreversible derivatives of thyrotropin-releasing hormone (TRH) with increased lipophilicity and resistance to cleavage by the TRH-specific serum enzyme, *Pharmaceutical Research*, **7**, 885–892.

BURTON, P.S., GOODWIN, J.T., CONRADI, R.A., HO, N.F.H. and HILGERS, A.R., 1997, *In vitro* permeability studies of peptidomimetic drugs: the role of polarized efflux pathways as additional barriers to absorption, *Advanced Drug Delivery Reviews*, **23**, 143–157.

CHRISTRUP, L.L., MØSS, J. and STEFFANSEN, B., 1996, Prodrugs, *Preservation of Pharmaceutical Products to Salt Forms of Drugs and Absorption*, edited by J. SWARBRICK and J.C. BOYLAN, pp. 39–70, New York: Marcel Dekker.

DELANGE, R.T. and SMITH, E.L., 1971, Leucine aminopeptidase and other N-terminal exopeptidases. *The Enzymes*, Vol. III, edited by P.D. BOYER, pp. 81–118, New York: Academic Press.

FREDHOLT, K., ØSTERGAARD, J., SAVOLAINEN, J. and FRIIS, G.J., 1999, α-Chymotrypsin catalyzed degradation of desmopressin (dDAVP) – influences of pH, concentration and various cyclodextrins, *International Journal of Pharmaceutics*, **178**, 223–229.

FRIEDMAN, D.I. and AMIDON, G.L., 1991, Oral absorption of peptides. Influences of pH and inhibitors on the intestinal hydrolysis of Leu-enkephalin and analogues, *Pharmaceutical Research*, **8**, 93–96.

FRIIS, G.J. and BUNDGAARD, H., 1996, Design and application of prodrugs, *A Textbook of Drug Design and Development*, edited by P. KROGSGAARD-LARSEN, T. LILJEFORS and U. MADSEN, pp. 351–385, Amsterdam: Harwood Academic Publishers.

GANGWAR, S., JOIS, S.D.S., SIAHAAN, T.J., VANDER VELDE, D.G., STELLA, V.J. and BORCHARDT, R.T., 1996, The effect of conformation on membrane permeability of an acyloxyalkoxy-linked cyclic prodrug of a model peptide, *Pharmaceutical Research*, **13**, 1657–1661.

GANGWAR, S., PAULETTI, G.M., SIAHAAN, T.J., STELLA, V.J. and BORCHARDT, R.T., 1997a, Synthesis of a novel esterase-sensitive cyclic prodrug of a hexapeptide using an acyloxyalkoxy promoiety, *Journal of Organic Chemistry*, **62**, 1356–1362.

GANGWAR, S., PAULETTI, G.M., WANG, B., SIAHAAN, T.J., STELLA, V.J. and BORCHARDT, R.T., 1997b, Prodrug strategies to enhance the intestinal absorption of peptides, *Drug Discovery Today*, **2**, 148–155.

GEARY, L.E., WILEY, K.S., SCOTT, W.L. and COHEN, M.L., 1982, Degradation of exogeneous enkephalin in the guinea-pig ileum: relative importance of aminopeptidase, enkephalinase and angiotensin converting enzyme activity, *Journal of Pharmacology and Experimental Therapeutics*, **211**, 104–111.

GRIFFITHS, E.C., 1987, Clinical applications of thyrotropin-releasing hormone, *Clinical Science*, **73**, 449–457.

GUDMUNDSSON, O.S., BAK, A., WANG, W., SHAN, D., ZHANG, H., FRIIS, G.J., WANG, B. and BORCHARDT, R.T., 1997, Prodrug strategies to enhance the permeation of opioid peptides through the intestinal mucosa, pp. 259–272, Copenhagen: Munksgaard.

HANSEN, J.F., 1991, The self-association of zinc-free human insulin and insulin-analogue B13-glutamine, *Biophysical Chemistry*, **39**, 107–110.

HARRIS, A.S., NIELSSON, I.M., WAGNER, Z.G. and ALKNER, U., 1986, Intranasal administration of peptides: nasal deposition, biological response, and absorption of desmopressin, *Journal of Pharmaceutical Sciences*, **75**, 1085–1088.

HESS, G.P., 1971, Chymotrypsin – chemical properties and catalysis. *The Enzymes*, Vol. III, edited by P.D. BOYER, pp. 213–248, New York: Academic Press.

HICHENS, M., 1983, A comparison of thyrotropin-releasing hormone with analogs: influence of disposition upon pharmacology, *Drug Metabolism Reviews*, **14**, 77–98.

HORITA, A., CARINO, M.A. and LAI, H., 1986, Pharmacology of thyrotropin-releasing hormone, *Annual Review of Pharmacology and Toxicology*, **26**, 311–332.

HRUBY, V.J., 1993, Conformational and topographical considerations in the design of biologically active peptide ligands, *Biopolymers*, **33**, 1073–1082.

HUMPHREY, M.J. and RINGROSE, P.S., 1986, Peptides and related drugs: a review of their absorption, metabolism, and excretion, *Drug Metabolism Reviews*, **17**, 283–310.

IVERSEN, E., 1988, Intra- and extravascular turnover of thyrotropin-releasing hormone in normal man, *Journal of Endocrinology*, **118**, 511–516.

JACKSON, I.M.D., 1982, Thyrotropin releasing hormone, *New England Journal of Medicine*, **306**, 145–155.

KAHN, M., 1993, Peptide secondary structure mimetics: recent advantages and future challenges, *Synlett*, 821–826.

KAHNS, A.H. and BUNDGAARD, H., 1991, Prodrugs of peptides. 14. Bioreversible derivatization of the tyrosine phenol group to effect protection of tyrosyl-peptides against α-chymotrypsin, *International Journal of Pharmaceutics*, **76**, 99–112.

KAHNS, A.H., BUUR, A. and BUNDGAARD, H., 1993, Prodrugs of peptides. 18. Synthesis and evaluation of various esters of desmopressin (dDAVP), *Pharmaceutical Research*, **10**, 68–74.

KASHI, S.D. and LEE, V.H.L., 1986, Enkephalin hydrolysis in homogenates of various absorptive mucosae of the albino rabbit: similarities in rates and involvement of aminopeptidases, *Life Sciences*, **38**, 2019–2028.

KIHLBERG, J., ÅHMAN, J., WALSE, B., DRAKENBERG, T., NILSSON, A., SÖDERBERG-AHLM, C., BENGTSSON, B. and OLSSON, H., 1995, Glycosylated peptide hormones: pharmacological properties and conformational studies of analogues of [1-Desamino, 8-D-arginine]-vasopressin, *Journal of Medicinal Chemistry*, **38**, 161–169.

KÖHLER, M. and HARRIS, A., 1988, Pharmacokinetics and haematological effects of desmopressin, *European Journal of Clinical Pharmacology*, **35**, 281–285.

KURTZHALS, P., KIEHR, B. and SØRENSEN, A., 1995, The cobalt (III)–insulin hexamer is a prolonged-acting prodrug, *Journal of Pharmaceutical Sciences*, **84**, 1164–1168.

KURTZHALS, P., HAVELUND, S., JONASSEN, I., KIEHR, B., RIBEL, U. and MARKUSSEN, J., 1996, Albumin binding and time action of acylated insulins in various species, *Journal of Pharmaceutical Sciences*, **85**, 304–308.

LEE, V.H.J. and YAMAMOTO, A., 1990, Penetration and enzymatic barriers to peptide and protein absorption, *Advanced Drug Delivery Reviews*, **4**, 171–207.

LEPIST, E.I., ØSTERGAARD, J., FREDHOLT, K., LENNERNÄS, H. and FRIIS, G.J., 1999, Stability and perfusion studies in of desmopressin (dDAVP) and prodrugs, in the rat jejunum, *Experimental and Toxicologic Pathology*, **51**, 363–368.

LUND, L., BAK, A., FRIIS, G.J., HOVGAARD, L. and CHRISTRUP, L.L., 1998, The enzymatic degradation and transport of leucine–enkephalin and 4-imidazolidinone prodrugs at the blood–brain barrier, *International Journal of Pharmaceutics*, **172**, 97–101.

LUNDIN, S. and ARTURSSON, P., 1990, Absorption of a vasopressin analogue, 1-deamino-8-D-arginine-vasopressin (dDAVP), in a human intestinal epithelial cell-line, *International Journal of Pharmaceutical Research*, **64**, 181–186.

LUNDIN, S., BENGTSSON, H.-I., FOLKESSON, H.G. and WESTRÖM, B.R., 1989, Degradation of [mercaptopropanoic acid[1], D-arginine[8]]-vasopressin (dDAVP) in pancreatic juice and intestinal mucosa homogenate, *Pharmacology and Toxicology*, **65**, 92–95.

LUNDIN, S., MØSS, J., BUNDGAARD, H. and ARTURSSON, P., 1991a, Absorption of thyrotropin-releasing hormone (TRH) and a TRH prodrug in human intestinal cell line (Caco-2), *International Journal of Pharmaceutics*, **76**, R1–R4.

LUNDIN, S., PANTZAR, N., BROEDERS, A., OHLIN, M. and WESTRÖM, B.R., 1991b, Differences in transport rate of oxytocin and vasopressin analogues across proximal and distal isolated segments of the small intestine of rat, *Pharmaceutical Research*, **8**, 1274–1280.

LUTHMAN, K. and HACKSELL, U., 1996, Peptides and peptidomimetics, *A Textbook of Drug Design and Development*, edited by P. KROGSGAARD-LARSEN, T. LILJEFORS and U. MADSEN, pp. 386–406, Amsterdam: Harwood Academic Publishers.

METCALF, G., 1982, Regulatory peptides as a source of new drugs – the clinical prospects for analogues of TRH which are resistant to metabolic degradation, *Brain Research*, **4**, 389–408.

MORIMOTO, K., YAMAGUCHI, T., IWAKURA, Y., MIYAZAKI, M., NAKATANI, E., IWAMOTO, T., OKASHI, Y. and NAKAI, Y., 1991, Effects of the proteolytic enzyme inhibitors on the nasal absorption of vasopressin and an analogue, *Pharmaceutical Research*, **8**, 1175–1179.

MØSS, J., 1995, Peptide prodrugs designed to limit metabolism, *Peptide Based Drug Design. Controlling Transport and Metabolism*, edited by M.D. TAYLOR and G.L. AMIDON, pp. 423–448, Washington, DC: American Chemical Society.

MØSS, J. and BUNDGAARD, H., 1990a, Kinetics and pattern of degradation of thyrotropin-releasing hormone (TRH) in human plasma, *Pharmaceutical Research*, **7**, 751–755.

1990b, Prodrugs of peptides. 7. Transdermal delivery of thyrotropin-releasing hormone (TRH) via prodrugs, *International Journal of Pharmaceutics*, **66**, 39–45.

Møss, J., Buur, A. and Bundgaard, H., 1990, Prodrugs of peptides. 8. *In vitro* study of intestinal metabolism and penetration of thyrotropin-releasing hormone (TRH) and its prodrugs, *International Journal of Pharmaceutics*, 66, 183–191.

Okumu, F.W., Pauletti, G.M., Vander Velde, D.G., Siahaan, T.J. and Borchardt, R.T., 1997, Effect of restricted conformational flexibility on the permeation of model hexapeptides across Caco-2 monolayers, *Pharmaceutical Research*, 14, 169–175.

Oliyai, R., 1996, Prodrugs of peptides and peptidomimetics for improved formulation and delivery, *Advanced Drug Delivery Reviews*, 19, 275–286.

Oliyai, R. and Stella, V.J., 1993, Prodrugs of peptides and proteins for improved formulation and delivery, *Annual Reviews Pharmacology and Toxicology*, 32, 521–544.

Palmieri, F.E., Bausback, H.H. and Ward, P.E., 1989, Metabolism of vasoactive peptides by vascular endothelium and smooth muscle aminopeptidase M, *Biochemistry and Pharmacology*, 38, 173–180.

Pauletti, G.M., Gangwar, S., Knipp, G.T., Nerurkar, M.M., Okumu, F.W., Tamura, K., Siahaan, T.J. and Borchardt, R.T., 1996a, Structural requirements for intestinal absorption of peptide drugs, *Journal of Controlled Release*, 41, 3–17.

Pauletti, G.M., Gangwar, S., Okumu, F.W., Siahaan, T.J., Stella, V.J. and Borchardt, R.T., 1996b, Esterase-sensitive cyclic prodrugs of peptides: evaluation of an acyloxyalkoxy promoiety in a model hexapeptide, *Pharmaceutical Research*, 13, 1615–1623.

Pauletti, G.M., Gangwar, S., Wang, B. and Borchardt, R.T., 1997a, Esterase-sensitive cyclic prodrugs of peptides: evaluation of a phenylpropionic acid promoiety in a model hexapeptide, *Pharmaceutical Research*, 14, 11–17.

Pauletti, G.M., Okumu, F.W. and Borchardt, R.T., 1997b, Effect of size and charge on the passive diffusion of peptides across Caco-2 cell mono-layers via the paracellular pathway, *Pharmaceutical Research*, 14, 164–168.

Rasmussen, G.J. and Bundgaard, H., 1991a, Prodrugs of peptides. 10. Protection of di- and tripeptides against aminopeptidases by formation of bioreversible 4-imidazolodinone derivatives, *International Journal of Pharmaceutics*, 71, 45–53.

1991b, Prodrugs of peptides. 15. 4-Imidazolidinone prodrug derivatives of enkephalins to prevent aminopeptidase-catalyzed metabolism in plasma and absorptive mucosae, *International Journal of Pharmaceutics*, 76, 113–122.

Saffran, M., Bedra, C., Kumar, G.S. and Neckers, D.C., 1988, Vasopressin: a model for the study of effects of additives on the oral and rectal administration of peptide drugs, *Journal of Pharmaceutical Sciences*, 77, 33–38.

Sawyer, T.K., 1995, Peptidomimetic design and chemical approaches to peptide metabolism, *Peptide Based Drug Design. Controlling Transport and Metabolism*, edited by M.D. Taylor and G.L. Amidon, pp. 387–422, Washington, DC: American Chemical Society.

Sayani, A.P. and Chien, Y.W., 1996, Systemic delivery of peptides and proteins across absorptive mucosae, *Critical Reviews in Therapeutic Drug Carrier Systems*, 13, 85–184.

Spatola, A.F., 1983, Peptide backbone modifications: a structure–activity analysis of peptides containing amide bond surrogates, conformational constraints, and related backbone replacements, *Chemistry and Biochemistry of Amino Acids, Peptides, and Proteins*, Vol. III, edited by B. Weinstein, pp. 287–300, New York: Marcel Dekker.

Szewczuk, A. and Kwiatkowska, J., 1970, Pyrrolidonyl peptidase in animal, plant and human tissues. Occurrence and some properties of the enzyme, *European Journal of Biochemistry*, 15, 92–96.

Taylor, M.D., 1996, Improved passive oral drug delivery via prodrugs, *Advanced Drug Delivery Reviews*, 19, 131–148.

Thompson, S.E. and Audus, K.L., 1994, Leucine–enkephalin metabolism in brain microvessel endothelial cells, *Peptides*, 15, 109–116.

Verhoef, J.C., Bodde, H.E., de Boer, A.G., Bouwstra, J.A., Junginger, H.E., Merkus, F.W.H.M. and Breimer, D.D., 1990, Transport of peptides and protein drugs across biological membranes, *European Journal of Drug Metabolism and Pharmacokinetics*, 15, 83–93.

VILHARDT, H., 1990, Basic pharmacology of desmopressin. A review, *Drug Investigation*, **2**, 2–8.

WANG, B., ZHANG, H. and WANG, W., 1996a, Chemical feasibility studies of a potential coumarin-based prodrug system, *Bioorganic Medicinal Chemistry Letters*, **6**, 945–950.

WANG, B., WANG, W., ZHANG, H., SHAN, D. and SMITH, D., 1996b, Coumarin-based prodrugs 2. Synthesis and bioreversibility studies of an esterase-sensitive cyclic prodrug of DADLE, an opioid peptide, *Bioorganic Medicinal Chemistry Letters*, **6**, 2823–2826.

WANG, B., GANGWAR, S., PAULETTI, G.M., SIAHAAN, T.J. and BORCHARDT, R.T., 1997, Synthesis of a novel esterase-sensitive cyclic prodrug system for peptides that utilizes a 'trimethyl lock'-facilitated lactonization reaction, *Journal of Organic Chemistry*, **62**, 1363–1367.

WILK, S., 1989, Inhibitors of TRH-degrading enzymes, *Annals of New York Academy of Sciences*, **553**, 252–264.

WEINBERGER, S.B. and MARTINEZ, J.R., 1988, Characterization of hydrolysis of [Leu] enkephalin and D-Ala^2-[L-Leu]enkephalin in rat plasma, *Journal of Pharmacology and Experimental Therapeutics*, **247**, 129–135.

12

Chemical and Pharmaceutical Documentation

KAREN FICH AND DEIRDRE MANNION

Danish Medicines Agency, Brønshøj, Denmark

12.1 Introduction

In this chapter the essential chemical and pharmaceutical documentation required in an application for a marketing authorization for a medicinal product in Europe (EU) will be outlined, with particular focus on the special requirements for products containing peptides or proteins as the active substance. The purpose of the documentation is for the applicant to prove a consistent quality and safety of its product, and to enable the authorities to be sure that the product does not give rise to any serious public health risks.

To assure a common level of the quality of medicines and an equal assessment of the dossiers throughout Europe, as well as to assist applicants in the preparation of the dossiers, several common rules and guidelines have been prepared in collaboration by the European authorities. During the preparation of guidelines it is also possible for the medicinal industry to comment on the proposed requirements. Recently a number of guidelines have been prepared by the authorities in Europe, US and Japan and the medicinal industry in the three regions, in a collaboration called the International Conference of Harmonization (ICH). The intention with these new guidelines is to make it possible to obtain marketing authorizations world-wide without having to produce three different dossiers.

However, it is important to remember that the guidelines are only guidelines, and are not legally binding. It is the responsibility of the applicant to provide satisfactory documentation for its specific product, and the right of the authorities to require further documentation if necessary, even if this is not particularly mentioned in a guideline. The administrative procedures and the overall requirements regarding the dossier are described in 'Notice to Applicants', and the specific requirements on several items are worked out in detail in the guidelines (European Commission, 1997).

Within the European Union, three possibilities of applying for a marketing authorization exist. In the central procedure, the application is assessed by two member states on behalf of all EU member states, and the marketing authorization is granted by the Committee for Proprietary Medicinal Products (CPMP). This procedure is obligatory for biotech products, and optional for other innovative products. The second procedure is the Mutual Recognition Procedure (MRP), where granting of marketing authorizations is based on mutual recognition of the assessment of the dossier by another EU member state. This procedure is obligatory from 1 January 1998, if marketing in more than one member state is to be achieved, and the central procedure is not used. The third procedure (national procedure) is only to be used when marketing in a single member state is wished for. However, the documentation necessary to obtain the marketing authorization is identical for all three procedures.

The content of the chemical and pharmaceutical part, part II of the dossier, will now be described in the order used in 'Notice to Applicants'. As it is impossible to cover all kinds of products, this description can only be in rather general terms, and therefore a list of the relevant guidelines for the particular section will be included at the end of each section for further and more detailed information. Further information on guidelines can be obtained on the Internet at *http://www.eudra.org/emea.html*, the homepage of the European Medicines Evaluation Agency.

12.2   Composition

In this chapter of the dossier a short outline of the composition of the product and the choice of container should be given.

The choice of composition of the medicinal product, including choice of excipients, especially if non-standard additives are used, and the possible addition of overages should be explained and justified, as well as the dosage form chosen. This part will often be rather short, as products containing peptides or proteins are usually formulated as injections, with only a few additives. The choice of the method of sterilization for products intended to be sterile should be discussed and justified, especially if a final heat sterilization is not possible or a non-standard method is employed. If the product contains preservatives this should be justified, and the efficacy of the preservative system should be proved, in accordance with *The European Pharmacopoeia* (European Department for the Quality of Medicines, 1997). Addition of preservatives to compensate for a less suitable manufacturing process, or low microbial quality of raw materials, is of course not acceptable.

The composition of the batches used in clinical trials should be shown, if different from the composition intended for marketing, and the consequences of such differences should be discussed. In the part 'Development pharmaceutics' any investigations and characteristics of the active substance and/or product leading to the choice of composition, pH, container, etc. should be referred to. Also, investigations on the suitability of the container, and the compatibility with excipients, container and other products, if relevant, should be described.

If the product is a modified release formulation, it should be explained how the modified release profile is obtained. An *in vitro* method for control of the *in vitro* dissolution profile should be developed, and a correlation of this method with *in vivo* data should be proved if possible. If it is not possible to obtain an *in vivo–in vitro* correlation, it should be demonstrated that the dissolution method can discriminate between batches with acceptable/unacceptable *in vivo* dissolution characteristics.

The bioavailability or bioequivalence with similar products should also be outlined in this chapter of the dossier, but this is rarely relevant for this kind of product.

Relevant guidelines for this section are:

- Development pharmaceutics and process validation

- List of allowed terms

- Quality of prolonged release solid dosage forms (even though this guideline is only about solid dosage forms, the general principles are also valid for prolonged release preparations for injection).

12.3   Method of manufacture

The manufacturing formula for the usual batch size should be stated here, followed by a detailed description of the manufacture of the medicinal product, including a detailed description of the sterilization process used for product and containers. A flow-sheet of the manufacturing process might be very helpful, especially if the process is complicated, or includes many steps.

The in-process controls carried out should be mentioned, together with the limits. Validation of the manufacturing process, including experimental data, and results of the investigations should be presented. Since peptide- and protein-containing products are often sterile products with a low content of active substance, special attention should be paid to validation regarding sterilization, bioburden and homogeneity of the final product. If part of the product, or the product itself, is stored during the manufacturing process, e.g. before filling, the maximum holding time should be stated, and justified by validation studies regarding microbial and chemical stability. The validation procedure should also ensure that the in-process controls, e.g. bioburden limits, are sufficient to ensure the quality of the product.

Relevant guidelines to this section:

- The use of ionizing radiation in the manufacture of medicinal products

- Manufacture of the finished dosage form.

12.4   Control of starting materials

12.4.1   Active substances

The content of this part of the file depends on the type of drug: whether it is a pure synthetic drug or a substance manufactured by fermentation or in cells by recombinant DNA technology, the differences primarily being in the description of the manufacture. The general information on the drug substance is the same for all types of substance.

Usually the section would be opened with a short statement of the identity of the substance, including nomenclature, i.e. International Non-proprietary Name (INN) and/or National Approved Names where relevant, laboratory codes and systematic chemical name(s) of the substance. A brief description of appearance and physical form should also be included, as well as information on structural formula, molecular formula and relative molecular mass if known. Information on stereochemistry should also be given here, if relevant.

One of the most important parts of the dossier is the description of the manufacture of the active substance. In this part it should be demonstrated that the manufacturing process will lead to a consistent product of as high a quality as possible, with a secure and well-known impurity profile. To ensure this, very detailed information on the manufacture and quality control of the active substances should be given.

Since there are some differences in the information required on the pure chemical synthetic processes and the biological processes, this section will be summarized in two separate parts, starting with the pure synthetic manufacture. For both types of manufacture, that name and address of all manufacturing sites should be stated.

The route of a chemical synthesis should be presented in a flow sheet, and a detailed description of each stage in the process should be given for a typical batch. This description should include information on starting materials, solvents and catalysts used, as well as the conditions where these are critical for the reaction. Where intermediates are isolated and purified this should be mentioned, and the final purification, including all solvents used, should be described in detail.

The scale of manufacture should be given, and yields at each stage might be helpful. If alternative routes of synthesis or alternative solvents or purification steps might be used, these should be described as well. The use of the alternative route should be justified, showing that the final quality of the product is not affected.

Specifications on all starting materials, solvents, catalysts and intermediates should be given. The specifications on starting materials and other compounds that significantly contribute to the molecular structure should include identity and purity tests and, where relevant, requirements on the maximum content of named impurities. Where the stereochemistry of the starting material is of importance, tests on the optical purity should be included.

When the active substance is produced in a biological process the necessary information is somewhat different, especially if recombinant DNA technology is employed. This will be only a brief outline, since the relevant information might vary significantly from product to product. The manufacturing process will often be a fermentation process, and the composition of the fermentation media, as well as conditions, should be described. The presence, nature and accepted levels of any microbial contamination in the culture immediately prior to harvest should be examined for each run.

The maximum permitted passage levels should be defined based on the stability of the host/vector system, and consistency and maintenance of growth and yield should be justified. Information on the monitoring of characteristics of the cell/vector system should be enclosed. If continuous culture production is to be applied, this entails special considerations regarding the stability of phenotypic and genotypic characteristics. Criteria for rejection of culture lots, as well as a clear definition of a batch, should be given.

The purification of the product calls for special attention. The methods used for purification should be described in detail, and the purification process should be validated. The validation should prove the capacity to remove unwanted products consistently and reproducibly, such as host cell derived proteins, nucleic acids, carbohydrates, viruses and other impurities.

Considering the starting materials, the same demands as mentioned above apply to the chemical substances in the culture media, but requirements on microbial purity should be included. Regarding the live starting materials, a different approach is needed. The basis of the control of cells is a well-defined seed lot system, a critical part being a full characterization of the seed material. Origin, storage, use and expected lifetime of the seed materials should be given, as should evidence on absence of potential oncogenes and infective agents such as viruses, bacteria, fungi or mycoplasma.

If recombinant cells are used, the host cell and expression vector should be described, including details on the origin and identification of the cloned gene, and the construction, genetics and structure of the vector. Details of the introduction and state of the vector in the host cell should be provided. Full details of the nucleotide sequence of the gene insert and of flanking regions should be presented, and all relevant expressed sequences should be identified. The sequence of the cloned gene should be confirmed at the seed lot level, and reconfirmed at least once after a full-scale fermentation cycle.

These are some examples of the type of data needed. Other kinds of information might well be relevant. It is the responsibility of the manufacturer and/or applicant to provide relevant, necessary and sufficient documentation to give a satisfactory description of its product.

Besides the information on the manufacturing process, other facts must be included in the file. These will be listed with some comments below.

Evidence of structure of the active substance should be given. This might include results of different kinds of spectroscopic investigations, elemental analysis, amino acid analysis, peptide mapping, isoelectric focusing, electrophoresis, circular dichroism, optical rotation, etc.; or it might be evidence from the route of the synthesis. For biological products, a wide range of both chemical and biological methods should preferably be used for the characterization. If it is not possible to give a full indication of the structure, the attempts made to prove the structure should be described. A physicochemical profile of the product should be given, with the relevant properties depending on the type of compound. The parameters could be pH, water content, solubility in different solvents, state of aggregation, hydrophobicity, melting point, boiling point, particle size or polymorphism, among others.

The development of analytical methods for routine control and their suitability should be discussed, especially in cases of unusual methods, and where biological methods are utilized. Methods of purity determination might be given as part of the impurity discussion instead.

Another very important part of the dossier, for obvious reasons, is the impurity discussion. This part should contain a comprehensive discussion on the possible impurities, their toxicity and level of content, and their influence on the safety profile and quality of the active substance. A list of all possible impurities related to the manufacture as well as degradation products should be provided, together with information on the content found in batches of the active substance and the analytical methods used for their detection. For active substances manufactured by biological methods, special attention should be paid to bacterial, viral and nucleic acid contaminants, cellular proteins or other impurities of host origin, and a wide range of analytical techniques should be used to demonstrate the purity.

The structure of the impurities should be determined if possible, or the attempts to identify the impurities should be mentioned. The maximum accepted levels of known and unknown impurities should be set. These limits should be as strict as possible, i.e. ppm level for protein contaminants, and justified by the amounts present in batches used in clinical and toxicological studies, or by separate toxicological studies on specific impurities.

In this part the content of other possible contaminants such as residual solvents should also be discussed, and limits proposed where relevant.

A specification on the active substance should be proposed, with sufficient requirements to ensure a consistent quality of the substance. The analytical methods used should be included, described in such detail that it will be possible for the authorities to carry out the tests.

The specifications should be based on results of elaborate analysis of an acceptable number of successive batches. Three batches will usually be acceptable for pure synthetic products, but often results of more batches will be necessary for substances manufactured by biological methods. The specification should contain requirements on identification, purity and assay/potency of the drug product.

Identification should be performed by at least two independent methods. For peptides and proteins, tests for the anticipated biological activity should be included, and physicochemical or immunological tests might be relevant. Verification of the identity by amino acid sequence analysis or peptide mapping might be necessary if other methods are insufficient for proper identification. The limits on the impurity content mentioned above should also be included in the specification, together with other relevant purity requirements such as tests for heavy metals, microbial contamination, endotoxins, or optical purity.

Requirements on assay and/or potency should be set, including lower and upper limits. For biological products such as peptides or proteins it is recommended, if possible, to correlate a biological potency test with and an accurate physicochemical assay, ensuring both an accurate content and the biological potency of the product. The potency should be reported in units of biological activity per ml or mg, based wherever possible on an appropriate national or international reference preparation.

Comprehensive validation studies on all analytical methods should be carried out, and the results reported, including raw data, to justify the suitability of the methods. Limits of detection and quantification of purity tests should be reported. Pharmacopoeial methods might be considered validated, and further validation might not be necessary or might be restricted to specificity.

The reference materials used in analysis of the drug substance should be described. The description should include information on manufacture and purification of the batch used as reference substance. The substance should be fully characterized regarding chemical composition, purity, potency and biological activity, and, where relevant and possible, full amino acid sequence. The reference batch should preferably have been evaluated clinically.

The section on the active substance should be closed with results of analysis of recent batches, normally at least three, to confirm accordance with the specifications.

It is possible to provide the documentation on the active substance using the European Drug Master File procedure. In this procedure the documentation is forwarded to the authorities in two parts, the active ingredient manufacturer (AIM) restricted part and the applicants part, also called closed and open parts respectively. The restricted part contains confidential information on details in the manufacturing process, and this part is sent to the authorities only. The open part contains all other information on the drug substance, and should be provided by the AIM to both the applicant and the authorities. The applicant should include the open part in the dossier, together with a so-called letter of access from the AIM, allowing the applicant to use the file. The AIM should enclose a statement that it will inform the applicant and authorities of any changes in the content of the Drug Master File.

Guidelines relevant to this section:

- Chemistry of active substances
- Requirements in relation to active substances
- European Drug Master File procedure for active substances
- Impurities in new active substances
- Production and quality control of medicinal products derived by recombinant DNA technology
- Quality of biotechnology products: Analysis of the expression construct in cell lines used for production of derived protein products
- Production and quality control of cytokine products derived by biotechnological process
- Production and quality control of monoclonal antibodies
- Validation of virus removal and inactivation procedures
- Validation of virus removal/inactivation procedures: Choice of viruses
- Analytical validation
- Validation of analytical procedures – definition and terminology
- Validation of analytical procedures – methodology.

12.4.2 Excipients

The quality of the excipients should also be mentioned. Many excipients are the subject of a monograph in the *European Pharmacopoeia* or the pharmacopoeia of a member state, and the documentation is then a question of verifying the accordance with the relevant monograph. For medicinal products intended for injection, it might be necessary to address the requirements on microbial purity.

If the excipient is not described in a monograph, an in-house specification should be enclosed, including relevant tests on identification, purity and physicochemical characteristics. New excipients, which have not been used in human medicines before, should be considered as new active substances and full documentation should be provided, especially regarding safety.

Relevant guidelines to this section:

- Excipients in the dossier for applications for marketing authorization of a medicinal product
- Limitations to the use of ethylene oxide in the manufacture of medicinal products.

12.4.3 Packaging materials

The packaging materials, including both glass containers and plastic containers, should be in compliance with the *European Pharmacopoeia*. The full composition of plastic containers for injections should be enclosed.

A specification, including identification of polymers in plastic materials, should be present. Suitability in relation to the actual product and compatibility of the container with the product should be demonstrated. If plastic containers are used as containers for liquid products, it should be demonstrated that additives do not leak to the medicine. The suitability and compatibility might be demonstrated as part of the stability investigations.

Guidelines relevant to this section:

- Plastic primary packaging materials
- Limitations to the use of ethylene oxide in the manufacture of medicinal products
- The use of ionizing radiation in the manufacture of medicinal products.

12.5   Intermediate products

This section is relevant only when tests are carried out on an intermediate product. However, it might sometimes be necessary to perform some of the tests on intermediate products instead of the finished product, especially when the intermediate product is stored before further manufacture. In these cases a specification on the intermediate product should be made, following the same principles as will be described for the finished product.

There are no guidelines that specifically describe requirements on intermediate products: in general, the requirements on the finished product are also valid for intermediate products.

12.6   Control tests on the finished product

A specification valid at the time of release of the finished product should be set up. The specification should contain a description of the appearance of the product, and pharmaceutical–technical tests relevant for the dosage form. These tests are described in dosage form monographs in the *European Pharmacopoeia*. Tests on identification of the active substance, assay and/or potency, and purity should be included. If the product contains preservatives, antioxidants or other stabilizing agents important for the stability of the product, tests for the identity and assay of these substances should be included in the specification. Requirements on impurities related to the manufacture of the active substance, which are not degradation products as well, and which are fully controlled in the active substance specification, can be excluded from the finished product specification if justified.

If the medicinal product is intended for reconstitution before use, e.g. powders for injection, relevant tests on the reconstituted product should be included. As for the active substance, a full description of the analytical methods should be contained in the specification.

The principles for setting up a specification on the finished product are exactly the same as described for the drug substance. The specification should be based on results of analysis of recent batches of the product, with as strict limits as possible, and justified by safety studies. Pharmaceutical–technical tests typically carried out on parenteral preparations are absence of particulate matter, and tests for sterility and endotoxins.

As for the active substance, all analytical methods should be fully validated. If the same methods are used for the active substance and the finished product, the validation in relation to the finished product need only deal with the specificity towards excipients. Standard methods for pharmaceutical–technical tests described in pharmacopoeias can usually be considered as validated.

In the same way as the active substance section, this section should be finished with batch analysis results on recent batches, to confirm accordance with the specifications.

Guidelines relevant to this section:

- Specifications and control tests on the finished product.

12.7 Stability

In this part of the dossier data should be submitted to document the stability of the bulk drug as well as the finished product when stored. Since the biological activity of peptides and proteins depends on the molecular confirmation, which depends on both non-covalent and covalent forces, these products are extremely sensitive to changes in the environment such as variations in temperature, oxidation, light, ion force and shear. It is often difficult to determine the effect of these variations on the drug substance or product. This makes it very important to design the stability studies carefully in order to gain necessary and realistic information and justification of the stability throughout the proposed shelf-life of the active substance and the finished formulation.

The major part of the stability studies is common to the bulk product and the finished medicinal product, and will be described in relation to the finished product. An exception is investigations on parameters directly connected to the dosage form of the finished product, i.e. stability of reconstituted solutions. Data on the stability of the active substance, also called bulk drug, should be included in the file, if the active substance is to be stored prior to further manufacture into the finished product.

The file should contain a detailed protocol on the stability studies containing all the necessary information to justify the proposed specification, shelf-life and storage conditions. The proposed specifications on the bulk drug should always remain the same at the time of manufacture and throughout the proposed re-test period.

The shelf-life specification on the finished product should be enclosed if different from the release specification, which will often be the case, as many biological/biotechnological products are subject to significant losses in potency. No guidelines provide direct recommendations on acceptable losses, and level of degradation products. Limits on content and purity tests should thus be justified on a case-by-case basis by the results obtained in the stability studies, as well as results from preclinical and clinical studies, and it should be demonstrated that the proposed specification is sufficient to ensure that the product will retain its clinical performance throughout the shelf-life.

To justify the proposed shelf-life and storage conditions, stability data should be provided on at least three batches representative for the manufacturing scale. Preferably, different batches of the bulk drug should have been used for the manufacture of the finished product batches. Data from pilot-scale batches might be acceptable at the time of submission, with the precondition that the first three production-scale batches will be placed in long-term stability studies. Such a concept should be discussed with the authorities prior to application. Where data on pilot-scale batches have been used for establishing the shelf-life, and long-term data on manufacturing-scale batches show that these batches fail the specifications, the authorities should be notified and appropriate actions agreed upon.

The storage period that should be covered by the stability data at the time of application depends on the proposed shelf-life. For products with a proposed shelf-life below six months, the amount of data should be determined on a case-by-case basis. For products with a proposed shelf-life longer than six months, at least six months' data should be present. The actual shelf-life will be established based on the real-time/real-condition data present, and it is therefore highly recommended that updated data be submitted to the authorities during the evaluation process, when available.

The testing frequency of the stability samples depends on the proposed shelf-life. For shelf-lives of one year or less tests should be carried out on a monthly basis for the first three months, and every three months thereafter. For shelf-lives longer than one year, analysis should be conducted every three months for the first year, with six-month intervals for the following year, and thereafter annually. Often data are to be submitted post-approval, and reduced testing might be acceptable if documented. If data exist that document such reduced testing, it is recommended to submit a stability protocol justifying longer testing intervals.

The samples in the stability studies should be packed in the container(s) intended for marketing. Where interactions between the container/closure system and liquid products cannot be excluded, the stability studies should be performed on packages in the inverted as well as the upright position, except for sealed glass ampoules.

When multiple-dose containers are used, it should be demonstrated that the closure is capable of withstanding repeated insertions and withdrawals, and that the product will retain its quality throughout the maximum in-use period specified in the product literature and labelling.

Where a product is packed in different fill volumes (e.g. 1 ml, 2 ml . . .), unitages (e.g. 10 units, 20 units . . .) or masses (e.g. 1 mg, 2 mg . . .), the matrixing or bracketing procedure might be used. However, this should always be discussed with the authorities prior to application. Matrixing is the systematic statistical design of the stability studies ensuring that all product strengths and package sizes are tested to the same extent, but only a fraction of the total number of samples are tested at any specific time point. At the subsequent time point another set of samples are tested. Matrixing should only be applied when it is confirmed by data that the samples tested are representative of all samples in the stability study.

Bracketing is the principle that only the smallest and largest container sizes are included in the stability study. This design can be applied when the same strength is packed in exactly the same container/closure system, but in three or more fill contents.

Samples should be stored at the real-time conditions proposed for storage of the product. However, it is highly recommended to submit data on accelerated testing as well. These data may give useful information on the degradation pathways and patterns, and might also provide guidance on which parameters are the most indicative of stability. Stressed samples will often be useful in the validation of the test procedures to be used in the stability studies, especially in relation to specificity and stability-indicating power. Further results of accelerated conditions might give some indications on the influence of accidental exposure to extreme conditions, e.g. during transportation. The accelerated test conditions should be defined in relation to the proposed real-time storage conditions, e.g. if the proposed storage condition is 2–8°C, 25°C will be considered as accelerated conditions.

Studies at high or low humidity should be included when relevant, i.e. for containers which do not offer proper protection against moisture or evaporation. When glass containers are used, this kind of investigation is rarely relevant.

The relevance of light testing should be discussed with the authorities from case to case.

To demonstrate the stability of the product, tests on relevant parameters should be carried out. As no single parameter is sufficient to characterize the stability of a biological/

biotechnological product, a multitude of tests that together will provide a stability-indicating profile should be proposed, assuring the identity, purity and potency of the product. The tests to be included in the profile will be product-specific.

The most important parameters will almost always be potency and purity. When the effect of the medicinal product depends on a defined biological activity, the potency should be tested in the stability study and reported in units of activity. The potency should be expressed against an appropriate reference material, preferably an international or national recognized standard. Where the biological activity depends on conjugation with an adjuvant or carrier, the dissociation from the second moiety should also be followed in the stability studies.

If direct characterization or quantification of the biological activity or the content of degradation products is not possible, appropriate surrogate tests should be considered.

Besides the extremely important parameters mentioned above, general parameters relevant for the specific type of product should be tested. Since most biotechnological/biological products are injections or powders for injection, requirements in relation to sterile products will usually be relevant. These requirements are described in the pharmacopoeia. Relevant tests could be appearance, colour of solution, reconstitution time, pH, moisture content, sterility or microbiological purity and content of additives, among others.

A wide range of methods should be employed to ensure a comprehensive characterization of the drug product or substance, and the degradation products. Some examples of such methods are SDS-PAGE, Western blot, peptide mapping, and different kinds of chromatography and electrophoresis. The choice of methods should be discussed and justified, and all methods should be validated. However, pharmacopoeial methods are in general considered as validated, except for specificity towards excipients and degradation products. The statistical methods that should be used for calculation and evaluation of the results are described in the ICH guideline on stability (see below).

If the product is intended for reconstitution or dilution in infusions, the stability of the reconstituted or diluted product should be assessed as well, for storage periods and conditions relevant to the in-use situation. The chemical, biological and microbiological stability should be addressed. The medicinal product should be labelled with precisely defined storage conditions, especially for products sensitive to freezing or elevated temperatures. The wording of labels and other product literature should be in accordance with national and/or regional requirements.

Guidelines relevant to this section:

- Stability tests on active ingredients and finished products
- Stability testing of new active substances and medicinal products (ICH)
- Quality of biotechnological products: Stability testing of biotechnological/biological products (ICH).

12.7.1 Expert report

An important part of the application is the expert report. This consists of a written critical appraisal of the application, highlighting the most important parts of the documentation. Deviations from guidelines should be discussed, and justified, and often the expert report will be an opportunity to elaborate the argumentation on difficult items or untraditional choices of formulation, tests or specifications. A comprehensive and really 'critical' part of the expert report is invaluable to the authority assessors, speeding up the assessment process. The tabular part should present the most important documentation in a tabular format, as the name suggests. Formats for the expert report are shown in the 'Notice for Applicants'.

The above outlines some rather general considerations on the documentation needed for the chemical–pharmaceutical dossier of an application for a marketing authorization. Due to the great variations in the group of biological and biotechnological products, this does not cover all possible aspects that should be discussed in the file, but only examples of the common documentation relevant for most products. As several items will have to be handled on a case-by case basis, it will often be useful to discuss such items with the relevant authorities prior to application, to avoid production of unnecessary information, misunderstandings or serious shortcomings in the file, leading to a prolonged evaluation phase.

References

European Commission, 1997, *The Rules Governing Medicinal Products in the European Union; Volume 2B: The Notice to Applicants. Presentation and Content of the Application Dossier*, Luxembourg: Office for Official Publications of the European Communities.
European Department for the Quality of Medicines under Europarådet (Council of Europe), 1997, *The European Pharmacopoeia*, 3rd edition and supplements, Strasbourg.

Index